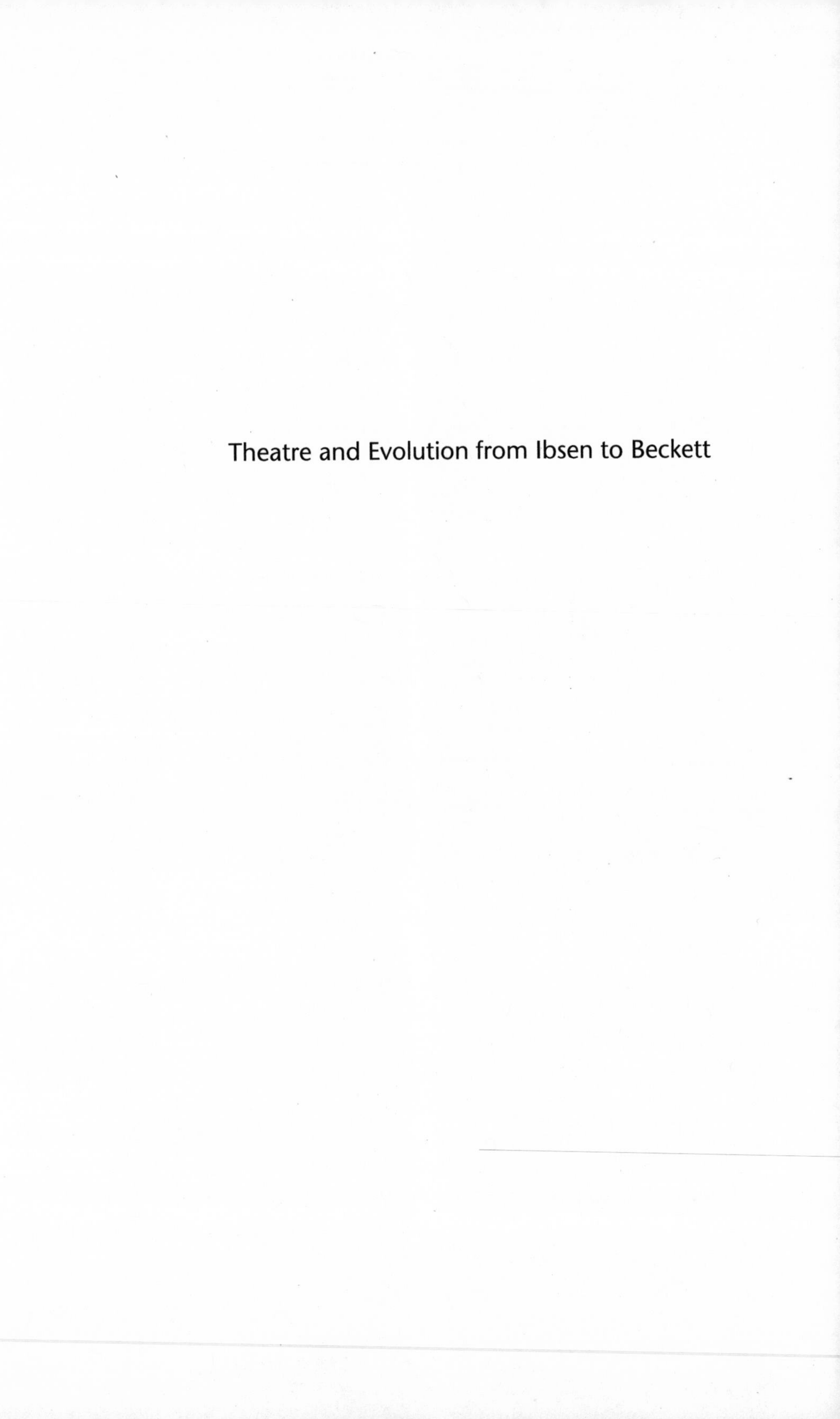

Theatre and Evolution from Ibsen to Beckett

KIRSTEN E. SHEPHERD-BARR

Theatre and Evolution from Ibsen to Beckett

COLUMBIA UNIVERSITY PRESS NEW YORK

Columbia University Press
Publishers Since 1893
New York Chichester, West Sussex

cup.columbia.edu

Library of Congress Cataloging-in-Publication Data

Shepherd-Barr, Kirsten, 1966–
Theatre and evolution from Ibsen to Beckett / Kirsten E. Shepherd-Barr.
pages cm
Includes bibliographical references and index.
ISBN 978-0-231-16470-2 (cloth : alk. paper)
ISBN 978-0-231-53892-3 (e-book)
1. Science and the arts—History. 2. Theater and society—History.
3. Evolution (Biology). 4. Darwin, Charles, 1809–1882—Influence.
5. Science—Social aspects—History. I. Title.

NX180.S3S54 2015
700.1'05—dc23
2014028479

∞

Columbia University Press books are printed on permanent and durable acid-free paper.
This book is printed on paper with recycled content.
Printed in the United States of America

c 10 9 8 7 6 5 4 3 2 1

COVER DESIGN: Noah Arlow
COVER IMAGE: Fiona Shaw as Winnie, in Samuel Beckett, *Happy Days* (directed by Deborah Warner, BAM Harvey Theater, January 2008). (Photograph by Hiroyuki Ito/Getty Images)

For Alastair

Contents

Preface

This project developed naturally from my book *Science on Stage* and from organizing a panel on Darwin and the Theatre for the international Darwin Festival in Cambridge in 2009. The book was originally going to be called *Darwin and the Dramatists*, a catchy title that paid homage to George Levine's wonderful book *Darwin and the Novelists*. But it soon became clear that this title would never reflect the true subject of the book, as theatre's engagement with evolution ranges far beyond "drama" (a too narrowly literary term) and far beyond "Darwin." In fact, it is nearly the opposite—a story of how theatre in an astonishingly wide range of forms took up specifically non-Darwinian ideas and transformed them in playwriting and in performance. As I was finishing this book, an eminent novelist and critic whom I had just met remarked, on hearing what I was working on: "Ah well, all artists are Lamarckians at heart!" To paraphrase Tennessee Williams, wouldn't it be nice if it were true. Instead, as I hope this book shows, theatre's engagement with evolution tracks an astonishing range of ideas in which all kinds of thinkers contribute, from Jean-Baptiste Lamarck and Herbert Spencer to Ernst Haeckel and Charles Darwin to Hugo de Vries and

Richard Dawkins, matched by an equally dazzling array of theatrical modes and performances. The title *Theatre and Evolution from Ibsen to Beckett* builds on and complements the pioneering work of Jane Goodall in *Performance and Evolution in the Age of Darwin*, expanding the scope of her work, which so brilliantly illuminated the ways in which nineteenth-century popular entertainment immediately engaged with evolutionary ideas. But, as Peter Allan Dale points out in his book *In Pursuit of a Scientific Culture*, it is not just about how scientific ideas are transformed in imaginative ways but about how art "discovers something rather than 'exploits' what others have discovered."[1]

Interdisciplinary work poses many challenges, not least to attain adequate knowledge of several different, and seemingly unrelated, fields. There are two main points to make about this problem. One is that, as Neil Vickers puts it, "interdisciplinarity is only valuable if one retains the academic standards and norms of one's own discipline." The other is the argument of (truly interdisciplinary) neuroscientist and former theatre director Stuart Firestein that the questions asked are more important than the answers given.[2] I hope this book presents not only answers to how theatre and evolutionary thought have interacted but also many new and intriguing questions.

Acknowledgments

This project was made possible through the generous help of many colleagues and institutions. It began with an invitation from Rebecca Stott to organize a panel on Darwin and the Theatre for the international Darwin Festival in Cambridge, England, in 2009. The ideas and the relationships formed there have been truly inspiring. A fellowship from the Leverhulme Trust 2011–2012 enabled me to complete the research and draft most of the book, and I am deeply grateful for this support. I also thank the Faculty of English at the University of Oxford for granting several periods of leave, funds for research assistance and travel to conferences, and help with permissions; I could not have completed the project without the Faculty's support and that of the John Fell Fund of the University of Oxford. St. Catherine's College has also provided invaluable support in the form of several periods of leave, generous financial assistance, and a uniquely interdisciplinary environment in which to work.

Many individuals have given generously of their time and advice. I am especially grateful to Stuart Firestein for enthusiastically supporting the book from its inception, to Martin Puchner for his astute reading of the entire manuscript, and to Sos Eltis, whose extraordinary knowledge of Victorian and modern drama has enriched my work; she patiently read and commented on drafts of chapters, and many of the plays I discuss in the book were her suggestions. Laura Marcus has given ongoing encouragement and support and I have benefited enormously from her expertise. Conversations with Jane Goodall anchored and clarified some of my ideas and stimulated the chapter on Beckett in particular; I have cherished these opportunities to learn from her. Jean-Michel Rabaté kindly read a draft of the Beckett chapter and offered invaluable suggestions, and Peter Fifield shared his wealth of knowledge and expert advice on many occasions. Other colleagues whose suggestions have helped develop and strengthen my project include John Holmes, Narve Fulsås, Sally Shuttleworth, John Bolin, James Secord, Will Abberley, and Chris Ponting. Discussions with Alastair Barr, Gordon M. Shepherd, and Gordon M. G. Shepherd have further enriched my understanding of a wide range of scientific fields and concepts.

Several colleagues have generously shared material with me, and I am especially grateful to Tore Rem, Eivind Tjønneland, Joanne E. Gates, Pietro Corsi, Michael Billington, and Elinor Shaffer. John Holmes, Helen Small, Michael Whitworth, and George Levine patiently read research proposal drafts and offered valuable advice. Sophie Duncan provided research assistance over several years, and I could not have completed this project without her skilful and insightful help. My undergraduate and postgraduate students have been a delightful and challenging sounding board for many of the ideas in this book, and I have learned so much from them, as I have from the ever-widening literature and science circle here at Oxford. The newly-established TORCH (The Oxford Research Centre for the Humanities) has also been invaluable in providing a home for Ibsen and Scandinavian Studies here at Oxford and strengthening my work in these areas.

Colleagues at St. Catherine's have given exceptional encouragement and support, in particular Bart Van Es and Jeremy Dimmick, Richard Parish in helping with French translations, and Barrie Juniper for sharing his wealth of botanical knowledge. College librarian Luda Gromova speedily procured every conceivable book I needed. The librarians in the Social Science Library, Oxford, have also been extremely helpful. My thanks also go to the archivists of the Fales Collection, who assisted me greatly with my research on Elizabeth Robins, and to the librarians at the British Library for helping me navigate the extraordinary resources of the Lord Chamberlain Collection of Plays.

I am grateful for the opportunities to share my work in progress in the Department of Drama and Performance at the University of Exeter; the Northern Modernisms conference in Birmingham; the Samuel Beckett: Debts and Legacies seminar series at Oxford; the University of Warwick's Beckett and the Brain project; the Samuel Beckett Centre at Trinity College, Dublin; the Ibsen Centre at the University of Oslo; the International Federation for Theatre Research conferences; and the Drama and Performance seminar series here at Oxford and other places. My thanks also go to Bernard Lightman and Bennett Zon for commissioning me to write a chapter on theatre and evolution for their book *Evolution and Victorian Culture* (Cambridge: Cambridge University Press, 2014); their comments have been invaluable for the opening chapters of my book. The editors of *Women: A Cultural Review* also deserve my thanks for their encouragement in publishing my article on James A. Herne and Elizabeth Robins.

My editor, Patrick Fitzgerald, has unfailingly supported this project, and I am hugely grateful for his enthusiasm and faith in the book. I am also grateful for the expert guidance and patient assistance of Kathryn Schell and for the astute and generous responses of the anonymous readers for Columbia University Press.

Living with this project for the longest time, though, day in and day out, have been my family, who have provided an unbeatable combination of encouragement, scientific and linguistic expertise, curiosity about the world, joy, and a sense of humor. I could not have written this without them: my parents, siblings, and siblings-in-law; my children Graham, Callum, and Gavin; and especially my husband, Alastair. This book is for him.

~

References to Elizabeth Robins's work printed with kind permission of Independent Age (Registered Charity No. 210729) (http://www.independentage.org).

I am grateful for permission from the editors to quote from selected chapters in Thomas F. Glick and Elinor Shaffer, eds., *The Literary and Cultural Reception of Charles Darwin in Europe*, vol. 3 (London: Bloomsbury, 2014).

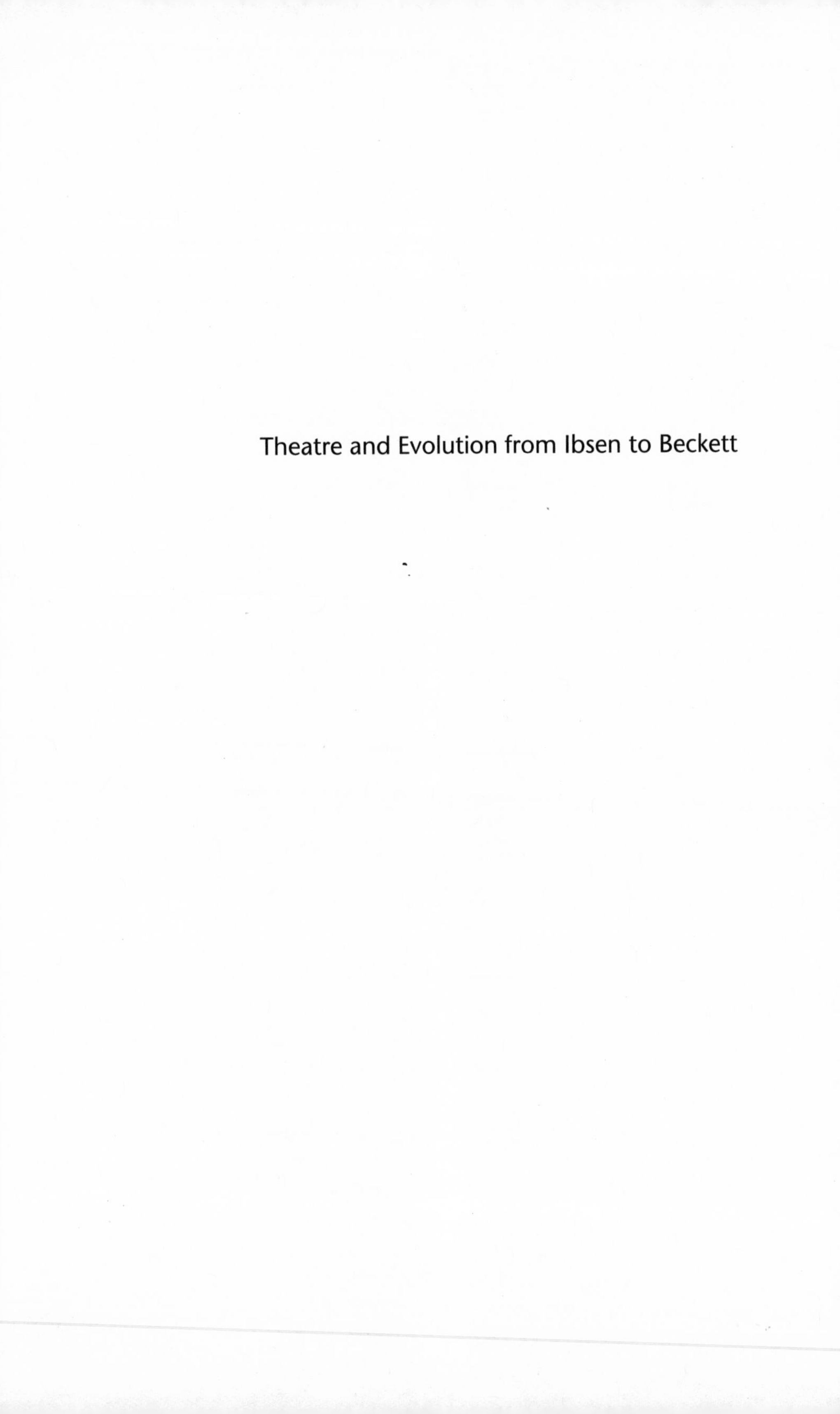

Theatre and Evolution from Ibsen to Beckett

Introduction

> *A play, after all, is a play, not a discussion-forum. It is the author's business, by means of a picture of life, to set us thinking, not to do all our thinking for us.*
>
> William Archer, introduction to *Alan's Wife*

An amused writer in the *Galaxy* in 1873 noted the "universal drenching" of literature and journalistic writing with Charles Darwin's ideas.[1] The concept of evolution caught the public imagination as no other scientific idea. There have been many studies of evolution in the novel (Gillian Beer, George Levine, Gowan Dawson, and many others); poetry (John Holmes); the visual arts; music; and other art forms, but apart from Jane R. Goodall's groundbreaking *Performance and Evolution in the Age of Darwin*, little consideration has been given to theatre in this context. This is a striking gap, given the popularity and variety of theatrical entertainment and the deep affinities between theatre and evolution, which explore the same basic questions: "What does it mean to be human? . . . to see (and be seen), to know (and be known), to relate to other humans?"[2] What is fundamental to evolution—"the relation of organism to organism . . . along with their relations to the wider shared environment"[3]—is also the basic stuff of drama and always has been. Theatre provides a particularly potent and fascinating example of how scientific ideas make their way into culture because of its combination of liveness and immediacy, kinetic human bodies in action, and time working on two levels ("real" and "theatrical" time) and because of its sheer variety, from mainstream drama to comedies to street theatre and the public lecture; from the expansive epic theatre of Bernard Shaw and Thornton Wilder to the spare minimalism of Samuel Beckett.

Yet, as the epigraph to this introduction indicates, more often than not theatre explores evolution to set the audience thinking, not to do its thinking for the audience (with the glorious exception of Shaw). This key distinction lies at the heart of this book. As one director puts it, theatre is a crucial, nonjudgmental space in which to explore ideas while the legislator's and the pope's backs are turned and where there is no committee on ethics.[4] It is easy to forget how central theatre once was, that it was once deemed so threatening that it prompted theatrical censorship laws in Britain lasting from 1737 to 1968. "Who shall deny the immense authority of the theatre, or that the stage is the mightiest of modern engines?" writes Henry James in *The Tragic Muse*. Despite theatre's position at what Joseph Roach calls "the center of civilized life," few studies have been done on the role of theatre and performance in intellectual history.[5] Looking at how theatre has engaged with evolution restores this sense of the cultural importance and role of the stage.

The playwright Henry Arthur Jones once wrote: "The dramatist has at the most two hours and three-quarters to portray all that is vital in the lives of some dozen characters, to portray what Nature takes some hundred of years to portray."[6] In this same way, the paradox is that one cannot literally *see* evolution taking place because it is such a slow, gradual process, so one might think that it would not be suitable for live performance; yet, theatre was quick to engage with it. As has been so fully demonstrated, fiction and poetry could explain and chart evolutionary processes through the use of narrative, descriptive language and the luxury of time.[7] In the theatre, the problem was how to depict something on stage without the luxury of narrative explanation, relying only on characterization, dialogue, the actors' bodies, and scenery, and constrained by the two-hour traffic of the stage.

There is no such thing as "Darwinian theatre," although a few plays refer directly to evolutionists by name, as in Shaw's *Misalliance* ("Read your Darwin, my boy. Read your Weismann") and *Back to Methuselah* ("Darwin is all rot") or Susan Glaspell's *Inheritors* ("Darwin, the great new man"). Some plays also physically place such figures on stage by calling for photographs or busts of them to be prominently displayed.

Although such instances are entertaining, the main purpose of this book is not only to catalogue theatrical allusions to evolution but also to show the more indirect, oblique engagement with the ideas themselves. Putting scientific ideas within a theatrical framework involves what Beer calls a "changing of contractual terms of belief."[8] Entire theories are not generally incorporated wholesale into novels, plays, poems, or paintings; instead, notes Michael Whitworth, "one finds fragments of imagery and vocabulary from science subjected to processes of condensation and displacement."[9] Theatre particularly embraced varieties of non-Darwinian evolution, from Lamarckism to Ernst Haeckel's monism to saltation. It did all of this indirectly, however, and it was by no means always the most scientifically accurate or faithful portrayal of evolution that was the most theatrically successful; instead, the stage often seemed to test scientific ideas in broader cultural contexts and with, as Goodall notes, "a cavalier attitude toward comprehension."[10] The danger of what George Henry Lewes termed "giving currency to vulgar error" was not a major concern of dramatists engaging with evolution. As Goodall puts it, "Show business intervened in the world of ideas with its own forms of expertise."[11]

A good model for the kind of interaction I explore in this book comes from the visual arts, in the exhibition "Endless Forms," organized for the Darwin celebrations in 2009, which showed how "science and the visual arts meet on equal terms, and the interplay between them is made clear."[12] The show is not about "cause and effect" but about "interplay," "resonance," "interchange," and the "two-way street" of mutual influence. Such terms pepper this book. This kind of influence has theatrical parallels in nineteenth-century performance modes as not only human nature (always the mainstay of theatre) but also the fascination for nature in all its "endless forms" became popular theatrical fare. Artists, novelists, poets, and playwrights were not passive recipients of scientific stimuli and ideas or merely reflective of them; they actively engaged with science and often transformed the ideas they encountered in the process, so that the artistic end product of that encounter was a different beast altogether, often questioning the very science on which it was supposedly founded.

Far from being a mere reflection of evolutionary theory, theatre often reacts *against* it. The "top-down" model challenged by important recent scholarship on the popularization of science by Peter J. Bowler,

Bernard Lightman, Gowan Dawson, Ralph O'Connor, Goodall, and others, whereby science is translated and disseminated through cultural mechanisms to a passive, receptive audience, simply does not apply here, for this is no story of benign assimilation: dramatists and theatre practitioners who dealt with evolution were often suspicious, doubting, and hostile to it even while deeply drawn to the new insights and possibilities it offered. *Theatre and Evolution* explores this often-deliberate wrong-headedness and what Beer has called the "creative misprisions" that can result when science meets art and literature. Theatre is no mere handmaiden to science, a means of transmitting its findings; theatrical encounters with evolution resist, challenge, and ultimately transform the ideas at their core, and my focus here is on how playwrights did this, complementing Goodall's focus on performance and extending the scope of her study through the twentieth century.

Thus, my study generally undercuts the kind of praise lavished, for example, by the young William Empson on Maurice Maeterlinck for adopting "one of the artist's new, important and honourable functions, that of digesting the discoveries of the scientist into an emotionally available form."[13] This "digesting" model breaks down quickly under the live conditions of theatre and in the hands of playwrights crafting work for an audience in the here and now. "Theatre is contrarian," notes playwright Simon Stephens.[14] In their uses of the biological sciences and evolutionary theory, playwrights often get it wrong, and this tendency of artists and writers has been noted by James Harrison; often, he writes, there is an "acceptance of evolution but dismay at Darwinism," specifically the blind, mechanistic aspect of natural selection.[15] In fact, it is those aspects of evolutionary theory that were its fallout rather than its main substance that tend to get their attention. Infanticide, rejection of maternal instinct—these are instances of the *questioning* of aspects and implications of evolution as they most fascinated and concerned playwrights. As Beer notes, "Scientific work always generates more ideas and raises more questions than can be answered solely within the terms of scientific inquiry."[16] She argues that "ideas cannot continue to thrive when locked away," and that when they escape the confines of the library or laboratory, they change, which is to the good, even though it may mean that they are misused or misunderstood. This risk is worthwhile from a broader perspective: the artistic transformation of science can pose "disturbing cultural questions" beyond the contexts in which the original ideas arose.[17]

Of course, one could argue that theatre stages evolution, whether consciously or not, simply through the common ingredients of most drama: a struggle between organisms for survival, the presence of competition and conflict (especially over mate selection), the human body signifying universal physiological processes, and the emphasis on how organisms respond to their environments. Theatre does indeed foreground the individual's relationship to his or her environment, within the ecosystem shown on stage. The audience accepts the physical limitations of the space and understands the metaphor of "all the world's a stage" framing what it is watching. Given the temporal and physical constraints noted by Jones, it is not surprising that theatre's engagement with evolutionary ideas tends to be metonymic, requiring techniques of extrapolation akin to Georges Cuvier's discredited law of correlations: a single individual stands in for or represents the species, a single fragment represents the whole. The same is true of ideas. Playwrights use ideas as starting points, and by its very nature, it is more difficult to contain ideas in the theatre; a script is just a starting point, embodied and given new meanings through each performance. Sometimes, notes the director Luca Ronconi, ideas *become* dramatic characters themselves rather than being conveyed *by* characters, or for that matter mediated through plot or biography.[18] Above all it is the presence of live bodies that gives theatre its strongest links to evolution; as Goodall puts it, theatre foregrounds "the evolutionary significance of the performer." The human body rather than the story became the shorthand for evolution on stage. So, instead of the novelist evoking eons of geological time and change, as Thomas Hardy is able to do in *Tess of the D'Urbervilles*, for example, the theatre relied on audiences recognizing in a single human body the processes and forces of evolutionary change.

But if all drama is fundamentally evolutionary, where do you draw the line? I adopt a more focused approach, looking at plays and performances that engage with evolution in ways that are not merely coincidental or loosely evolutionary but that map onto a specific theme or motif under debate, such as extinction or sexual selection. I do this because tracing the complex history of this engagement means understanding theatrical works as manifestations of the larger "cultural embeddedness of science."[19] Given the popularity of theatre, it is likely that more people encountered evolutionary theory through theatrical performances than through other art forms, especially during the late nineteenth and early twentieth centuries. At the same time, not only evolution specifically, but also biology more generally had an innate popular appeal partly through its direct relevance to everyday human life and partly through its accessibility. It could be fairly easily grasped (and misunderstood, as often happened) and could be found readily in nature just by looking around the garden or going to see fossils in a museum. By contrast, the emerging field of modern physics for many years "had no practical impact on people's lives" until the first atomic weapon was used in 1945, and although it was "available to literary writers as a storehouse of images, concepts, and thought experiments that could be put to various uses, the new physics was inherently suited to describing unfamiliar and uncanny states of subjectivity."[20]

In this age of public engagement, the question of how science is disseminated, as well as how aware we are of its hegemonic power, is becoming increasingly important. "Ways of viewing the world are not constructed separately by scientists and poets; they share the moment's discourse."[21] Many of the playwrights mentioned in this book had profound interests in science, far from dilettantism, such as August Strindberg, Maeterlinck, and Anton Chekhov. New research has uncovered valuable examples of performance influencing science, for example, in David Ferrier's experiments with making monkeys flinch.[22] Science has often borrowed from theatre for useful metaphors and analogies. Recent examples of this borrowing include Nessa Carey's *The Epigenetics Revolution*, which uses the dual modes of drama as text and performance as an analogy for how epigenetic changes are registered, like pencil marks and Post-it notes on a script; and Robert Crease's *The Play of Nature*, which argues that "scientific experiments and theatrical performances have deep, striking similarities. . . . The theatrical analogy allows one to develop not only a comprehensive approach to experimentation but also an adequate language in which to speak of it philosophically."[23] Analogies also abound between theatre history and evolutionary science in their respective methodologies of reconstructing past events. Joseph Roach likens historians of performance to paleontologists digging in a fossil record: it is a "discipline devoted to the reconstruction of a once vital phenomenon now characterized by its total extinction. The evidence has forever receded from the thing itself," so we extrapolate from "a few old bones" to imagine "the creature's disposition and behavior, its natural habitat and local adaptations."[24]

In doing the research for this book, I tried to find as much evidence as possible regarding how the plays under discussion were performed and received, but this evidence is limited, and my discussion remains necessarily reliant on texts of plays. I do, however, treat all plays as dual entities consisting of both text and performance. Throughout the book, I try to balance the need to place plays within their historical and scientific contexts with the recognition that works of art are, in an important sense, autonomous.[25] I am unable to go into the kind of depth I would like on some of the plays under discussion, each of which merits a fuller consideration—but I hope this will spark some interest for others to investigate further, especially finding more evidence of how these plays were acted.

In terms of the book's scope, I make no claims to complete coverage. Rather than a comprehensive survey, I offer interconnected studies of representative examples that can be looked at in some depth. Each example is suggestive of distinct currents and patterns in the way theatre has encountered evolution. The book is largely chronological, with the first three chapters centered on the nineteenth century, followed by three chapters primarily on the period 1900 to 1920, although inevitably there is some overlap and the borders between these "periods" are fluid. After a chapter looking at theatrical engagements with evolution from about 1920 to 1950, I conclude with a chapter on Beckett and with an epilogue summarizing the main tendencies of more recent evolution-themed plays and performances and taking stock of what the book has suggested is important in this kind of theatre. Just as period boundaries are fluid, so are geographical ones when it comes to ideas. For the sake of scope and coherence, I focus mainly on British and American drama, but I include relevant cross-cultural examples when they illuminate how theatre engaged evolution in a distinctive and important way. Such geographical spread is necessary because evolutionary thinking was international and because ideas cannot be simply contained.[26]

Elin Diamond notes that "artistic careers, when they are long and complex, continually demand new critical framing,"[27] and at the core of this book are several key playwrights: Henrik Ibsen, whose profound engagement with evolutionary theory is just beginning to be understood; Shaw, whose "Creative Evolution" provides an alternative (and, as Tamsen Wolff argues in her recent study of eugenics and theatre, not altogether desirable one) to natural selection; James A. Herne, sometimes called "America's Ibsen," whose lifelong interest in evolution manifests itself in his plays; Elizabeth Robins, actress and playwright whose work deals trenchantly with questions about gender that are central to human evolution; Glaspell, whose plays *The Verge* and *Inheritors* offer pioneering theatrical treatments of plant genetics signifying larger concerns about human evolution; Wilder, who brings evolution to the stage in a modern, Brechtian mode; and Beckett, whose theatrical work is a complex, sustained, and mutating engagement with evolutionary ideas. Dozens of other playwrights (both well-known and less-canonical ones) also feature in the discussion. Many of them presciently raised concerns about the negative impact of human activity on our climate well before the crisis of global warming. Yet the dominant narrative of theatre history still gives little space to such developments.

Theatre and Evolution builds on the momentum of the Darwin celebrations of 2009, which focused both scholarly and popular attention on evolutionary theory as never before and spawned a wealth of publications presenting new ideas about the man and his work, yet also narrowed the range of our understanding of evolutionary theory in concentrating so exclusively on Darwin and the "Darwin-o-centrism" of so much scholarship on nineteenth-century culture.[28] Ideas about evolution go back to Aristotle.[29] It is simply not the case that Darwin's "dangerous idea" exploded out of the blue onto an inherently conservative Victorian scene; it was one of many voices in the discourse on evolution, and certainly one of the most prominent, but by no means the only valid one. What was new through Darwin's work was the idea of descent with modification: species are not fixed but change through natural selection and sexual selection, the determining forces in the processes of evolution, with the irrefutable implications for humans that all the evidence from the plant and animal worlds showed. And it was "new" in the way it quickly became Social Darwinist. Goodall argues that Joseph Roach, Daniel Dennett, and others overstate Darwin's impact as a paradigm shift and a "dangerous idea," and she points to critics like Bram Dijkstra and Michael Ruse, who argue that far from being shocked and scandalized, Victorian society was predisposed toward evolutionary ideas.[30]

Darwinian evolution, argues Peter Bowler, was a "catalyst that helped to bring about the transition to an evolutionary viewpoint within an essentially non-Darwinian conceptual framework."[31] By *non-Darwinian*, he means "preserving and modernizing the old teleological view of things."[32] The rediscovery of Gregor Mendel's work in 1900 led to the founding of the study of genetics, but these developments took decades, and it is essential to recapture what people thought before the Modern Synthesis around 1930 when genetics finally confirmed natural selection as the key mechanism of evolution and showed how it worked, as doing this can "reveal many of the factors that still shape the nonscientist's image of evolution."[33] Theatre reveals this image as it was taking shape through a complex interaction of plays, performances, audiences, and critics. Lamarckism held a particular attraction for artists and writers well into the twentieth century: Spencerian ideas as they took shape in "social Darwinism" proved irresistible; Haeckel's mystical materialism as well as his recapitulation theory continued to garner support. In short, Darwin's was just one of a range of competing theories, and *Theatre and Evolution* shows the staying power of non-Darwinian ideas in the theatre long after their supposed demise in scientific circles. The so-called eclipse of Darwinism (from about the 1880s to the 1920s) coincides with some of the most strikingly original, creative, and profound theatrical engagement with key aspects of evolutionary thought. There is now an exciting timeliness to this endeavor, as the field of epigenetics suggests a possible validity to once-discredited ideas about what factors bring about evolutionary change.

So enduring is the "culture shock narrative" of Darwin's impact that even so careful and scholarly a work as Richard D. Altick's *The Shows of London* is conflicted about this, claiming that Darwin came like a "thunderbolt" into Victorian culture (a term echoed by Michael M. Chemers in *Staging Stigma*), yet also saying that well before *Origin* the concept of evolutionary theory was widely discussed and the foundations already laid for Darwin's ideas.[34] This is not just about getting the facts right about Darwin's impact; it is also about letting go of our stubborn fondness for stages, epochs, watersheds, and other forms of "'punctual or decisive change'" that characterize the Kuhnian "revolution narrative" of science.[35]

Two aspects of evolution have particular relevance for theatre. First, there is often a too-easy equation of evolution with degeneration, as if the two are synonymous. I do not see Ibsen, Beckett, and other playwrights treating evolution in this way; the picture is far more complex than that.[36] Although there certainly were fears across European cultures about reversion and regression, climaxing in the degeneration *furore* caused by Max Nordau in the 1890s, there was equally profound belief in progress emanating from a Darwinian worldview. Darwin once wrote in a letter to George Henry Lewes that "if we could collect all the forms which have ever lived, we shd have a clear gradation from some most simple beginning."[37] He was troubled by the idea of "brutal reversion" as he put it, but distanced himself from eugenic proposals like those of Francis Galton. In fact, Darwin often characterized evolution as progressive. *On the Origin of Species* ends with the claim that "all corporeal and mental endowments will tend to progress towards perfection," and his *Autobiography* asserts the belief that man "in the distant future will be a far more perfect creature than he now is."[38] This is in the teeth of all the evidence showing the blindness of nature and the randomness of natural selection. Natural selection "is not perfect in its action, but tends only to render each species as successful as possible in the battle for life with other species under wonderfully complex and changing circumstances."[39] Yes, there is much suffering in the world, but "if all the individuals of any species were habitually to suffer to an extreme degree they would neglect to propagate their kind"; this clearly is not the case.

Second, as knowledge about human evolution grew, it foregrounded gender and raised issues concerning the roles of women that had lasting societal impact. Bowler writes that sexual selection was still largely ignored by scientists in the 1950s, and that it "has become important for perhaps the first time since Darwin proposed the idea."[40] But theatre immediately took up sexual selection as a central and productive idea with great audience appeal. This is partly because of the conditions of theatre, whose raw materials are bodies and emotions, not just words in a script. Wilder once wrote that "a laugh at sex is a laugh at destiny. And the stage is peculiarly fitted to be its home. There *a* woman is so quickly All Woman."[41] This sense of what happens to the female body in the eyes of the spectator, what she could signify from an evolutionary perspective, underpins many of the plays discussed in this book and helps to explain why gender is at its core.

Darwin and other evolutionists were building on an internalized, unquestioned depiction of nature itself as female; animals and plants functioned according to gender stereotypes, got married, were seduced, and so on, in a system laid down by Linnaeus and Erasmus Darwin.[42] The miniature dramas within this conceptual framework replicated gender stereotypes of the submissive, passive female, the fallen woman, the heroic male. The Victorians inherited this tradition of seeing nature as gendered and anthropomorphizing the relations between the different sexes within the plant and animal kingdoms. Although of course to a certain extent a product of his age and its view of women, Darwin had great sympathy with the plight of women, not least because his father's experience as a physician gave him particular insight into women's lives: "He often remarked how many miserable wives he had known."[43]

Recent work has looked beyond the stereotypical views Darwin expresses about women's inferiority in *The Descent of Man* and shown how thoroughly his startlingly progressive theory undermines them.[44] Already at the turn of the century, Mona Caird and other women writers were quick to see how aptly suited Darwin's work was for feminism by freeing it up from the notion of gender essentialism to focus instead on the constant process of change and flux at the heart of evolution. We still refer to "Mother Nature." The animal kingdom's division based on those with mammary glands and those without illustrates this tendency to ascribe to other living things gendered roles that have more to do with social mores than with biology. This system, and the gender roles embedded within cultures, has been of primary concern to playwrights, many of whom directly challenged these assumptions in their works, in both mainstream and popular theatre.[45] In exploring gender, I pay particular attention to the actress in case studies of Robins, Eleanora Duse, and Thorndike that are spread throughout the book.

I have outlined what I attempt to do in this book; prior to moving on, I want to mention briefly some of the things I am not doing. Joseph Carroll, Brian Boyd, and others have argued that literature plays a key role in human evolution, claiming, for example, that storytelling makes us fundamentally human and distinct from animals.[46] One has to wonder why, as Seamus Perry puts it, we need to "ennoble" art by thinking of it as "an evolutionary advantage."[47] The literary Darwinists give scant attention to theatre, perhaps because it does not fit into their idea of storytelling, despite the fact that "theatre is so deeply embedded in human culture in some form or other it is probably a human universal," writes William Flesch.[48] Literary Darwinism reintroduces a teleological element to the discussion, as if we are constantly evolving toward ever-higher forms and literature merely assists in this process. It fails to account for a far more salient candidate for theatre's evolutionary "functionality": not the verbal element of drama, but its performance side.[49] In a generally nonnarrative form, interiority must be conveyed through other means—facial expression, voice, movement—putting the focus on what is around the character as much as what is inside. What happens when actors simulate and mimic, and what cognitive processes are engaged (e.g., empathy), have been the subject of much investigation lately, including areas such as mirror neurons.[50] This seems to me far more fruitful as it has little to do with design and teleological development. It also requires much more attention than I can give it in this book.

Likewise, the concept of the drama as a living organism that *itself* evolves, and its role in the evolution of human culture, is not a concern here. If anything, theatre consistently and deliberately flouts adaptation to its environment, for example in the "self-annihilating" productions of avant-garde theatre, with its short-lived experiments and fleeting groups that seem antievolutionary, self-destructing rather than self-preserving; to create new forms, theatre continually attacks itself, threatening its own extinction.[51] This would seem to go against the principles of evolution. Obviously, there are many other kinds of theatre that do survive and thrive, but theatre historiography still tends to privilege the subversive, often counterevolutionary and ephemeral productions that effected lasting change while themselves dying out. Many scholars have objected to this romanticizing of theatre as "primarily a site of subversion," so that the historiography is full of "narratives of resistance in which theatre triumphs over the social evils of conservatism and conformity."[52] This book intervenes in the debate through the selection of a range of plays that do not fit neatly into either category.

In this final section, I briefly sketch some of the main evolutionary ideas and thinkers referred to in this book. Given the wide availability of this information, this discussion is not intended to be comprehensive but simply to provide some necessary background for the rest of the book.

In the early part of the nineteenth century, William Paley's natural theology (his "argument from design") and the notion of the fixity of species prevailed. The only way to account for the complexity of organisms in nature and their fittedness to their environments was to assume a Creator who had made them. Prominent scientists like Robert Owen fiercely upheld these ideas even in the face of growing evidence to the contrary. In 1844, the sensational best seller *Vestiges of the Natural History of Creation* suggested close evolutionary links between humans and animals, but the author's natural theological position kept design in the picture.[53]

Vestiges opened the way for further evolutionary ideas to be disseminated and discussed and indeed raised a much greater outcry than *Origin of Species* (1859), which, if anything, helped to resolve rather than create a crisis, in James Secord's analysis.[54] In *Origin*, Darwin put forth his "long argument" for natural selection but avoided dealing with the question of human evolution. The connection between primates and humans had already been established, however; in 1832, in volume 2 of his *Principles of Geology*, Lyell had suggested the evolution of man from the orangutan, and the concept of ape ancestry made its way into the popular imagination through an array of satirical treatments in both print and performance.[55] It was not until *The Descent of Man* (1871) that Darwin addressed the issue of human evolution fully, and that book caused a much greater scandal than *Origin*, which on its first publication was actually well received and did not generate the kind of frenzied response one often supposes because of events like the famous debate between Thomas Henry Huxley and Bishop Samuel Wilberforce in Oxford in 1860.[56] As Bowler puts it: "The antiteleological aspects of Darwin's thinking prized by modern biologists were evaded or subverted by the majority of his contemporaries," by quietly assuming some degree of agency and purpose behind seemingly random adaptations.[57]

Darwin combined his own expert observations of the natural world with several key influences: Alexander von Humboldt's biogeographical approach, Charles Lyell's uniformitarianism as elucidated in *Principles of Geology*, and the work of Thomas Malthus on population control. Darwin got from Malthus the insight that "the tendency of any population to breed beyond the available food supply would result in a constant 'struggle for existence.'"[58] This, Darwin realized, meant that "those individuals best adapted to the environment would do better in the struggle than their rivals and would thus survive and transmit their characters to future generation."[59] From Lyell came the insight that species have been constantly changing and indeed being replaced by new ones at a steady pace "ever since the oldest fossiliferous rocks were formed, will surely continue in the future, and is, therefore, going on right now, though too slowly to be noticeable."[60] He also drew on Jean-Baptiste Lamarck's idea of the inheritance of acquired characters, one of the most popular and enduring ideas of the nineteenth century and well into the twentieth. Even now it is enjoying something of a return through epigenetics, showing that certain environmental influences can be passed on to subsequent generations.

The dominant model of evolution in Darwin's time was "developmental," put forth mainly by Herbert Spencer years before *Origin* and emphasizing the "orderly, goal-directed, and usually progressive character of evolution." This developmental model continued to dominate late nineteenth-century thought.[61] Spencer accepted the mechanism of natural selection, came up with the term *survival of the fittest*, and extended these concepts to humankind, but his "development" hypothesis emphasized progress through an ever-increasing heterogeneity; he supplied "some of the same reassurance" as "a divinely ordained teleology."[62] Spencer's ideas had a widespread impact in Britain and America in the late nineteenth century and the first few decades of the twentieth century, and many playwrights in my discussion were heavily influenced by him. Spencer was "the common man's Darwin, proffering the intellectual certainties of Darwinism on a cosmic scale, without Darwin's scientific rigor" to the extent that "his popularity soon outstripped Darwin's." Where Darwin appealed to the "harder, clearer intellects. . . . Spencer appealed to the multitudes."[63]

Darwin maintained that "new species originate in conformity, not to any law of adaptation to conditions nor to any law of progress, but rather to the law of common descent. Within any natural grouping of species there has been an irregularly branching adaptive divergence, in which new species have arisen from pre-existing varieties, the main agent in the conversion of varieties into species being a *natural selection of hereditary variants in the struggle for existence*."[64] Natural selection had nothing to do with animal will or with a Creator, with divine intent or design, or with a progression toward perfection; if species progressed they could just as easily decline.

It is common to think of Darwin's ideas supplanting those of Lamarck, who inherited the Enlightenment assumption of man as the pinnacle of life forms.[65] He believed that nature had a plan and a goal (teleology), yet also that nature goes through unpredictable changes induced by the environment. Lamarck emphasized adaptation (although he did not use this term) and thought of it as an interaction between organism and its circumstances. He was also uniformitarian and gradualist, against Cuvier's catastrophism; instead, he believed in the "gradual accumulation of small changes." In this, and in his opposition to natural theology's theory of design, he and Darwin think alike. Where they strongly diverge is in the degree of importance accorded to environmental factors and in the possible role of the will in evolution, although as a central plank in Lamarckism, articulated in Shaw's preface to *Back to Methuselah* (1921), this has been distorted by historians of science because Lamarck's discussion of will amounts to only seven pages, a tiny faction of his principal work, *Zoological Philosophy*.[66]

Lamarck is most famous for the idea of the "inheritance of acquired characters" by which environmental impact on an organism could effect changes that could be passed on (such as the familiar example of the giraffe getting a longer neck by successive generations reaching for higher and higher leaves). By this logic, one might expect changes wrought on an organism to be inherited, but in the late 1890s, August Weismann spectacularly discredited this idea by showing that chopping the tails off mice had no effect on the tail lengths of successive generations. But the inheritance of acquired characters retained an interest for many biologists, and Darwin himself accepted it to a certain degree, actually increasing the role he ascribed to it in his works over time rather than diminishing it.[67] The key point here is that because the mechanism of inheritance of acquired characters occurs in both Darwin and Lamarck, it is not accurate to describe a theory that accepts the inheritance of acquired characters as non-Darwinian.

Lamarckism continued to exert wide appeal for decades into the twentieth century (and in fact is now experiencing something of a vindication through epigenetics). Lamarckism gained ground because it could more easily accommodate religion, "since direction could be translated into theological terms as providence," and because it offered, "as Darwinism did not, firm scientific support for the efficacy of education in improving humanity . . . [and] muting the initial harshness of the Darwinian challenge to established truths and thus facilitating its general acceptance."[68] American scientists tended to combine the inheritance of acquired characters with natural selection in their approach to evolution.[69] This is true across other fields as well; Freud, who greatly admired Darwin, yet remained "an unrepentant Lamarckian to the end of his days" because natural selection, which depended so completely on chance, seemed not to support his own psychological theories.[70] Natural selection and the inheritance of acquired characters were complementary.

The idea of descent with modification through the mechanism of natural selection is identified as Darwin's, but it should really be called the Darwin–Wallace theory since they codiscovered it. Alfred Russel Wallace's manuscript formulating natural selection forced Darwin to rush his own findings into print after twenty years of hesitation, completing the manuscript of *On the Origin of Species* in less than a year.[71] George Beccaloni suggests that Wallace was eclipsed because, in the late nineteenth and early twentieth centuries, natural selection became unpopular as an explanation for evolutionary change; most biologists posited alternatives such as neo-Lamarckism, orthogenesis, or mutation. Wallace continued to make key contributions to evolutionary thought well into the twentieth century; he is sometimes neglected because of the more mystical tone of his later work.[72] Such neglect is similar to the fate of the German biologist Haeckel, whose groundbreaking work in morphology and embryology and as a disseminator of Darwin's ideas has been overshadowed.

More people learned about Darwin through Haeckel than through Darwin's own writings.[73] Haeckel's hugely successful book *The History of Creation* (1876), expressing a Lamarckian interpretation of evolution that did not support natural selection, continued to be reprinted until 1926. He never accepted natural selection as the sole explanation of the formation of new species. Instead, his Darwinism "concentrated on the struggle between rival species or subspecies."[74] Haeckel's reputation suffered from allegations of plagiarism in the use of illustrations in one of his books and more lingeringly from the erroneous and widespread notion that he was a proponent of scientific racism and that his ideas led to Nazism; this has been convincingly refuted by Robert Richards.[75] Throughout the last two decades of the nineteenth century and well into the twentieth, Haeckel was a highly public scientist, frequently lecturing to large audiences and being discussed at length in the press. His many contributions have sometimes been forgotten, such as his coining of the term *ecology* and his correct prediction of where to find one of the "missing links" of human evolution in Africa in the 1890s. He is perhaps best known now for his biogenetic law, "phylogeny recapitulates ontogeny," which posited that an individual organism's development from embryo to adult retraced the path of its genetic heritage in accelerated fashion.[76] He also argued that in species development, the lower forms of mental activity had always existed, while the higher forms were recent in evolution.

Haeckel's combination of evolution and recapitulation remained hugely popular and "powerfully convincing" for decades.[77] This idea fell from grace in the early twentieth century, challenged by genetics and the availability of increasingly powerful microscopes to test it, and in 1932 the English embryologist Gavin de Beer finally put the nail in the coffin of recapitulation in his scathing *Embryos and Ancestors*.[78] Haeckel also retained an "almost mystical belief in a purposeful nature."[79] This was articulated most fully in his theory of monism, as expounded in *The Riddle of the Universe* (1901), positing the unity of all matter in a single substance throughout the universe, rather than the dualism of matter and energy, mind and body. It was a modern version of an old idea, in effect uniting Benedict Spinoza with contemporary science to argue that the physical world and the human mind were contained within one system, and it proved popular. In 1905 Haeckel founded the Monist League, which continued until 1933 with chapters in many countries. Widely derided by scientists, monism nonetheless attracted a passionate following, particularly among artists, writers, and theatre-makers, and it was a deep influence on many of the playwrights discussed in this book.

What remained elusive throughout the nineteenth century, and helped to feed the artistic imagination as much as the scientific, was an adequate explanation for *how* the modifications brought about by the pressures of natural selection were actually passed on. One theory put forth was pangenesis, espoused by scientists from Darwin to Hugo de Vries, who published *Intracellular Pangenesis* in 1889. Pangenesis posited that the entire organism is involved in heredity, the parent transmitting "gemmules" (tiny units of heredity) to the offspring. This had a pronounced Lamarckian element, and it did not explain how certain traits were inherited. Weismann's theory of germ plasm was of definitive importance in positing a distinction between "germ cells," which contain heritable information, and "somatic cells," which do not but which carry out other, nonreproductive functions. The two kinds of cells operate separately in that the information passed through the germ line is not affected by the somatic line; a kind of wall exists between them, known as the Weismann barrier. The rediscovery of Mendel's work in 1900, made available through the efforts of de Vries and colleagues, confirmed Weismann's work and led to the establishment of modern genetics through the work of William Bateson, Thomas Hunt Morgan, and others. This was eventually combined with Darwinian natural selection and population studies to create "the Modern synthesis," announced by Julian Huxley in 1942 but worked on before then by many leading scientists.

One of these scientists, Ernst Mayr, also wanted to correct a misconception about the seeming randomness and blindness of nature. Survival, "the essence of the whole concept of natural selection," he writes, "the ability to contribute to the genetic content of the next generation, is not at all a matter of accident, but a statistically predictable property of the genotype." He claimed that science has abundantly demonstrated that "natural selection is a direction-giving force, within the limitations of the evolutionary potential set for a given species by its genotype," rendering absurd the "glib claim that Darwinism expounds the production of perfection by accident, the rule of 'higgledy-piggledy,' as Samuel Butler called it."[80] Butler, Shaw, and many of their contemporaries had trouble with this particular aspect of Darwinian thought, refusing to accept that nature leaves everything to chance, insisting on some degree of human agency. In *Origin*, Darwin refutes the notion of chance as a "false" view.[81] The urgent and ongoing sense of debate about this issue is an important part of the cultural landscape of this book.

Two other aspects of evolutionary theory relate specifically to the plays discussed in this book. One is the eugenics movement, which I discuss in chapter 5 and intermittently throughout the book as it forms an important part of the landscape of ideas with which theatre was engaging. Wolff's *Mendel's Theatre* is indispensable reading for anyone interested in the relationship between theatre and eugenics, and my work is indebted to her pioneering scholarship. Second, the interaction of theatre with evolution reflects a key shift brought about by *The Descent of Man* and *The Expression of the Emotions*, which together had the effect of focusing inordinately on animal development as synonymous with evolution—defining evolution mainly through the behavior of mammals and birds. These are, of course, most liable to anthropomorphism. Botany and marine and insect life receive short shrift in the popular reception of evolution from about the 1870s onward, even though it was through Darwin's work on barnacles, orchids, and earthworms; Mendel's experiments with peas; and Morgan's work on fruit flies that we eventually got to the modern understanding of evolution. This makes those rare plays that eschew furry mammals in favor of marine life or plants—like Hubert Henry Davies's *The Mollusc* or Glaspell's *The Verge* or *Inheritors*—all the more significant and give a more complex picture of how theatre has engaged with evolutionary ideas.

In later editions of *Origin*, Darwin wrote of theatricality as mere trickery: "The wonderful manner in which certain butterflies imitate, as first described by Mr. Bates, other and quite distinct species" is like acting, employing "mimicry" and "counterfeiting." He asked why "certain butterflies and moths so often assum[e] the dress of another and quite distinct form: Why, to the perplexity of naturalists, has nature *condescended to the tricks of the stage*?"[82] I want to use this damning view of theatre as a starting point to explore how evolution found its way into the public imagination through the stage's many "tricks," showing both the deep engagement between theatre and evolution and the distinctly questioning stance of the former toward the latter that begins to emerge and that becomes one of its defining characteristics.

1

"I'm Evolving!"

Birds, Beasts, and Parodies

Victorians lived in a cultural landscape in which "curiosity was a great leveler."[1] Every type of theatrical performance was on offer, from melodramas (and their even more spectacular cousins, the equestrian melodramas) to pantomime to music hall to fairground shows, circus, *tableux vivants*, and public exhibitions. These shows cut across class divisions and catered to all kinds of audiences, a thriving theatrical eclecticism that lasted the entire century.[2] The combination of entertainment and education was a particularly potent one in an age of reform and progress. New spaces like zoos and museums were opening up all the time where the public could spectate biology to its hearts' content. Spectating by its very nature implies a predominantly visual experience, and "Victorian audiences and scientific performers still regarded seeing as the paradigm of knowing."[3] Public lectures and scientific experiments that were so popular in the Victorian period presented science as something to gaze at and that can provoke admiration, awe, and wonder. Theatricality fed directly into the popularization of science; for example, Henry Neville Hutchinson wrote in *Extinct Monsters* (1892) that he was attempting to depict "the great earth-drama that has, from age to age, been enacted on the terrestrial stage, of which we behold the latest, but probably not the closing scenes."[4] Thus science, popular culture, and performance continually converged, from the crew of the *Beagle* giving "pantomime names" to the Fuegians whom Robert FitzRoy colonized in the 1830s (Fuegia Basket, York Minster, Jemmy Button, and Boat Memory) to the dramatic dissection of a lobster by Thomas H. Huxley in 1860 to show that adaptive modification had "unity of plan, diversity in execution."[5]

Theatre had a decisive role to play in the public taste for biology. Plays about everything from the "birds and the beasts" to physiognomy and galvanism testify to the public appetite for biologically based entertainment.[6] This extended well beyond Britain.[7] The period saw a growing conceptualization of human life and interactions, origins and future, as "biologized." Alfred Russel Wallace urged natural history museums "to design exhibits such as dioramas that would show the intricate relations between life forms in each region."[8] Peter Morton charts the impact of "the vital science" of biology on Victorian culture; Margot Norris explores "biocentrism" throughout the literature of the period; and Angelique Richardson notes the "increasingly biologized" discourse of the late nineteenth century around the maternal aspect of femininity.[9] Highlighting this biocentric appetite can help to dismantle some of the myths about the Victorians. For example, in the enduring cliché of a society "obsessed with respectability and therefore gratifyingly shockable, fixed in an archaic worldview preached to them from the pulpits of every church," we have sustained a "simplistic fiction" that obscures "the complex social and cultural dynamics" of the period.[10] Looking at how theatre engaged with evolutionary ideas confirms this and helps to bring back some of those complex dynamics that have been obscured.

The key point about theatre's engagement with evolution in this period is its extraordinary freedom, described by Jane Goodall as gleefully exploding the "parameters of enquiry" that science so strictly controlled and exploring scientific ideas "without a map, exercising the freedom to invent as well as observe," showing an "eager receptiveness to new ideas from the realms of science, a fascination with their implications and an alertness to changing directions of speculation."[11] Evolution was just one of many scientific domains that caught the popular imagination, including medicine, astronomy, physics, chemistry, microbiology, psychology, anthropology, and geology. But because of its particular relevance for human life and for the lives of animals as well, not to mention its questioning of religion, it went straight to the heart of concerns shared by many audience members across all sections of the theatre-going public. For example, many Victorian writers and artists worried about where Darwinian evolution left the individual will or effort and what happened to an overall sense of purpose. They often turned to a more comforting "Lamarckian auto-teleology operating at the level of the individual organism (shades of Samuel Smiles's *Self Help*, 1859), and some overriding principle or force to provide an assured and universal upward thrust."[12] Even those writers and artists attracted to Charles Darwin's ideas and accepting of natural selection had a difficult time with the often-"crass" nature of evolution "which gives no assurance that the *worthiest* shall survive."[13]

This is the crux of the development of naturalism in the late nineteenth century, which profoundly influenced theatre as well as literary forms like the novel. Naturalism gave expression to the key concern raised by evolutionary thought: What shapes us as human beings? Is there anything we can do to push against the formative influences identified by the Positivists as race, milieu, and moment? Are we completely determined by our environment? In his seminal *Le Naturalisme au theatre* (1878), Emile Zola crystallized the assumptions of Positivism and called for theatre that would unflinchingly show the truth and reality of life. It was almost inevitable that this would make naturalism synonymous with the underbelly of life, dramatizing how the seamier, more primitive aspects of human nature always get the better of us, an idea epitomized in plays as various as August Strindberg's *Miss Julie* (1888), Gerhart Hauptmann's *Before Sunrise* (1889) and *The Weavers* (1892), and Maxim Gorky's *The Lower Depths* (1901). Theatre lent itself perfectly to this enterprise, capturing the shaping power of the environment on the individual by being, in effect, a controlled "experiment" (Zola's own term for what he was trying to do in his stage adaptation in 1873 of *Thérèse Raquin*): presenting specially constructed environments and revealing their effects on a few individuals. His naturalist manifesto proclaims this desire to "bring the sweeping movements of truth and experimental science to the theatre," building on the impetus given by "the new scientific methods . . . subjecting man and his actions to exact analysis, concerning itself with social circumstances, milieu and physiology."[14] This, combined with the rapidly developing field of psychology, led to a deep interest in depicting the darker forces unconsciously shaping our lives, impulses often beyond our control and with intensely theatrical potential for new depths of characterization.

Naturalism is thus of central importance to my explorations in this chapter of specific plays, playwrights, and scenography. But my discussion also builds on, and complements, the existing scholarship on performance modes *outside* mainstream theatre, such as freak shows and exhibits of commercially displayed peoples in the newly founded zoological gardens of cities like Paris, London, Berlin, and Copenhagen.[15] These exhibits of exotic "natives" were often used by anthropologists to take biometric measurements that would taxonomize the different races, giving scientific credence to sometimes-spurious methodologies.[16] Rae Beth Gordon notes that in 1877, Africans were installed in the Jardin d'acclimatation in "villages nègres," exhibited behind a fence.[17] Nigel Rothfels discusses the hugely popular exhibition of Fuegians in Paris and Berlin in 1881, quoting one contemporary's observations about how the crowd of spectators responded to this practically naked group appearing in "such complete naturalness that they had first to be given bathing suits": for the Parisians, "precisely this abundance of naturalness met with all the more approval. The draw there was . . . massive, and it was just as large in Berlin where the Fuegians came next, for there the public flattened the barricades and it was necessary to retain security guards to keep order."[18] The attendance figures for these shows are staggering; more than fifty thousand people visited the Paris exhibit on one Sunday, and a special stage had to be erected for the Berlin show to protect the Fuegians from the wild "rush of the public."[19] Yet what they came to see was ordinary domestic life, not drama. People seemed satisfied with "simply gazing" at these "primitive" people.[20]

Consider this in relation to Darwin's passing comment in his *Autobiography* that "the sight of a naked savage in his native land is an event which can never be forgotten."[21] This comment, made just a few years after such exhibits, contrasts the "real thing" with its tawdry imitation; although the people on display may be real, their environment is simulated, so the audience cannot possibly feel what the reality of those on display would actually be like. These hugely popular human exhibits of so-called savagery coincided with the development of theatrical naturalism ushered in by Zola and given fullest expression through André Antoine's productions at the Théâtre Libre in the 1880s. Antoine's production of *Ghosts* was in the same year as the Fuegians' exhibit in Paris. Indeed, theatres themselves sometimes served as exhibition sites: in 1908, the Théâtre Antoine mounted an exhibit of skulls representing human evolution to coincide with the discovery of a new missing link.[22]

In another performance-related development, the popularization of science through such diverse means as geological panoramas and the public lectures of Huxley and other scientists shows theatre and science borrowing from each other.[23] Peter J. Bowler points out that "by the 1870s, the reconstruction of life's ancestry had become a major scientific industry, which in turn became the layperson's image of what evolutionism is all about."[24] The performing-science strand of theatrical engagements with evolution contrasts with the imaginative engagements of playwrights and directors and may indeed help to explain their contrariness, their outright hostility to science at times. These two developments are, I believe, closely connected, the one feeding into the other and radically changing the very concept of nature in the audience's imagination.

As all these examples show, the public performance of science was overwhelmingly orchestrated and carried out by men. The scientist doing a public experiment or giving a lecture was "the face of the idealized Victorian public gentleman"; hence, Victorian audiences would have identified successful scientific performance as "unambiguously male."[25] (Notable exceptions to this include Sonya Kovalevskaya and Clémence Royer, both of whom gave public lectures on science.) That scientific performances tended to be male reinforced the alignment of science with masculinity. By contrast, the actress became, in Goodall's terms, the most provocative signifier of evolution by the sheer suggestive power of her body on stage, often questioning rather than reinforcing gender assumptions and roles. My discussion of nineteenth-century theatre and evolution therefore concludes, in chapter 2, with a brief look at the phenomenon of the female actor in relation to evolution. This is in illuminating contrast to the focus here on two key issues: the stage's engagement with the abstract notion of geological "deep time" and concomitantly an increasing interest in the inner lives of animals.

Natural History, Deep Time, and Wild Nature on Stage

It was difficult to comprehend a gradualism that even scientists sometimes rendered in hundreds rather than millions of years.[26] If the stage could not literally show such slow processes, it could at least suggest them, and it seems reasonable to speculate that Victorian theatrical spectacle built on the new interest in geology brought about by Charles Lyell's *Principles of Geology* in the 1830s, finding ways to suggest "deep time" through breathtaking landscapes and natural disasters.[27] Even before Lyell, one of earliest theatrical engagements with evolution was Lord (George Gordon) Byron's *Cain* (1821), "the original paleontological drama," which shows the influence of contemporary thinking about geology.[28] It is an important template for the kind of epic treatment of evolutionary history on stage that Thornton Wilder later attempts in *The Skin of Our Teeth* (discussed in chapter 7). Byron has Lucifer telling Cain that he will take him on a journey to show him "the history / Of past, and present, and of future worlds."[29] This play was regarded as a closet drama and remains unperformed, despite developments in staging that could have enabled production. Large-scale venues and sophisticated machinery could display vast natural tableaux, registering an increasing curiosity about the natural world and the inherent drama of small humans pitted against great natural forces. Productions frequently re-created waterfalls, rivers, lakes, frozen ice floes (one of the hallmarks of Harriet Beecher Stowe's *Uncle Tom's Cabin* in the 1850s), even watery caves (Dion Boucicault's *The Colleen Bawn*, 1860). The popular equestrian melodrama *The Cataract of the Ganges* in the 1820s, for example, offered stunning scenes that included stampeding horses, burning forests, and vertiginous precipices for the heroine to (nearly) fall off. This may well have been the kind of show Darwin took his cousins to see at Astley's in 1838, one of his rare references to theatregoing.[30] Wild nature was very much present in theatrical forms that lay outside the interiors of domestic drama.

Alongside this interest in conceptualizing wild nature and geology as drama came an interest in depicting animals, not merely as decorative objects but as characters; one of the pioneers of ethology, Charles George Leroy, wished to have "the complete biography of every animal."[31] In 1859, Étienne Geoffroy Saint-Hilaire re-created ethology as "the branch of biology devoted to the study of animals in their natural habitats."[32] The field became increasingly defined by the belief that "animals possess specific, innate 'characters' which can be understood, often by direct analogy with human character, on the basis of prolonged and sympathetic observation."[33] The idea that animals could have characters that might be analogous to human personalities, that they might have both an inner life and an existence worthy of being narrated, further links evolution to the stage because "animals were seen as actors in their own social worlds, and indeed the analogy with drama was extremely common," with one ethologist describing pigeons' sounds and gestures before a fight as an elaborate "pantomime" and another likening bird-watching to being a spectator at a play.

This analogy with drama gave rise to new imaginative possibilities by directing attention to what specific behaviors might signify or mean. In 1901, one ethologist referred to "the wonderful drama . . . of bird-life" that seemed almost comical in its heightened formality, a key connection here to the dramatic art of mime. How far were animals' ritualized gestures and acts of expression giving "insight into the feelings of the performer"?[34] Darwin's *The Descent of Man*, especially its descriptions of bird behavior, foregrounded "the drama of . . . birds' lives," how evocative their actions were, how charged their courtship rites. Julian Huxley "came to see the elaborate movements involved in the birds' lengthy 'courtship' as the gestures and postures of two performers caught up in some sort of evolutionary play," whose "plot" Huxley was trying to discern.[35] Indeed, so taken was Huxley with his bird-watching that he rhapsodized it as "the foundation of a real science . . . [and also] a sport" whose participants will find themselves "not only experiencing the sportsman's thrill, but also the intellectual interest of the detective piecing together the broken chain of evidence, and the human feelings of a spectator at the play."[36]

Birds are hardly the same as humans, of course, but their behavior as observed and catalogued in *The Descent of Man* presented an irresistible example for many of Darwin's contemporaries. Huxley discerned "a rudimentary 'ethical process' in the social systems of birds and animals which formed the starting point for moral evolution."[37] In *Man and Superman*, Bernard Shaw uses them to make a point about human evolution: "The birds, as our friend Aristophanes long ago pointed out, are so extraordinarily superior, with their power of flight and their lovely plumage, and, may I add, the touching poetry of their loves and nestings." Why then, he asks, would Life, "having once produced them . . . start off on another line and labor at the clumsy elephant and the hideous ape, whose grandchildren we are?"[38]

Anthropomorphizing animals as performers with individualized characters is a long-standing tradition, and a potent example comes from the parade of a cameleopard, one of two baby giraffes given as gifts by Egypt to the king of France and the king of England.[39] The *Morning Post* of London gives a clear indication that its arrival in Paris was a sensationally choreographed public performance: "It was accompanied by an escort of twenty-five gens-d'armes. . . . A wagon, containing several other animals sent by the Pacha to the King, preceded the cortege, in which we noticed M. Geoffrey [*sic*] Saint-Hilaire, who, forgetting the care of his health in his anxiety for the interests of science, had constantly accompanied it till he approached within a few leagues of Paris."[40]

Note the presence of a celebrity scientist, the esteemed zoologist and founder of ethology, lending the performance the authority of science. The cameleopard was all the more breathtaking because it was the first living specimen the public had seen of a giraffe—hence the pomp surrounding its arrival and the almost-loving descriptions of "its elegant head . . . its long neck gracefully [rising] . . . its well set large black eye filled with mildness and joy." The anthropomorphism is notable: this is a superior animal because it is so wonderfully endowed with the best *human-like* qualities (elegance, grace, rising above everyone, well-set eye beaming with benevolence). The reporter notes that since its arrival "more than 10,000 persons have been to view it."[41] Another account confirms this anthropomorphized sense of benevolence and good will: "It looks upon the crowds that press forward to admire it with pleasure and tranquility."[42] Such vivid reports were fodder for the theatre. A few months later, in October 1827, an extravaganza called *The Giraffe; or, The Cameleopard* was performed at Sadler's Wells. Theatrical interest was fueled by constant commentary in the press keeping the public, not only in London but all over Britain, informed about the welfare of the giraffe.

An entirely different kind of show featured humans impersonating animals, typified by the pantomime *Birds, Beasts and Fishes; or, Harlequin and Natural History* (1854), which involved a range of animals including bear, salmon, trout, jackal, and lion interacting in entertaining ways, controlled by Dame Nature and culminating in a spectacular scene in the Zoological Gardens.[43] Although it might seem a far cry from instances like the cameleopard of public encounters with real animals, the show builds on the popularity of the newly opened zoos and natural history museums in this period, feeding the public curiosity about animal-human interaction and the perception of fluid boundaries between them.[44] The final scene of *Birds, Beasts and Fishes* was a staggering transformation: the entire width and breadth of the stage was made into an aviary and bowers with a large square cage right in the middle, "magnificently gilt and covered with what represents gems of every hue" as well as bejeweled peacocks' wings, all set against a backdrop of gold. The scene ("which it is utterly impossible to do justice to in description," writes one reviewer) progressed with constant, enchanting transformations, each one meeting with a "volcanic shock of applause."[45] The *Era* wrote that "parents and guardians can pat their youngsters on the head, and take them to see the City Pantomime, if only to improve them in natural history."[46]

Nineteenth-century theatre (particularly the "illegitimate" stages, such as in Buffalo Bill's hugely popular Wild West shows) made regular use of dogs, horses, and other live animals. American playwright James A. Herne partly paid the bills for loss-making productions like *Margaret Fleming* through popular shows like *The Country Circus*, succinctly described by his biographer as "rot" that drew hordes of people to see it on Broadway in 1891. This "menagerie of dogs, cats, baboons, and monkeys"—which can be seen as "an early example of environmental theatre"—drew ladies in evening dress, evoking the well-to-do spectators of ethnological exhibits so popular a few years earlier.[47] Iwan Morus emphasizes the active, constitutive role of the audience at such events and their resistance to a reduction "to a common type. If they were really all the same, then they are neither interesting nor particularly helpful as explanatory tools. Audiences constituted themselves as the public through their active participation in performances. They went there avidly to see and be seen as part of the same publicly declarative act of belonging."[48]

Freaks and Missing Links

This declarative act extended to the freak show. Victorians were obsessed with freakery, "human curiosities," and the idea that there might be living "missing links" representing intermediate species on the path from early to modern humans.[49] Even Darwin's pioneering use of photographs of the mentally disturbed in his book *Expression of Emotions in Man and Animals* suggests their freakishness and alienation from the reader. Normally taboo themes like monstrosity, sex, violence, insanity, and degeneracy were made more respectable when cast in evolutionary discourse.[50] "Nearly every critic writing on freaks" has pointed to the Victorian period as "central in the establishment of freak shows and in the evolving understanding of 'freaks' as a social construct. Indeed, it was in 1847 that the term developed its contemporary association with human anomaly," write Marlene Tromp and Karyn Valerius.[51] Yet Heather McHold argues that the focus has been too much on early and midcentury freakery; "modern European historians have largely ignored the cultural history of late-Victorian freak shows," and in fact far from losing interest, "audiences continued to flock to freak shows well into the 1890s."[52]

The Hottentot Venus exemplifies the fluid boundaries between modes of performance in this period; she was both a "freak show" and an anthropological display, an important early example of the phenomenon of "commercially displayed peoples." Posters for shows such as those at 6 Leicester Square regularly advertised displays like "Men Monkies" and "The Earthmen" (three-and-a-half-foot-tall men "who burrow under the earth, subsisting upon insects and plants"). Female physical extremity like grotesquely large buttocks (steatopygia) or excessive facial hair was a particularly big audience draw, and not only in Britain: Gordon convincingly documents an aesthetic of ugliness as central to the Parisian avant-garde and intricately linked to a fascination with the pathological implications of human evolution.[53] Likewise the ape-like Krao (a hirsute girl touted as "Living Proof of Darwin's Theory of the Descent of Man") and Jocko; at the time, there were even multiple versions or spin-offs of these original shows.[54] Hirsute women in particular seemed most powerfully to embody this hybridization, as the popularity of the famous bearded woman Julie Pastrana in America in the 1850s attests.[55]

The question of human origins, especially after publication of *The Descent of Man*, was explored on stage in such performances of human anomalies and missing links, often in comic mode and in ways that reflect and engage with the hot pursuit of fossil evidence for missing links in real life. The sensational discovery of *Pithecanthropus erectus* may partially explain the revival in the 1890s of interest in staging plays dealing with transitional forms, as in the farce *The Missing Link* (1893), which revolves around the simple and well-worn stage device of a man in a chimp suit, disguising himself and fooling an eminent scientist and then winning the scientist's daughter.[56] The play opens with a blood-curdling scream; the servant Robert staggers on stage holding a bloody kitchen knife and tells Maude that he has just killed a big monkey belonging to Maude's father, the eminent zoologist Professor Boggles, because it has bitten him. She says that her father will be very angry, as he has had that monkey specially sent from Central Africa. Robert goes off to bury the monkey in the garden, and the predictable farce of mistaken identities unfolds when Maude's lover, Frank, of whom the professor disapproves, dons a monkey suit and pretends to be the newly arrived monkey, Pongo.

This trifling piece embeds some profound questions about evolution, race, humanity, and the concept of equality within its farcical format. To begin, the line between humans and apes constantly breaks down as the play questions who really is the ape. Professor Boggles is "a scientific old gentleman of a decidedly monkeyish type of countenance." Boggles soliloquizes his gleeful anticipation of meeting Pongo and reads aloud the letter describing him as

> *'a most wonderful ape, just captured in the virgin forest here—he may prove to be the missing link between man and monkey.' What a chance for me. In my latest book I contended that the ancient Britons believed that Adam was an Ourang-Outang and Eve a baboon. I shall welcome this Pongo with delight and watch his simian intellect as it buds into manhood. Who knows? He is perhaps a relation of mine! Perhaps the same blood flows in our respective veins! Oh! Science! Science!*[57]

The "same blood" shared by man and monkey is a persistent theme in the play, and one that Boggles puts to Maude: "Tell me child, candidly—how would you like to have a chimpanzee for a relative? . . . or a baboon, or an ourang-outang, a wild man of the woods. Call them what you will they are our poor relations." Then the farce takes this scandalous idea further with the shocking suggestion of interspecies sex. Boggles sees the monkey (Frank) at Maude's feet; he asks whether the Missing Link has spoken, and she says that yes, he has professed his love for her. In a hilarious tête-à-tête with the monkey, Boggles tries to dissuade him from marrying Maude, using arguments like, "You understand that I cannot reasonably entertain the idea of hearing myself called grandfather by a brood of little apes! No!" It is a fascinating example of theatre's elasticity: the same idea framed in a more serious mode (as Edward Albee does in his play *The Goat; or, Who Is Sylvia?*) becomes taboo.

The concept of the missing link was also used rather cleverly by feminists. In her book *Facts and Fictions of Life* (1893), the American campaigner and popular lecturer Helen Hamilton Gardener argues that "the marks of the human race, and the real physical characteristics which distinguish us from the animals, are feminine peculiarities." She says that men have always regarded women as a kind of "missing link"[58] between them and the animals, but a physiologist in Germany suggests it

> is the other way around—comparisons of the bone structure of the pelvis in men, women, and gorillas indicate that "the pelvis of woman is a new type which has appeared on the earth. Until now we have sought in vain for that animal which shall complete the chain between us and animals. It is striking: the narrow, high pelvis of the man is more ape-like than that of the woman. If the assertion is correct that the upright gait (on two feet) is the mark of distinction, and the noblest one for man, then woman certainly possesses the advantage of a pelvis particularly suitable for upright walking.[59]

Deflating Darwinism

In Shaw's *Man and Superman* (1903), Mrs. Whitefield complains that "it's a very queer world. . . . Nothing has been right since that speech that Professor Tyndall made at Belfast."[60] John Tyndall's address in Belfast in 1874 represented a watershed for the way science was perceived within broader culture, and theatre registered the shift well before Shaw's reference to it. Since Huxley's debate in Oxford with Samuel Wilberforce in 1860, there had been an entente cordiale between science and religion, and science had enjoyed a rise in cultural position. But Tyndall's address came after other shocks to the system, such as Darwin's *Descent of Man* (1871) and the passage of the Education Act of 1870.

The result was that Tyndall was attacked: following his address, he was seen as an "aggressive, dishonest, devious, and distinctly un-British materialist. Even in the 1870s, the charge of materialism was a serious one. It grouped Tyndall together with lower-class atheists, casting aspersions on his status as a member of the intellectual elite. Moreover, Tyndall became a symbol of everything that was wrong with modern science and scientists in general."[61]

The Christian establishment seized this opportunity to limit scientists' cultural role and to mount "a sweeping indictment of modern scientific culture and of the cultural organs which had facilitated the spread of scientific materialism."[62] In terms of longer-term repercussions, Tyndall's address caused discussion in the periodical press about "how scientists created an illusion of authority"; some writers felt that "scientists were too prone to lavish praise on one another, a strategy that helped bolster their authority. But to these critics, it merely revealed the superficial nature of modern science."[63] They noted Tyndall's effusive praise of Darwin as an example.

This shift in the public perception of science meant that alongside theatrical productions that seemed to celebrate the richness of the natural world, or the commercially displayed peoples exhibits that traded on the authority of science to justify their human exploitation, there were counter-evolutionary forces in the popular theatre and these were developing fast.

Based on Alfred, Lord Tennyson's *Princess* (1847), William S. Gilbert and Arthur Sullivan's *Princess Ida* (1884)—in which male wooers dress up as women to infiltrate an all-female college—ridicules the idea that humans are descended from primates. Song number 15 (sung by Lady Psyche) follows Gilbert and Sullivan's tried-and-true formula of putting topical references into their works, here lampooning Darwinism. It tells how "A Lady Fair, of lineage high / Was loved by an Ape, in the days gone by" and has the refrain "The Ape was a most unsightly one." The Ape's efforts to woo the lady by buying fine clothes and by shaving, docking his tail, bathing, and other attentions to his personal toilet having failed, he "christen'd himself Darwinian Man!" But the maiden, "with a brain farseeing," saw through this: "Darwinian Man, though well-behav'd, / At best is only a monkey shav'd!" The song taps (and simultaneously feeds) the same vein as "missing link" shows and *Punch* cartoons lampooning man's lowly origins. The song (and the opera as a whole) skewers both feminism and Darwinism. The Rabelaisian atmosphere of cross-dressing and gender-bending is firmly dissolved through the restoration in the end of the status quo: marriages all around, women put firmly in their place, and humans and animals decidedly separate.

Carolyn Williams notes the central role this type of song plays in the Gilbert and Sullivan repertoire. Like the fable of "The Magnet and the Churn" in *Patience* or "The Junction Song" in *Thespis*, the "Darwinian Man" song in *Princess Ida* is a "crux," "a little embedded allegory of the work as a whole."[64] In *Princess Ida*, an opera with an "evolutionary argument," the Darwinian Man song provides an "interspecies scenario" by posing an explicit question: "Can men imitate (and therefore learn to be) women? Can the lower classes 'rise in the social scale' by dressing up and imitating their betters? The song answers these questions in the negative; in this little fable, biological essence trumps socialization. Gender and class are biologically determined," because as the song puts it, the Ape remained "the apiest Ape that ever was seen."[65] It is a "monstrous error" for an ape to love a fair maiden; the natural order must not be disturbed.[66] Whereas in Tennyson's *Princess* "culture emerges from nature to drive further evolution," notes Williams, in *Princess Ida*, "nature is all."[67]

Tennyson's poem was, as Rebecca Stott points out, an "experiment with conversational and narrative form, an attempt to use the poem not only to tell a story, but in this case to dramatize a series of contemporary debates about politics, gender and education."[68] The poem is full of radical ideas about the future of women in particular, seeing evolution as "politically utopian for the Princess, . . . [who is] convinced that natural laws enhanced by social regulations and human laws will ensure the progressive improvement of the race."[69] Progress, hopeful endings, improvement all around: this is exactly what people were seeing in popular theatre as well. And the poem also delivers sex, as it "eroticizes change." If everything on earth is perpetually, slowly transforming, if "nothing is static or fixed, if slippage is the state of nature, then sex is the means by which new future comes into being."[70] Gilbert and Sullivan defang this radical idea through their ridicule and demystification of evolutionary science generally. The operetta, which incorporates conventions of pantomime such as cross-dressing, music, and the glorious restoration of the status quo, also has fun with the idea of instinct. Psyche, Melissa, and the three young men sing a song with lines like:

My natural instinct teaches me
(And instinct is important, O!)
You're everything you ought to be,
And nothing that you oughtn't, O!

In this song, they celebrate the biological naturalness of heterosexuality and indeed "the power of biological sex to trump any social attempt to disguise or deny it." The opera stages a debate between nature and culture in which nature triumphs.[71] Everything is predetermined; we may cross-dress and behave in ways that defy our gender, but it is all in vain—nature will win out. This completely misconstrues the relationship between Darwinism and essentialism, missing entirely the way in which, as Angelique Richardson has shown, Darwin's theory refutes the idea of fixity and essence.[72]

Parodying Evolution

Robert Buchanan's *The Charlatan*, which played for a full two months at the Haymarket in 1894, is a rare example of a play that uses the word *evolution* specifically in relation to contemporary science to make a satirical point. It therefore provides valuable evidence of a particular use or interpretation of evolution as it entered popular culture. In *The Charlatan*—described by one reviewer as "a most unworthy rubbish"[73]—Buchanan uses evolution as a source of fun; there is no attempt at intellectual engagement, no scientific content, and the completely superfluous use of the term indicates its faddishness (similar to the "Darwinian Man" song in *Princess Ida*). The self-absorbed, decadent Mervyn Darrell's self-proclaimed "evolution" becomes a running gag, with constant variations on his catchphrase "I'm evolving," puncturing a weighty and controversial scientific concept by attaching it to the most vacuous character in the play.

Darrell embodies the modern spirit (he embraces every new trend, from Aestheticism to Radicalism) and is drawn as comical for that and for his utter self-obsession. His cousin Lottie says that "self is your study," and he proclaims it "a great, indeed the only one. How to evolve. How to be!" She then asks if he is "evolving at present?" and says it is surely "a dreadfully uncomfortable process." When he proclaims that he does not believe in society, morality, duty, respectability, or any other "bourgeois superstition," but only in himself and "in my right to evolve in my own way," she remarks, "Isn't that a little selfish?" He replies, "Certainly. Self is the only reality!" Buchanan explicitly links Darwinism to lack of humor, gloominess, morbidity, and total self-absorption. Above all, it is boring and exasperating for Lottie, who sets about to change Darrell. "Just look at him! Do you know what he's doing? He's—he's evolving!" She tells him to "go away. . . . Anywhere. Go, and evolve."[74]

This subplot becomes more interesting when he proposes marriage to her because she is his second cousin (i.e., they already are connected, so that is good for "posterity"—an unusual interpretation of the workings of heredity) and because "you are nice-looking." He says, "Let us forget all the other phenomena of existence and evolve together." She says no because he has never done anything with his hands—painted a picture, written a book—and besides, "You're dreadfully fond of somebody else," and holds a mirror before him saying: "Adore it—look how it's evolving." In act 4, Darrell presents Lottie with a nosegay he has gathered but then notices a snail in it, and says as he removes it: "What a contradiction Nature is! Always by the side of the Beautiful, she places the Execrable!" He and Lottie then talk and he confesses to a fighting streak—once he had a fight with a bargeman in Oxford that gave him a black eye for a week: "Between ourselves, I don't get on with some of the culture apostles at the university, and sometimes—don't despise me—I almost think that another set to, such as I had with the bargeman, might do me good." Darrell, it turns out, is actually virile, and Lottie immediately says that she would marry him "if I really thought you a *man*." As he lifts her up in his arms, she exclaims admiringly: "How strong you are!"

Why bother with this "rubbish" play at all? Its dialogue attests to the extent to which evolutionary terminology and discourse could figure in such a wide range of theatrical modes, even at the lightweight and flimsy end of the spectrum. It shows the degree to which the word *evolution* had accumulated specific semantic meaning and cultural resonance. It also shows how playwrights could be highly selective in their engagement with evolution. Buchanan uses "evolution" as a comic handle and a means to sketch character with a broad brush, but the very fact that he can do this indicates the potency of associations the term held by this time. Buchanan is not just parodying the effeminate Aesthete and affirming the manly man (with a little help from Lottie, who does the proposing). The context is self-consciously the theory of evolution, with female sexual selection as the defining evolutionary force, and the women (and everyone else looking on) are in the helpless grip of it.

This was not in fact Buchanan's first effort to incorporate evolution into his theatrical work. In September 1890 he had (with Fred Horner) adapted Alphonse Daudet's *La Lutte pour l'existence* (*The Struggle for Life*, 1889) for production at the Avenue Theatre with George Alexander as Paul Astier. Daudet's play had just been brought to London in a production in June that year, so audiences and critics would have had either direct or indirect exposure to its central premise and its characters. It received mixed reviews. The *Pall Mall Gazette* flatly pronounced it "directly at variance with the tastes of the public," but the *Times* critic praised it as an improvement on Daudet's "unfortunate" original because the adapters "have done him the signal service of cutting out his 'Darwinism,' root and branch," leaving a "somewhat gloomy, but quite unpretentious," drawing-room tragedy without the "Darwinian counterblast."[75] The critic informs us that Darwin's name is not even mentioned in the dialogue and that, "relieved of its sham science, the piece is now at least a perfectly inoffensive production." The critic then cleverly analyzes the play as a specimen of evolution. Its title, for example, is "merely the trace of a disused faculty which has survived the evolutionary processes of adaptation; it is the rudimentary tail, so to speak, which serves to remind us of the morphological changes effected in M. Daudet's original scheme."[76] The main character, Paul Astier, is a "monster."[77] An "easy, cynical, polished heartless man of the world," he represents the "heartlessness" of Darwinism, with his "detestable, but quite incontrovertible, doctrine that the weak must give way to the strong."[78] One critic felt that Buchanan and Horner substantially improved on the original when they turned this notion on its head, refuting the "theory that the strong always destroy the weak, the latter sometimes in their turn rising in self-defence and destroying the strong."[79]

Clearly, the critical responses to this adaptation of Daudet's play provide valuable evidence of how evolution was popularly understood during this period, and two things in particular stand out. One is the way in which "survival of the fittest" became reductively, imprecisely used and indiscriminately applied. For example, it becomes a drama of willed and cruel destruction of the weak by the strong rather than an impartial act of adaptation. The idea of the weak going to the wall does not necessarily mean that they are being literally destroyed by the strong; they could just naturally be weeded out, superseded, die. The other thing that stands out in these reviews is the tendency to ascribe purpose and agency to the forces of nature. One critic says that "instead of the 'Survival of the Fittest,' we have the triumph of the basest."[80] The repeated emphasis on "heartlessness" and baseness in the context of Darwinism is especially revealing, indicating that in the popular conception of evolution, the mechanism of natural selection was understood as *purposefully* cruel, not just blindly and randomly so. It shows how common it was for those writing about evolution to slip into anthropomorphism: Nature is "heartless" and cruel; "Nature never forgives."[81]

2

Confronting the Serious Side

More serious attempts to incorporate evolution thematically into plays were taking place, and many of these hinge on the relationship of people to their environments. Downing Cless has argued that by the mid-Victorian period, with the development of domestic drama, settings move indoors: "Nature does not disappear, . . . but it is distanced—what's outside the window or what's down the stream."[1] While appealing in its simplicity, there are significant exceptions to this claim. Henrik Ibsen's plays do feature people talking intensely in rooms, but they also emphasize and indeed rely on their natural settings (*Ghosts* with its ceaseless rain and remote fjord just outside the big picture window, the mountainside setting of *Little Eyolf*, the apocalyptic avalanches of *John Gabriel Borkman* and *When We Dead Awaken*). As will be shown in chapter 3 in my discussion of Ibsen's engagement with evolution, these environments directly shape the action; they are not just "down the stream." One of the playwrights who exemplifies this emphasis on environment is James A. Herne, who, influenced by Ibsen, brought nature even more directly on stage, often in astonishing ways.

Shore Acres (1893)—"the play that made a million dollars"[2]—requires a live horse on stage and (though not simultaneously) a raging storm that takes up the entire third act and presents a challenge to the set designer: "At intervals waves dash against the window" of a lighthouse on stage, "the thunder crashes, the sea roars, the lightning flashes," and the scene moves to the exterior of the lighthouse and "an expanse of wild, storm-tossed waves, with the lighthouse . . . rising from the rocky coast."[3] The scene immediately following opens with heavily falling snow and howling wind. These scenic demands are striking and show nature as inherently dramatic: it is center stage, not merely a backdrop.

What distinguishes Ibsen's and Herne's treatment of nature from earlier natural spectacle is the prominent framing of their plays within evolutionary discourses. Herne was steeped in Charles Darwin and Herbert Spencer, particularly a watered-down version of the latter, a vaguely defined sense of man's eternal struggle with his environment being the determining factor in his life, and an insistently progressive view of human evolution. These evolutionary interests were shaped by Hamlin Garland, one of the founders of American realism, who quoted Darwin and Spencer in his notebooks, showing a particular interest in Darwin's comments on the brute nature of man, our "animal origin," and Spencer's comments in *First Principles* on evolution as "an integration of matter and concomitant dissipation of motion."[4] Herne was also influenced by Ernst Haeckel and Francis Galton and would tirelessly discuss "the constitution of matter [a possible nod to Haeckel's monism] and Spencer's theories of evolution" with Garland.[5] In short, evolutionary theory permeates his work, though sometimes it seems "squeezed through the Darwin-Spencer wringer."[6] It also came from Helen Hamilton Gardener, the southern suffragist novelist and activist whose comments on the "missing link" I quoted in chapter 1. An early Herne biographer describes her as "one of the most fiery radicals in the country, the friend of Susan B. Anthony, Elizabeth Cady Stanton, the eminent alienist Spitzka," renowned for her "sensational writing and lecturing on such subjects as women's rights, atheism, heredity, and insanity." For many years, Gardener was "Katharine's [Herne's] closest friend."[7] She lectured widely on heredity and evolution, showing particular concern for women's place in society and in relation to nature. Her lectures were published in magazines and collected in book form, and one of her most popular was "Sex in Brain," in which she challenged the assumption of female brains being inferior because they are smaller than men's. She also questioned the notion that woman's role is to bear unlimited children, anticipating Eugène Brieux's dramatization of this concern a decade or so later.[8]

A recent assessment questions how "fiery," radical, or "sensational" Hamilton was, pointing out that she actually belonged to the less-strident militant strand of early American feminism, "quietly" lobbying Congress and meeting with President Woodrow Wilson.[9] This would be consistent with the tendency of many militant suffragists to distance themselves from biological issues such as reproduction to avoid alienating the public from the main objective of their cause, which was to get the vote. Yet she was hardly reticent on such subjects, and her friendship with Katharine Herne, who so directly shaped not only the role of Margaret Fleming but also all of Herne's other plays, is a vital link that has hardly begun to be fully explored. Katharine had a striking "intellectual enthusiasm," wrote Garland, never happier than when discussing "the nebular hypothesis . . . atomic theory . . . the inconceivability of matter . . . Flammarion's super-sensuous world of force, Mr. George's theory of land-holding, or Spencer's law of progress."[10] She also followed "the latest scientific theories of child raising."[11] One can only speculate about the intellectual rapport she had with Gardener, but it is highly probable that their ideas fed on one another and, in turn, exerted an influence on Herne's plays, particularly his elastic conception of parental and sexual roles.

In *Shore Acres*, for example, Helen, described as "the modern girl," is already aware that the books she reads are "going to get me into trouble. . . . [*amused*] Why, the other day I was trying to tell Father something about evolution and 'The Descent of Man,' but he got mad and wouldn't listen."[12] Her religious and traditional New England farmer father, Martin, opposes Helen's love for Sam Warren, a free-thinking doctor who also reads Darwin and Spencer, which Martin derides, offended at the idea (reiterated several times) that "my grandfathers was monkeys."[13] When Martin rejects "interference from . . . Darwin" and confronts Sam about his religious beliefs, Sam tries to explain: "Do you hear those insects singing? . . . Well, that's their religion, and I reckon mine's just about the same thing."[14] Before we even meet Sam, he is described as studying "frogs an' bugs an' things," liable to sit in the middle of the road "watchin' a lot of ants runnin' in an' out of a hole," and lecturing in the schoolhouse on "evolution as he called it."[15] Evolutionary theory is not merely paid lip service, as in Robert Buchanan's *The Charlatan*, but takes center stage from the start of the play, though one reviewer wished for "much less dissertation on Darwin."[16] By making the anti-Darwinian character so unpleasant and narrow-minded, Herne steers audience empathy firmly toward the attractive and open-minded young Darwinians.

Shore Acres shows an overarching interest in the issue of adaptation and reverses the standard practice of naturalist plays that show humans beings defeated by the harsh conditions around them. Rather than struggle to fit in and survive in a hostile environment, Sam and Helen elope and go west, where they prosper. Far from being crushed by their circumstances, they seek new ones and flourish, and this is emphasized repeatedly by the stage directions regarding Sam's new "big warmhearted manner."[17] The play lives out the myth of the regenerative American West, but rather than disappear into it, the couple brings its constructive forces back east to effect a transformation in that negative environment. Though only seventeen, Helen successfully asserts her "Spencerian individuality and right to live independently of social mores."[18] Martin also changes in the end through his meeting with the next generation, his granddaughter, who turns his manner from dour and resentful to mild and gentle. His apology to Sam is met with another repeated idea: "You didn't quite understand me, that's all," "You folks around here didn't understand fellows like me, that's all."[19]

Spencer's popularity with Victorian dramatists indicates how dominant a role he played in bringing evolution to people's consciousness, how accessible his version of it was. During the so-called eclipse of Darwinism in the last two decades of the nineteenth century, Spencer's star was at its highest, coasting a combined wave of social Darwinism and eugenics. Dramatists—especially those with a moralizing bent—found this appealing and easy to translate into theatrical material. As Donald Pizer has noted, Spencer was attractive because his ideas were so adaptable; Spencerian evolutionary theory was "like a large apple shared by several boys," who each take "a bite here, another there, but no one swallows it whole."[20] Political and literary reformists alike found Spencer a means of challenging the existing order. Social Darwinists needed "to show that the social order in some way mirrored the natural order."[21] Yet, what was the natural order? One constant in all the permutations and varieties of social Darwinism was the view of "the laws of nature as both beneficent and malign, as something to be emulated and as a force to be feared, as both a model and a threat."[22]

From the age of eighteen, exactly at the time when he first started going to the theatre, Henry Arthur Jones began reading Spencer.[23] This was around 1871, when Darwin's *The Descent of Man* was published. In 1878, Jones confides: "I am still reading hard; all my spare time in the day and sometimes half the night. I am now approaching the end of Herbert Spencer's system of philosophy. It has been a hard nut to crack, but I wanted first of all to get a good groundwork of the latest science to build upon. And Herbert Spencer must not merely be read; he must be learned."[24] For all his insistence on its importance, Jones never says exactly *what* is so significant in Spencer's philosophy. Concrete clues, however, may be found in his play *The Dancing Girl*.[25]

Drusilla Ives is a dancer who has fled her stultifying Quaker life on an island off the southwest coast of England for a career dancing in London, where she is the mistress of the Duke of Guisebury. He owns the land on the fictitious island of Endellion where Drusilla's village is founded; its citizens are his tenants. She returns in disguise to visit the village, and he follows her there. He discovers that he has neglected the land and the community so badly that a group of its men has joined a voyage to the Antarctic in desperation to find a better life for their families; the boat is now believed lost. Meanwhile, some of those left behind are trying in vain to erect a breakwater against the imminent flooding that will no doubt destroy their village. The duke sees all of this, repents his wasted life of luxury in London, and secretly resolves to kill himself after giving a final farewell feast for all his friends in honor of Drusilla so she will be deemed respectable and can continue to teach dancing in their homes after he is gone.

Sybil Crake, a lame young woman who is the daughter of the duke's steward and land agent and whom he rescued many years ago from an accident in which she was crushed beneath a horse, finds out his plan of suicide and manages to intercept him just as he is about to drink the vial of poison. Sybil is a kind of Jenny Wren figure—like Jenny, the lame young doll maker in Charles Dickens's *Our Mutual Friend*, she hobbles about acting as *raisonneuse* to the other characters who confide in her, and she is full of wisdom and sage advice.

The final act, set two years later, shows the duke a broken man, prematurely aged, who now devotes himself to building the breakwater for the village. We learn that Drusilla has died of a fever in New Orleans, where her dancing had won her the adoration of the public. To the very end, she continued to flout the wishes of her stern Quaker father, but he forgives her posthumously at the urging of Sybil. The duke joins forces with Sybil, whom he realizes he loves. Meanwhile, the boat from Antarctica miraculously returns with all men safely on board, and Drusilla's angelic sister Faith has married local boy John (who had once passionately loved Drusilla) and they now have a baby—establishing the new generation and the hope for the future.

Theatre censorship laws meant that all plays intended for production had to be submitted for licensing to the Lord Chamberlain. Frequently, they were refused licenses, or he would wield his "blue pencil" and delete references that were deemed objectionable, usually of a sexual, moral, or political nature.[26] Jones refers to Spencer twice in the original version of *The Dancing Girl.* This licensing copy of the play, held in the Lord Chamberlain Collection of Plays in the British Library, contains the following lines in the fourth and final act:

> SYBIL: . . . *You know they teased me about reading Herbert Spencer the other day*—(Guise nods) *I've found out something.*
> GUISE: *What?*
> SYBIL: *That he teaches exactly the same thing as Dante. Dante says "in His will is our peace." Herbert Spencer says "You must bring yourself into perfect agreement with all these great laws around you. You must, or you'll get crushed." And that's what you've had to do. You have obeyed.*[27]

However, the published text of the play does not contain these lines; all mention of Spencer and of adapting to one's environment have been excised.[28] Between the first production of the play and its publication, then, the references to Spencer were dropped, even though, as Doris Arthur Jones recalls, they seemed to go down well (particularly with Spencer himself):

> Among the many letters of congratulation and praise received by my father, none gave him keener pleasure than a letter from Herbert Spencer asking him to go and see him. *The Times* criticism of the play referred to

> the lines where Sybil Craig [*sic*], in speaking of Herbert Spencer, says, "I've found out." Guisebury, "What?" Sybil, "That he teaches exactly the same thing as Dante. Dante says, 'In His Will is thy peace,' Spencer says, 'You must bring yourself into perfect agreement with your environment or get crushed!'" Herbert Spencer was very pleased at this quotation from his teaching, and H. A. J. derived the keenest pleasure from the talk he had with the great man. . . . My father said constantly, "Any clear thinking I've done I owe to Herbert Spencer."[29]

Such direct interaction between a playwright and an evolutionist is rare. Again, though, there is little indication here of what exactly it was in Spencer's thought that appealed to Jones and that he might have put in his plays; was it just the "survival of the fittest" idea? Adapt to one's environment or be crushed? This was hardly new and hardly identified solely with Spencer. *The Dancing Girl* also refers explicitly to the "principle of selection." In act 3, when the aristocrats are visiting Guisebury for his last supper, Lady B complains: "I'm not squeamish, Guise, but really society is getting too mixed!" and Guise responds: "It is mixed, but so it will be bye and bye . . . whatever principle of selection is adopted."[30] This is intriguing: What exactly are the different principles of selection Jones is alluding to, and is he allowing for natural selection, at a time of much skepticism and hostility to it? The play at least suggests such a possibility through its multiple versions of the "adapt-or-perish" theme—not only Drusilla's story but also the depiction of the encroaching sea. This story beneath the fallen woman narrative is arguably far more interesting and has much more contemporary relevance for us now, showing the fight to save the coastal village, to defend itself against the sea, and the voyage to Antarctica. In this last regard, Jones must be responding to the feverish race to explore Arctic regions that gripped so many nations in this period and is bound with imperialism—Britain and Norway in particular leading the races to the Poles.[31]

The Dancing Girl kills off the "fallen woman," the standard procedure for plays featuring this scandalous female type, and it demonstrates the impossibility of women's position in society by showing how few options there are for anyone wishing to deviate from the norm of getting married and having children. The highly talented dancer who openly renounces marriage, children, and religion must die. Yet the play also features Regy, a man literally on the run from exaggeratedly calculating society matriarchs and their daughters pursuing him for the "trap" of marriage. So, while on the one hand Jones depicts in negative terms the woman stalking her prey, on the other he deplores the woman who turns away from that role. What, then, should women do? What shall they be? As in William S. Gilbert and Arthur Sullivan's *Princess Ida*, the answer is to maintain their traditional roles as wives and mothers; anything else would be absurd. In his play *The Case of Rebellious Susan* (1894), produced at the height of the "new woman" craze, Jones broaches (and rejects) the idea that the two-sex state of human evolution is flawed. Elaine states that she and her fellow feminists will correct nature, leading Sir Richard Kato to ask how they will do that: "By changing your sex? Is that what you ladies want? You are evidently dissatisfied with being a woman. You cannot wish to be anything so brutal and disgusting as a man and unfortunately there is no neuter sex in the human species."[32]

There is another reason why it is instructive to compare the licensing version of *The Dancing Girl* with its published version. In the former, Faith, Drusilla's sexually pure, pointedly named younger sister, reinforces the natural role of woman. Toward the end of the play, when she has become a mother, Faith declares that birth is "such a miracle. I think and think for hours, and the more I ponder, the more I cannot understand how such a thing can be. . . . I wonder if other mothers have the same thoughts that I have—I suppose they have—and yet it seems as if nobody but me could have such a sweet secret."[33] This episode was cut from the published version, which thus offers no answer to the question of what is the viable role for women. It also suggests that Jones may well have been more attuned to the discourse and debate on issues like biological determinism and gender essentialism than he let on since he removed this pointed reference to women's natural role as mothers. This would indeed present a more complex side to him than we have previously seen. By his own admission, Jones was a champion of "what is called bourgeois morality," an anachronistic sentiment at a time when most of his contemporaries in the theatre were attempting to chip away at that.[34] The scholarly consensus is that Jones was always hampered by his stuffiness, though given what we know about his intellectual interests it is going too far to say that "he failed to introduce any deep thought [into the theatre] . . . , for deep thought was beyond him." The very attempt to create a play around an idea—even if Jones's idea of evolution was so simplified—gave impetus to the stage to shake off "the innocuous and stupid farces which hitherto had been the rage."[35]

The Drama of Extinction

Much as he emphasized "origins," one of the things that Darwin, and for that matter Charles Lyell, Thomas Malthus, and evolutionary thinkers generally, brought home with great force was the reality of life ending, and not only the deaths of individuals or whole groups, but also of the sun and hence all life. As Cyril D. Darlington put it in 1969, echoing Darwin to the same effect in *Origin*, "The ultimate destiny of man . . . is probably extinction."[36] Georges Cuvier in 1796 had argued that past catastrophic events had obliterated whole species. Lord (George Gordon) Byron's *Cain* (1821) dramatizes this concept, referring directly to the fierce debate between the Cuvier catastrophists and the gradualists. The sense of "deep time" brought about by findings in geology lent greater force and reality to the idea of extinction and gave it a fascination that no longer exists; what was, for Darwin in the *Origin*, a natural phenomenon to be "marvelled" at is now, for us, a sad catastrophe too often of our own making.[37] This fascination with extinction is bound with the Victorian interest in seeing living fossils and transitional forms on display, a pastime that simultaneously highlighted the robustness and the fragility of species. Darwin argued that extinction "has played an important part in defining and widening the intervals between the several groups in each class. We may thus account for the distinctness of whole classes from each other—for instance, of birds from all other vertebrate animals—by the belief that many ancient forms of life have been utterly lost."[38]

Darwin's views on extinction changed, however. By the time of his autobiography, he is bemoaning the finality of extinction and finding, as Bernard Shaw would do, a cruel paradox in the thought that ever-improving humankind is "doomed to complete annihilation after such long-continued slow progress."[39] Such terms as *complete annihilation* (Darwin, *Autobiography*) and *a universal winter* (Thomas H. Huxley's phrase) abound in the writings of Darwin and his contemporaries, presaging a bleak Beckettian view of the world as obliterated, entropied, with humankind at its last moments before extinction. The phrase *universal winter* signals as well a convergence of key scientific ideas of the nineteenth century: evolution and entropy, biology and physics.

But extinction could furnish dramatic material. Dramatists seize on the idea of extinction as a current event, happening in real theatrical time, showing the last of the species at its end point (for example, in Thornton Wilder's *The Skin of Our Teeth*, discussed in chapter 7). They can thus focus on the drama (and trauma) of "sudden" extinction, and in fact, they create this counter-evolutionary idea for the stage; dinosaurs apart, most extinctions happen over a long and gradual process, a slow death. Dramatists convert it into something over which we have some power, rather than the passive Darwinian sense of organisms (and hence entire species) being at the mercy of genetic and environmental factors beyond their control, thus having no say in whether they go on or go extinct.

One of the ways in which extinction plays out on stage in the nineteenth century is through the idea of the end of the family line. Henry James, François de Curel, and St. John Hankin all depict this, and it is often tied to broader concerns about the national stock and a dwindling birth rate.[40] De Curel's play *Les Fossiles* (1892) deals with extinction, although it was more notorious for its depiction of incest.[41] The "fossils" of the title are the outmoded, decadent, and corrupt nobility. In the climactic scene in act 3 between father and son (and with sister Claire listening outside the door), the duke admits he slept with Robert's wife at Chantemelle before she married Robert—but the audience has already heard this shocking revelation from Helen's own lips, earlier in act 3, when she tells Robert that for two years his father essentially had his way with her.[42] The duke says that he has engineered Robert and Helen's marriage to perpetuate the family line. The weak and ill Robert dies, and the final scene of the play takes place in the family morgue, where, gathered around his body, they read out his will and learn his express wishes that the future Duke of Chantemelle, his son, be a modern man in the most profound sense of the word: someone capable of dying for his ideas and not a fossil facing backward and steeped in the Revolution, the age of the guillotine.[43] It is also revealed that sister Claire has promised Robert never to marry and to remain forever with Helen and the baby.

In one of the play's key speeches, Robert draws on the metaphor of two distinct, competing natural environments to point to the human condition. The forests and the sea hold equal attractions for Robert, who loves them both, and sees in each two vastly different "pictures of humanity." Will we "advance in unison like the waves," rushing together toward the shore "without clashing"? Or will we be like trees that grow so massive they "strangle everything"?[44] The unusually long monologue allows the extended metaphor to develop, working out the intricacies of the forest versus sea motif and its implications with regard to class, nobility, attitudes, inherited beliefs, and so on. The metaphor remains, however, solely verbal, simply an elaborate description, whereas this same contrast between forest and coast would in Susan Glaspell's hands be strikingly enacted through the staging of her one-act play *The Outside* (1917).

Strictly speaking, there is no guarantee that the family line of "fossils" will die out in de Curel's play, but extinction is used more symbolically than literally and it suffuses the play; Robert invokes the ideas of his forebears, the need to die among those ideas and the memories of his youth, surrounded by the honor and the name of his ancient family. It all sounds a lot like memes and replicators rather than genetics and heredity. This fits with Ibsen's sense of heredity, which commentators like Max Nordau and Max Beerbohm derided for its lack of medical accuracy. In fact, *Ghosts* was produced at the same theatre by Antoine at almost exactly the same time as *Les Fossiles*, and the two plays do share the theme of incest, the promiscuous father overflowing with joie de vivre and the son doomed to die by disease. But Ibsen is not concerned with the dying out of an aristocratic family line; in fact, the implication is that mating with Regina would bring welcome "new blood" into the Alving family. This shows the distance between Ibsen's more contemporary evolutionary vision and that of de Curel and, for that matter, one of Ibsen's great admirers who also treats extinction: Henry James.

The idea of "fanatical attachment to the nobility of race; the necessity of its life and strength by every and any means"[45] is also a concern of James's *Guy Domville*, which addresses celibacy and its annihilating consequences.

Guy has "an ancient name"; the Domville family is "one of the two or three oldest in the kingdom!"[46] One of the lines in the play (repeated twice), "I am the last, my lord, of the Domvilles," is famous in literary history for having elicited from an audience member the retort "I bloody well hope so!" The play was such a resounding failure that it put James off playwriting for good. Yet considering this neglected work in light of evolutionary motifs gives it new meaning. In *Guy Domville*, the extinction of the line is, in a sense, enacted for the audience. Guy's decision to enter the priesthood, in full knowledge of his status as the last in his family line and that celibacy will therefore spell its extinction, is the hinge for all the action of the play. It is also deeply anti-Darwinian given what Darwin writes in *The Descent of Man* about celibacy as a "senseless" practice valued by the ultra-civilized for the quality of self-control that it requires.[47]

Extinction is not, in fact, an immediate threat in *Guy Domville* due to the revelation of illegitimate children, just as in *Les Fossiles*. This is presented as wholly positive in *Les Fossiles*, but in *Guy Domville*, these offspring are of a lower class and therefore unacceptable, despite perpetuating the family genes. The "numerous progeny" left by Guy's recently deceased and unmarried relative are dismissed by his friend and self-appointed adviser Lord Devenish as "not worth speaking of" and "a pack of village bastards," even though Guy immediately refers to them as "my family."[48] Technically, they do not exist, and Guy is informed that he is the next in the line of succession to the family home: "the heir of your kinsman, the last of your name," a phrase that will of course be echoed in the last moments of the play.[49] Lord Devenish tells Guy he must not enter the priesthood because he has a duty to "your position—to your dignity—to your race." Guy evidently absorbs this rhetoric because he then ignores the existence of the "village bastards," saying a few moments later in conversation with Mrs. Peverel: "My cousin is dead—*there are no other kin*—and I'm sole heir to the old estate. . . . I'm sole of all our line, I'm sole of all our name."[50] Not only does this contradict what he and the audience have just learned, but from a scientific point of view it is nonsense; the Domvilles themselves are hardly in danger of extinction, only their aristocratic name. Interestingly, none of the reviews of the play acknowledges the existence of these illegitimate offspring.

In fact, James's idea of extinction harks back to an earlier usage, more in keeping with Cuvier than with Darwin, when in Britain the term "was mainly linked to the history of landed families: a line becomes extinct and with it the family name and the succession of property and practices," as Gillian Beer observes. In the history of science, the implications of extinction were at first familial; in eighteenth-century Britain, for instance, "the term 'extinction' was mainly linked to the history of landed families: a line becomes extinct and with it the family name and the succession of property and practices."[51] Beer notes that in *Origin of Species* Darwin expanded the idea of family away from the exclusiveness of "pedigrees and armorial bearings" to embrace all "the past and present inhabitants of the world." This meant that "instead of 'special creations' all organic beings are now 'lineal descendants of those which lived long before the Silurian epoch.' He turns the metaphor of the great family around in a way that dignifies all species: all are members of the oldest of all families," so that they seemed to Darwin to become "ennobled," thus radically redefining the concept of "nobility" through biology: "For Darwin this inclusiveness and continuity is the 'grand fact' he has uncovered: 'The grand fact that all extinct organic beings belong to the same system with recent beings, falling either into the same or into intermediate groups, follows from the living and the extinct being the offspring of common parents.'"[52]

Guy Domville and *Les Fossiles* reflect this shift in thinking about kinship and nobility, as well as being about extinction (abstractly and symbolically more than literally). The fact that James's play is set in 1790 might have obscured this direct link by making it feel remote and irrelevant compared with the contemporary society dramas then in vogue on the London stage, such as Oscar Wilde's *The Importance of Being Earnest.* Similarly, de Curel depicts a family line that is six hundred years old (how Robert hates the sight of the moldy old house, "la façade rébarbative de Chantemelle").[53] In addition, James's focus on religion—linking extinction with the enforced celibacy of the Catholic priesthood—further removes the play from an overtly biological context.

Yet the biological language of the play places evolutionary concerns up front, despite the religious theme. This is hardly surprising, since throughout his career James mined evolution for metaphors; his appreciation of Ibsen as "evolved," for example, and the geological imagery with which he recalled the impact of stage adaptations of Dickens "in the soft clay of our generation," like "the wash of the waves of time."[54] *Guy Domville*'s treatment of extinction can be read as part of a wider discourse on evolution. James sets up the situation so that Guy explicitly chooses extinction, despite the full possibility of continuing his line. Lord Devenish applies eugenic pressure: Guy must remember that he is a gentleman, and that character is a "treasure" that his philandering kinsman soiled.[55] James depicts Lord Devenish as a threatening, mysterious agent of change that redirects an entire life course. Guy asks him, "Who *are* you, what *are* you, my lord . . . ?" and likewise asks Mrs. Peverel: "Who *is* he, Madam—*what* is he, that he comes here to draw me off?"[56] It is like the enigmatic power of the Rat Wife in Ibsen's *Little Eyolf*, a play of the same year as *Guy Domville* by a contemporary playwright James deeply admired. Indeed, like Eyolf, Guy seems passive, too easily led. He veers from feeling in act 2 an obligation to breed ("to do my duty to my line") to utterly rejecting by the end of act 3 this sense of reproductive duty.[57] Guy's final speech in act 3 makes no evolutionary sense: he loves Mrs. Peverel, and she loves him; he knows that he could happily produce more little Domvilles with her; yet he says he must renounce everything and enter the Church. He will thwart sexual selection (Mrs. Peverel has clearly chosen him as her mate). Guy is deliberately non-adaptive, and as Darwin shows, failure to adapt leads to extinction not only of the individual but also of the species.

But set this behavior in the context of Huxley's statement in *Evolution and Ethics*:

> The practice of that which is ethically best—what we call goodness or virtue—involves a course of conduct which, in all respects, is opposed to that which leads to success in the cosmic struggle for existence. In place of ruthless self-assertion it demands self-restraint; in place of thrusting aside, or treading down all competitors, it requires that the individual shall not merely respect, but shall help his fellows; its influence is directed, not so much to the survival of the fittest, as to the fitting of as many as possible to survive. It repudiates the gladiatorial theory of existence.[58]

Guy's altruism (sacrificing his own love to help his rival elope), his celibacy ("the rigid rule of [my] life, is to abstain" from marriage/sex),[59] and his willed extinction all go against Darwinian principles. Yet in every respect, Guy's behavior matches Huxley's description, and it begins to make sense from an evolutionary standpoint that includes a moral dimension. The play maps Guy's necessary progress toward the fact that even if he does not keep the line going he still plays an evolutionary role, just as the very fact of extinction plays a role in the perpetuation of life.

Critical responses to *Guy Domville* were generally attuned to the Jamesian voice but puzzled over the theme and dramaturgy. However, the reviews are not at all as damning as theatre legend would suggest. Shaw opined that the play was not bad, just out of fashion, and he praised its "rare charm of speech."[60] Arnold Bennett wrote that although sometimes "tedious," the play contained some excellent scenes and beautiful writing. The first act was the best, he felt, "natural, impressive, and studded with gems of dialogue—gems, however, of too modest and serene a beauty to suit the taste of an audience accustomed to the scintillating gauds of Mr. Oscar Wilde and Mr. H. A. Jones."[61] H. G. Wells found the whole thing too "delicate" to get across the footlights unscathed; one needs bold, broad brush strokes. He also found the second act tedious. He mentions that we need to see more clearly "Guy's growing disgust with life . . . a disgust that forms the key to the third act."[62] Perhaps this links to the fact that, as John Stokes shows, 1895 was still more or less the peak of the "suicide craze" and the fashionable "tired of life" ennui of Decadence.[63]

Victorian thinkers grappled with a fundamental problem articulated here by Huxley: where morality fit into the new picture of the natural order. This had also dogged Malthus, one of the great inspirations for Darwin. As a later edition of Malthus's *Essay on the Principle of Population* (1890) points out, the response to the first edition of the book (published anonymously in 1798) prompted Malthus to "soften some of the harshest conclusions" for the next edition. "He not only introduced his new idea of moral restraint and postponement of marriage till it could be afforded, but he showed that, on the whole, civilization was capable of alleviating the pressure of population, and had done so sensibly in modern times."[64] Both Malthus and Darwin found they had to spell out the moral implications of work that stood perfectly well without that, simply because it was deemed too bleak. This continued throughout the responses to evolution. Leslie Stephen argued that in human evolution it was not individual modifications but social transformations that constituted adaptive change, and moral evolution reflected this fact.[65] Both Huxley and Stephen "wished to present human evolution as a story of moral progress, while remaining haunted by the implications of the naturalistic process that had made this evolution possible—selection."[66]

Putting Darwinian ideas to moral use meant taking a step toward not only Spencer but also eugenics; taking it upon themselves, as Gideon Lewis-Kraus puts it, to "help the herd thin itself out" in the service of a higher humanity. Fortunately, "we now understand that just because a gene is selfish it doesn't follow that a person should be."[67] *The Struggle for Life,* Buchanan and Frederick Horner's adaptation of Alphonse Daudet's *La Lutte pour l'existence*, though written well before the findings of Gregor Mendel came to light and modern genetics was developed, takes just this line, reductively equating Darwinian evolution with pure selfishness. Recent studies are arguing the idea of cooperation as the driving force of evolution rather than competition, just as Peter Kropotkin had done in his seminal work *Mutual Aid* (1903). At the heart of cooperation and altruism lies the capacity for empathy and identification. This in turn is central to the art of acting. In the final section of this chapter, I discuss briefly how evolutionary thought relates to the theory and practice of acting, beginning with the actor's main tool: the body.

Evolution and Acting

The science of acting has received a great deal of scholarly attention, much of it linked to cognitive processes and the larger context of evolutionary theory, and a full exploration of this connection lies beyond the scope of this study. The discovery of mirror neurons has had particular relevance to theatre and performance studies.[68] In addition, the growing field of biosemiotics links biology to theatre through their shared status as sign systems. Here, my concern is with how nineteenth-century understandings of acting hinge on the performer's body and what it came to signify in light of knowledge about human evolution. Jane Goodall writes that the body is the most visible site of the realization of Darwinian ideas, not plays that reflect them: "An evolutionary view of the human was one that foregrounded embodiment, and the performing arts have in common the body as their primary instrument of communication, so that physiognomy, sexuality, energy, expression and mobility—important theoretical concerns for the evolutionist—are integral components of the performer's work."[69]

Acting could even be seen as embodied recapitulation: actors as representatives of humanity in growth and development, enacting universal stages. This challenges the idea that one cannot represent evolution on stage because there is not enough time. Joseph Roach likewise refers to the human body as "an evolutionary text" and to the plays of the New Dramatists like Ibsen and Shaw as, "like the palimpsest of the Darwinian body, revelatory of motivation and desire in deep subtextual layers, demanding of the actors a corresponding density of psychological impulse."[70] By the late nineteenth century, the actor's body is no longer under a rigid and closed system à la Denis Diderot and eighteenth-century acting theory but is, like evolution, unpredictable, a "tangled bank" of emotions and gestures that cannot be codified and systematically deployed. For many, that body is also an avatar bearing the stamp of all evolution in its form and features, and actors from Henry Irving to the cabaret performer Loie Fuller exploit this. Irving created the role of Mathias in *The Bells* in 1871, the same year as *The Descent of Man* and a year before *Expression of the Emotions* was published, and the role was linked instantly to contemporary science in its suggestion of both atavism and mesmerism.[71] Darwin's investigations suggested the presence of atavistic traces in the actor—a mere facial expression (perhaps Irving's wildness is an example of this) could suggest the animalistic essence that lurks beneath the most advanced human, showing "the survival of this evolutionary past in the present."[72] Since Irving revived this role repeatedly for the next thirty years, its link to Darwin would have been accentuated further.

Along with the impact of other factors such as the rise of psychology, Darwin's ideas inspired the newly physical emphasis in acting well before Constantin Stanislavski and psychological realism. Studies of the development of naturalistic acting by Roach, Rose Whyman, Lynn Voskuil, and others have sparked debates about how far Darwin's *Expression of the Emotions in Man and Animals* revolutionized the actor's art.[73] Using subjects from the margins of civilized society (the insane and the racially different, and also infants), Darwin's investigations sought to "strip the face of its civilized mask of convention and reveal a language of expression that derived from the struggle for survival and various forms of adaptation to environmental or physical forces."[74] In effect, the experimental emphasis shifted from studying the norm (the object of traditional physiognomy, which sought to establish an ideal) to "a science of deviant faces."[75] Darwin showed not only that the lines between the normal and the abnormal are blurred when it comes to emotions, but that the expressions of our emotions are organic processes, universal (innate, not learned) and brought about through evolution.[76] In addition, Ivan Pavlov's work on conditioned reflexes—which revealed that "the individual's behaviour is a constant series of interactions with the environment"—gained currency in Stanislavski's lifetime.[77]

The idea of a universal language of expression was of course prevalent in the theatre, for instance in the commedia dell'arte in Europe, in Johann Wolfgang von Goethe's *Rules for Actors*, and in the eighteenth-century English theatre with its acting manuals demonstrating codified gestures. It penetrates well into the twentieth century as a code that playwrights and actors sometimes used and that audiences recognized; a stage direction in James M. Barrie's *What Every Woman Knows* (1908), for example, simply indicates that the Comtesse "assumes the pose of her sex in melodrama"—actors and audiences alike would have known what this meant.[78] Jerome K. Jerome pokes fun at this in his book *Stage-land*. Goodall warns against overstating the impact of *Expression of the Emotions* on theories or practice of acting at the time, questioning the idea that it single-handedly triggered "a paradigm shift in the theory of acting," as the complexities warrant further exploration.[79] Bruce McConachie, for instance, argues that Stanislavski drew not so much on Darwin as early behaviorist ideas "that have long since been abandoned."[80] Whatever the finer points of the debate, Darwin's book did show the similarities between human and animal gesture and facial expression, suggesting that "grimaces and extravagant gestures are particularly clear vestiges of our animal past."[81] The significance for the actor is clear: acting is linked to an "instinct to imitate," an involuntary mimicry.[82]

The type of mimicry in nature that we know most about is visual: "mimetic resemblance," or camouflage. Mimicry was discovered shortly after *Origin* was published. It was a particularly elegant illustration of adaptation, and it shows how natural selection favors mimetic patterns across the board (in both artificial and natural kinds of environments). There is both conspicuous mimicry (bright colors) and the more subtle kind (blending in). Occasionally, mimicry can lead to the development of a new species, as when a butterfly species that was red becomes blue and white.[83] But mimicry does not simply mean simulation or imitation. In nature, it is "the parasitic or mutualistic exploitation of a communication channel. More plainly, the term describes the situation in which one organism gets the better of another organism (known as the dupe) by looking, smelling, sounding, or feeling like something else."[84] Here, science draws explicitly on the language of Renaissance theatre with its easily fooled dupes and gulls, but these direct theatrical analogies are not flattering; they make theatricality synonymous with trickery, echoing Darwin's aforementioned allusion to the "tricks of the stage."

In essence, the mechanism of mimicry raised a key question: How could one be natural while being artificial if all nature was a form of performance? George Henry Lewes memorably observed that natural acting (as opposed to the stylized systems of representation that had gone before) was an attempt "to catch nature in the act." He also characterized human beings as innately theatrical: "We are all spectators of ourselves."[85] This tension between the artificial and the natural is captured in the work of the Italian actress Eleonora Duse, one of the great interpreters of Ibsen. Duse specialized in "pain and the representation of pain," and she expressed the painful emotions of embarrassment and shame by famously blushing at will.[86] Shaw, for example, makes a big deal of the blush in his review of both Duse and Sarah Bernhardt's performances as Hermann Sudermann's Magda.[87] Blushing is usually associated with emotion rather than physical pain, and this deepens the mystery surrounding her uncanny ability, noticed by contemporary critics, to manufacture a blush (and to reverse the process, going suddenly pale).

Duse's apparent ability to blush and pale at will goes directly against Darwin's claim in *Expression of the Emotions* that "we cannot cause a blush . . . by any physical means,—that is by any action on the body," as blushing comes entirely from the mind. "Blushing is not only involuntary," he goes on, "but the wish to restrain it, by leading to self-attention, actually increases the tendency."[88] Blushing at will appears frequently in stage directions of contemporaneous plays; Hauptmann calls for three different actresses to blush spontaneously several times in his play *Lonely People* (1891), when Miss Mahr twice "changes colour," when Kitty goes "red," and when Mrs. Vockerat's "colour changes."[89] Maeterlinck, who began his playwriting career in the 1890s, asks for blushing in his play *Betrothal* (1921): the silent white-shrouded woman's "colour comes and goes," and there are also several references to her spouse blushing and paling.[90]

Shaw put it into his plays as well. He calls for Gloria "blushing unendurably" at the end of act 2 of *You Never Can Tell* (1897), but "she covers her face with her hands and turns away," giving the actress an easy way out of the physiological challenge. In *Man and Superman*, Octavius produces "an eloquent blush" and then immediately runs off.[91] As with so many of Shaw's excessively discursive stage directions, the instruction to blush seems to serve a narrative purpose; it may be more for the reader than for the actor. But, given his admiration for Duse's blush, we may fairly surmise that his playwriting draws directly on what he has witnessed an actor do on stage and deeply admired. Roach notes that the fastidious Shaw was "clearly embarrassed" by Duse's body but "couldn't stop writing about it." This is why his assessment of her "equivocates between zoocentrism and extreme idealism—the divine special creation of womankind."[92] Shaw describes Duse as "ambidextrous and supple, like a gymnast or a panther," yet distinct from the animals by that "high quality" of the human, eschewing "explosion of those passions which are common to man and brute"; behind every gesture is "a distinctively human idea."[93] But this is more than just the "asexual's ambivalence about the animal body. Shaw wants to separate women from beasts as well as mind from body."[94] Susan Bassnett writes that it was precisely this defiance of categorization—Was she a new kind of female? Was she beast or human? Was she natural or artificial?—that entranced critics, who noted how she "portrayed inner struggle, how she rejected make-up and corsets and tinted hair in favour of natural physical decay and change, how she emphasized the pause and the infinitesimal movement rather than the wide gesture. . . . [they all seem to suggest that] Duse was insisting on representing a femininity that had nothing to do with artifice . . . [or] with feminism."[95] Not until the "third phase" of her career, after about 1909, did Duse espouse feminism, beginning to play women "who celebrate their Otherness and who impose their will upon the world around them. Woman may still suffer, but she survives and is a source of life and energy."[96]

Women's bodies on stage carried atavistic connotations, signifying the primordial and the hidden: the emotions on display might only scratch the surface of a deep inner life. And this possibility of concealment of what Roach calls the "iceberg" beneath the merest look or gesture gave a new kind of power to the actress. Sos Eltis points out that from the 1890s onward, actresses increasingly show emotion through its suppression, acting crying by suppressing it rather than sobbing hysterically. Mrs. Patrick Campbell was famous for blowing her nose instead of crying.[97] Firsthand accounts of Duse playing Nora Helmer note this lack of emotional ostention: when "the misfortune happens [Duse] makes no desperate attempt to resist it, she gives no hysterical cry of fear, as a meaner soul would do in the struggle for life," this last phrase echoing Spencer and Darwin. "Duse's Nora hastily suppresses the first suggestion of fear."[98] Elizabeth Robins's unpublished play *The Mirkwater* has Felicia utter an "inarticulate cry."[99] Gerhart Hauptmann has Mrs. Vockerat "forcibly repressing her excessive emotion" in *Lonely People*.[100] In act 1 of *Hedda Gabler*, finally alone on stage, Hedda crosses to the window, raises her clenched fists "as if in rage", and yanks open the curtains, wordlessly conveying her deep frustration. Which comes first, these gestures of suppression by actresses, or playwrights inscribing them into their plays? And how do these instances relate to the broader discourse on human evolution?

Duse's acting hinged on restraint; she used "more contained gestures" rather than grand ones.[101] She emphasized separate parts of the body rather than giving a notion of wholeness as in the grand gesture style of acting.[102] Luigi Rasi, a fellow actor, thought that the basis for Duse's acting was creating *la faccia convulsiva* (the distorted face), normally associated with mental illness.[103] This resonates with the distorted faces of the mentally ill subjects in Darwin's *Expression of the Emotions* and the empty, forward-facing stare that accompanied Robins's "autistic gesture" as Hedda Gabler. Duse likewise used her eyes to speak: "Her tormented eyes screamed at us in silence."[104]

Rhonda Blair, in her study of acting and cognitive science, mentions Duse's blush as evidence of an actor using what Sanford Meisner calls "true emotion." For Meisner, Duse's blush in her performance as Magda was "'the epitome of living truthfully under imaginary circumstance, which is my definition of good acting.'"[105] It is surprising that in a book whose starting point is the biological basis of the actor's art there is no attempt to understand *how* Duse achieved this feat. One possible route might be through a hypnosis-like "suggestible state" in which "stimuli contrary to an impulse or command are ignored," and consciousness functions as normal but without the inhibiting element of self-doubt: in other words, the subject under hypnosis can "blush on command because she doesn't doubt that she can. She tells her brain that she has received some stimulus that demands a blush, and the brain sends the appropriate signals to the body, regardless of whether or not the stimulus exists."[106] It is a kind of neurological trickery.

Darwin was evidently fascinated by the phenomenon of the blush; he devotes an entire chapter of *Expression of the Emotions* to "Self-Attention, Shame, Shyness, Modesty: Blushing," and it begins with the statement that "blushing is the most peculiar and the most human of all expressions. Monkeys redden from passion, but it would require an overwhelming amount of evidence to make us believe that any animal could blush."[107] He also claimed that blushing was exclusive to humans, although on his voyage with the *Beagle* he observed the changing color of an octopus and likened it to blushing.[108] After extensive discussion, Darwin concludes that it is attention to personal appearance rather than one's moral conduct that causes blushing.[109] He seems conscious not only of the paradox of performed concealment associated with shame but also of its fruitfulness for the artistic imagination. And, although blushing does not necessarily depend on an audience—humans can blush in solitude or even in the dark—it is still due to the *thought* of what others think of us, projecting ourselves performing "acts" in front of others.[110] Long before mirror neurons were discovered that would help to explain this phenomenon scientifically, Darwin concludes from this that it is other people's "close attention" to our bodies, in particular our faces, that triggers blushing.

Earlier studies posited that blushing was God's way of encouraging sympathy between people. Thomas Burgess, in *The Physiology or Mechanism of Blushing* (1839), wrote that "such adaptation and harmony of arrangement as here evinced, could never be the effect of chance; on the contrary, in every link of the chain which combines all the organs engaged in the production of [blushing], there is a palpable evidence of *Design*."[111] Darwin uses the blush as evidence *against* creationism, since it makes little sense: "Those who believe in design, will find it difficult to account for shyness being the most frequent and efficient of all the causes of blushing, as it makes the blusher to suffer and the beholder uncomfortable, without being of the least service to either of them."[112] Nor would this account for the universality of blushing, regardless of race, which Darwin discusses at some length, and for the fact that "almost every strong emotion, such as anger or great joy, acts on the heart, and causes the face to redden."[113]

Darwin describes the underlying physiological changes that happen during emotions associated with blushing: the heart beats rapidly and breathing is disturbed, which "can hardly fail to affect the circulation of the blood within the brain."[114] Bassnett suggests that Duse was able to recognize, simulate, and successfully manipulate such physical states within her own body and could manufacture the blush through a "simple technical device of holding her breath and keeping the tension in the chest which would cause a rush of blood to the face and head."[115] The apparent ease and speed with which she could change color belies the difficulty of controlling such complex physiological and emotional processes. One critic described how Duse's "cheeks went from blush to pallor with incredible rapidity," and others commented on her suddenly going pale as Nora in the tarantella dance (surely a feat, given the frenzy of physical activity that dance demands).[116] Again, Darwin's observations are pertinent, suggesting that "in some rare cases paleness instead of redness is caused under conditions which would naturally induce a blush. For instance, a young lady told me that in a large and crowded party she caught her hair so firmly on the button of a passing servant, that it took some time before she could be extricated; from her sensations she imagined that she had blushed crimson; but was assured by a friend that she had turned extremely pale."[117] In this case, the young lady's mortification stemmed from her being publicly embarrassed and was registered on her face in a way that even she herself was not aware of, or had misunderstood, showing the strong causal link between blushing or paling and the perception of the self as performer.

Duse's controlled blushing and paling seems to defy science even while it harnesses it, and with such extreme control that she could distinguish different mechanisms for these two physiological effects based on mental power and concentration. She developed this organic technique when there are other, easier ways to express shame (the most common emotion associated with blushing, and with the fallen woman, as Magda is); as Darwin points out, people more often avert their heads or look downward, their "eyes wavering or turned askant," than show a change of skin color when ashamed or embarrassed.[118] These are actions redolent of acting manuals of the eighteenth and nineteenth centuries, no doubt still found in melodramas in Duse's time; it is significant that she avoids them. Darwin devotes several paragraphs to "movements and gestures which accompany blushing."[119] He notes "the strong desire for concealment" that accompanies shame, which is paradoxical for an actor in full public view, yet is completely consistent with the trend toward showing emotion by its suppression.

Duse's blush was thus a potent signifier for its knowing audience—this one gesture showed the desire for concealment due to public shame, the evidence of guilt, but also the feat of the actress who after all in a realistic play is pretending to be unaware of an audience watching her—bypassing, or even hijacking, the normal trigger for a blush ("close attention") to induce it herself. This was part of her overall, ongoing experimentation with how to show emotion nonverbally and through the attempt to suppress it.

James Agate describes Duse's Ellida in Ibsen's *The Lady from the Sea* in 1923 as "pure passion divorced from the body yet expressed in terms of the body."[120] How is this Cartesian paradox achieved? Duse illustrates how acting at this time is in transition from more stylized nineteenth-century modes to deep psychological realism; her acting embodies this transition, and the key point is that audiences hold these two seemingly opposed styles in their horizon of expectations at the same time.[121] Duse's acting was "based on fluidity, on a physicalizing of dynamic internal processes"; Luigi Pirandello noted how Duse's technique was "a technique of movement. A constant, gentle flowing that has neither time nor possibility of stopping, and certainly not of crystallizing itself into predetermined behaviour."[122] The link to evolution here is pronounced: the strong Bergsonian overtones in this description of Duse's "fluidity" and "flowing," signaling a new concept of time in acting, combine with the Darwinian emphasis of showing the body's "dynamic internal processes."

What is fascinating about the critics' responses to Duse is the paradoxes they raise. While they are obsessed with capturing every detail of Duse's acting to show that it is unique, novel, and completely idiosyncratic, they also claim that she is acting a type: the modern woman, "with all her complaints of hysteria, anaemia and nerve trouble and with all the consequences of those complaints . . . [her acting repertory] consists of a complete collection of that sort of abnormal woman with all their weaknesses, quirks, unevenness, all their outbursts and languors."[123] The blame for the increased theatrical attention to this hysterical type was often, and repeatedly, directed at Ibsen, whose work explored evolutionary themes more broadly in ways that bring new meaning to his plays, not least through his unprecedented representation of women.

3

"On the Contrary!"

Ibsen's Evolutionary Vision

> *[Ibsen] accepted what so many of his contemporaries could not bring themselves to accept from the newer biology, man's unprivileged position in the evolutionary process.*
>
> Brian W. Downs, *Ibsen: The Intellectual Background*

Henrik Ibsen's works address humankind's "unprivileged position" brought about by "the newer biology," as he consistently probes our struggle with this demotion. More specifically, he explores evolutionary mechanisms such as artificial versus natural breeding, sexual selection, and adaptation. In his letters and speeches, his drafts and notes to his plays, as well as in the plays themselves, Ibsen reveals a sustained interest in evolutionary ideas and these are often surprisingly broad, touching on a wide range of aspects beyond the issues for which he is so well known, such as his treatment of heredity and his portrayals of women. In 1906, a *New York Times* obituary noted Ibsen's awareness of both Charles Darwin and Ernst Haeckel, "reflected in many of his works with telling power," but until this claim was echoed in a recent biography of Ibsen, little interest was shown in the connection between Ibsen and Darwin, let alone other evolutionary thinkers.[1] This chapter explores the extent of Ibsen's engagement with evolution through close analysis of his writings, building on recent work in a wide range of disciplines (history of science, biography, drama, eco-criticism) that has been done on this subject both in Norwegian and in English.[2]

Brian Downs argues that we cannot fully comprehend Ibsen's development without a sense of the history of ideas of his time, because he was not some "sudden, causeless phenomenon . . . but stood well in the stream of the ethical, religious, political and sociological thought of his time and, besides making notable contributions to it, took a lively interest in many of its aspects," including science. Downs is not, however, "presenting him as the mere product of" these tendencies of thought.[3] As we will see with Beckett, we are dealing with an oblique and indirect engagement, not clear-cut influence, but all the more thorough for that, going far beyond the Darwinian window dressing in plays like Robert Buchanan's *The Charlatan.* As Michael Meyer notes, Ibsen's engagement with the most progressive and radical ideas compelled audiences and readers "to rethink their basic concepts of life. . . . [It was] like reading Darwin or [Karl] Marx or [Sigmund] Freud."[4]

We are unsure of what exactly Ibsen read due to his many smokescreens and disclaimers of influence. We can confidently assume at least indirect contact with Darwin's works through Suzannah Ibsen, who was "a great reader . . . ploughing through the year's books and recommending to Ibsen what he should read." The first thing the family did when they arrived in a new town was to locate the library, and Sigurd would regularly visit it with a basket to exchange books.[5] Most scholars suggest an indirect, perhaps secondhand, reading of Darwin; this might account for the lack of precision in it.[6] This is both frustrating and liberating: it links him firmly to Darwin and other evolutionary thinkers but not in ways that are exact, detailed, and thus constraining. Eivind Tjønneland sees Ibsen as adopting a full-scale, "totalizing" evolutionism at some point around 1879—the time of *A Doll's House* and the beginning of his "social problem" plays, though as I will argue, *The Pillars of Society* (1877) already has strong elements of evolutionary thought in it.[7]

Ibsen's engagement with evolution is not limited to thematic motifs, however, but extends to structure, methodology, characterization, and staging. Both Jane Goodall and Tamsen Wolff notice the sense of "deep time" in his dramaturgy—how he manages to stage the past in the present, "conveying the deep histories behind critical events in the present and . . . suggesting the atavistic elements in human personality."[8] Yet Wolff notes that in *The Wild Duck* (1884), Ibsen "throws the whole notion of revelation [his usual method of retrospective arrangement through present revelation of formative past events] into question since the play grants no single, unencumbered moment of clarity."[9] This seems consistent with Darwinian thought: the idea not only of entanglement and randomness, but also of no single causal, and therefore teleological, events leading to one particular outcome. In this play, at least, there is no determinism.

Certainly by the end of the nineteenth century Darwinism and evolution were part of the "literary tool kit" of naturalist writers.[10] Yet Mathias Clasen, Stine Slot Grumsen, Hans Henrik Hjermitslev, and Peter C. Kjærgaard caution against reading Darwinism where none exists or seeing authors as picking up whatever "cultural debris" happens to be lying around. So the influence might be, as Inga-Stina Ewbank puts it, "almost incalculably diffusive." Darwin might be for Ibsen similar to what Dickens became: more of a "catalyst" than a source—"a catalyst in self-directed experiments: texts to learn from *and* to react against."[11] But for other Ibsen scholars the Darwinian connection is clear and decisive. Brian Johnston shows how Ibsen uses "supertext" to expose the "widest-ranging human drama" within local, limited events and patterns, and "this merging of the particular within the universal, this immersing of the individual within the history and discourse of the species, was Ibsen's aesthetic response to the intellectual world he inherited, in which the Hegelian idea of our cultural evolution as a species was linked to the Darwinian account of our biological evolution."[12] Robert Ferguson asserts that as a modern dramatist, Ibsen operated "in the philosophical space left by Darwin's discoveries, keenly filling the space—this vacuum—with determinist theories of behaviour."[13] With regard to *Brand* (1866), for instance, Ibsen "may have lost his own faith; but the social visionary in him read the signs of a new age dawning in the sensational success of *On the Origin of Species*."[14] Pointedly, Ibsen drops a reference to fossils into *Peer Gynt* (1867) directly after—and distinctly undermining—a reference to God creating the world.[15] Around the time of writing *Rosmersholm* (1886), "in the continuing absence of God Ibsen had been cultivating Darwin, feeling a spiritual need for that sense of biological determinism that allowed him to replace original sin with an equally implacable genetic heritage."[16]

In fact, there are surprisingly many references to evolution in Ibsen's writings. For example, in 1887, he proclaims in a speech at a literary banquet: "It is said that I have been prominent amongst those who have helped usher in a new era. On the contrary: I think the era we are currently in could just as accurately be described as a conclusion, and that something new is being born. In fact I think that natural science's teaching about evolution is also relevant to life's spiritual elements."[17] As Clasen and his coauthors put it, in this speech Ibsen publicly declares himself a Darwinist.[18] The speech not only reveals a striking vision of a new era brought about by evolutionary theory, but also uses the English word *evolution*, not the Dano-Norwegian "arternes udvikling." The speech goes on to describe a synthesis of ideas, as poetry, philosophy, and religion will "melt together," obliterating old forms and categories to create a single new "life force" that we can as yet only dimly imagine. Although it expresses an important idea for Ibsen—that of synthesis of concepts, forms, and laws—commentators on this speech tend to emphasize the first part with its striking reference to evolution. No commentator on this speech has, however, noted its remarkable similarity to a passage in Robert Chambers's *Vestiges of the Natural History of Creation* that likewise foresees a time of synthesis, when all systems will melt into one: "Thinking of all the contingences of this world as to be in time melted into or lost into the greater system, to which the present is only subsidiary, let us wait the end with patience."[19] It is also important to note here Ibsen's reference to the life force he sees dawning, resonating with the then-popular theory of vitalism.

Ibsen again uses the word *evolution* in a letter to Georg Brandes in 1888, this time applying it to himself. He writes that we can no longer be satisfied with the society in which we live: "The national consciousness is in the process of becoming extinct and will be replaced by a racial consciousness. I for one have gone through this evolution."[20] Ibsen's letters contain several references to natural history and science, including a curious allusion to extinction. In a letter of 1883, he says that art forms, like species of animals, can become extinct, and that verse drama will meet the same fate as the dodo, "of which only a few individuals remain down on an African island." It is worth noting the odd but also incorrect mention of the dodo; the last reported sighting of one was in 1662, so its extinction was a well-established fact by the late nineteenth century. Ibsen's mention of the dodo is resonant; it was a cultural marker not only of extinction but also of evolution more broadly and above all of the devastating consequences of human invasion of natural habitats.[21]

Clearly, then, Ibsen is engaged with evolution, from applying it as a general term to his own artistic approach to invoking specific aspects of evolutionary thought such as extinction. However, it is through gender that Ibsen exerts the most profound impact on the subsequent treatment of evolution in drama, opening the way for so many playwrights to explore women's roles within the "natural order." Ibsen pins his hopes for human progress and evolution on women: "A Pastor Manders will always provoke some Mrs. Alving into rebelling. And just because she is a woman, she will, once she has begun, go to great extremes."[22] But Ibsen is quite specific about what he means by woman's evolutionary power. It is this "extreme" capability combined with her maternal capacity: "It is up to the *mothers* through hard and slow work to awaken a conscious feeling of culture and discipline. This must be created in humans before we can bring the people forward. It is women who shall solve the human question. As mothers they will do so, and *only* as mothers. Herein lies the great task for women."[23]

What is "the human question," and why is it linked to women and specifically to mothers? Is it solely spiritual, or could it be something similar to what Bernard Shaw insists is needed: the state should subsidize women bearing children (he suggests a salary of £10,000 per year) to make motherhood "a real profession as it ought to be"?[24] And if Ibsen hails mothers in this way, why are his plays so full of thwarted mothers, abandoned or dead children, and generally distorted versions of motherhood? Paradoxically, the two qualities seem to cancel each other out—Ibsen's plays repeatedly show women behaving in extreme ways, yet with little space for motherhood. In *The Master Builder* (1892), Mrs. Solness's dolls symbolize her dead babies and her lost maternal vocation. Nora in *A Doll's House* leaves her children to be brought up by their nurse. *Ghosts* (1881) ends with Mrs. Alving forced to choose between watching her son die and helping him end his life. Rita in *Little Eyolf* (1894) prefers her husband to her child, her sexual desire helping to bring about Eyolf's disability. Ella Rentheim in *John Gabriel Borkman* (1986) loves only her surrogate son, and in act 2 she pointedly declares herself incapable of the maternal feeling that supposedly springs up in any woman when confronted with a needy child: "When a poor, starving child came into my kitchen, frozen and crying and begging for a bit of food, I let the serving girl look after it. Never felt any desire to take the child to my bosom, warm it by my own oven." Far from being resolved by scientific developments, the questioning of motherliness as innate in women is exacerbated by them and is being explored by Ibsen and, as will be discussed further, dozens of other playwrights who follow suit.

Already in his draft for *The Pillars of Society*, Ibsen alludes to the status of women as the pressing social issue that needs to be addressed and praises women for their power, which is still frustrated and unchanneled. "Our society is a society of bachelor-souls," says Bernick, "we do not see the woman."[25] In the final version of the play, a subtle but significant shift has the proto–New Woman Lona Hessel say these lines, making them an accusation rather than an observation: "Your society is a society of bachelor-souls; you don't see the woman."[26] Bernick agrees with her. His last statement is that "it is you women who are the pillars of society," though Lona rejects such pat claims and the enormous responsibility this places on women as she speaks the play's final lines: "That's a frail wisdom. . . . No, my dear; the spirit of truth and freedom—*that* is the pillar of society."[27] Ibsen is already working out exactly in what way women are to be the key to evolution, and he is aware of the immense responsibility this places on them. How will his vision of women's roles be different from that of his contemporaries, with the expectation that women both stand on the pedestal and raise men to their level?

The notes to his next two plays expand on his concern with women as the primary means of moral evolution, as he raises the problem of the sexual and legal double standard (*A Doll's House*) and the abjection of women generally (*Ghosts*): "These contemporary women, mistreated as daughters, as sisters, as wives, not raised with respect to their talents, kept away from their calling, deprived of their inheritance, their souls embittered—it is these who are the mothers of the new generation. What will be the result?"[28] In particular, argues Ross Shideler, *Pillars* and *A Doll's House* ask "what will happen to male authority in a world governed by evolution? What will happen to the patriarchy when women demand to be treated as equals? . . . Ibsen will stage the questions on an international scene."[29] Shideler identifies in Ibsen and August Strindberg a systematic undermining of male authority within the family and of the family itself as a stable unit and the cornerstone of society. These playwrights write Darwin into their plays through their depiction of "family conflicts between women striving for equality and men trapped in historically defined notions of masculinity. Usually, the disruption in the family comes from a woman whose 'biocentric,' independent, and sometimes newly discovered sexuality forces her to reject a weak or interfering husband."[30]

Ibsen's plays show individuals, especially women, striving to adapt and the chosen unit of human evolution—the nuclear family—simply failing to function. It is not only individuals, or marriages, that fail, but also whole families: the Helmers, Werles, Alvings, Rosmers, Allmers, and Solnesses. And, since families are the main unit of the tribe, or the group, Ibsen seems to be saying something fairly damning about human evolution so long as it is based on this flawed unit. What does he suggest in its place? A return to natural, rather than socially constructed, forms of life? This is what Jens Peter Jacobsen, citing Haeckel, believed: "A complete and sincere reversion to nature and the natural circumstances is required," rather than the current "barbarism" brought about by "our entire social and moral organisation."[31] Ibsen dramatizes this question of how the constructed social order sits in relation to the "natural order."

This problem of how to reconcile the social with the natural as they increasingly grow apart in the modern world informs Ibsen's engagement with evolutionary thought. He perceives a natural order to which human organization and will are subject, yet his characters often realize this too late, hence the feeling of doom, futility, and pessimism that critics often see in Ibsen's plays. Death, the random succumbing to this order, is "so meaningless. So utterly meaningless. And yet the order of the world requires it."[32] This is an important indication of Ibsen's Darwinian understanding of evolution: seeing the natural order as meaningless, nature as blind and random. In the draft of *The Wild Duck*, Ibsen has Gregers sarcastically refer to "the order of nature," the pretense at normal family life that he thinks his father wants to present to the world to avoid Mrs. Sörby looking like a loose woman:

GREGERS: *"Why, it's in the order of nature—"*
WERLE: *"Yes, it ought to be the order of nature, Gregers—"*
GREGERS: *"Oh, I haven't a rag of belief either in nature or in its order."*[33]

In *An Enemy of the People* (1882), when Dr. Stockmann realizes that even the liberal press will oppose him, the editor retorts, "'That's the law of nature. Every animal has to fight for survival, you know.'"[34]

As these examples show, Ibsen frequently refers to natural laws, and in one case, it is linked to contemporary scientific advances. In a "notice" tentatively dated 1879 by his biographers, he begins by noting that "contemporary natural scientists have more and more come to recognize that phenomena in their areas of scientific expertise actually rest upon a very small number of natural laws which, as research and the resulting knowledge progresses, constantly diminishes; and that it is likely that we will someday be confronted with the discovery that there is actually only one such law—if there really exists any at all."[35] Here again, Ibsen is invoking the notion of synthesis (all of nature's laws will gradually synthesize into one all-encompassing law), also present in his dream of returning to an original oneness with nature (something Darwinism might lead to). He seems to go against the movement of nineteenth-century science itself from generalized natural philosophy to increasing diversification and specialization into discrete fields of knowledge. Yet, might he be alluding specifically to evolutionary theory as the "one such law"? The tentative date of 1879 would have been about the time he encountered Darwin's writings and perhaps also Haeckel's.[36]

Given the strong connections that are emerging between Ibsen and Darwin, let us explore how Ibsen might have encountered his works and what other evolutionary thinkers he may have soaked up along the way. Ibsen was living in Dresden (1868–1875) when Darwin's works began being translated into Danish and being discussed in the newspapers and journals of Denmark and Norway, to many of which the Scandinavian Club subscribed; Ibsen was a member and went there regularly to read the papers. The titles the club subscribed to most likely included the Danish weekly *Illustreret Tidende* (*Illustrated Times Journal*), which as early as February 1860 had published the first brief review of *On the Origin of Species*.[37] He also interacted with the Danish writer and naturalist Jacobsen, who, along with the Norwegian naturalist and folklorist Peter C. Asbjørnsen, was one of the main conduits for Darwin's writings in Scandinavia at the time, although Clasen and coauthors point out that Darwin was already known to Danish scientists, journalists, and members of the clergy. Jacobsen makes a striking

Danish parallel to Asbjørnsen in his interdisciplinary interests and accomplishments as creative writer, journalist, and scientist. Jacobsen's translation of *Origin* put Denmark ahead of Norway in producing a translation of this work: it "also became the most important one for Norwegian readers, until a Norwegian edition appeared at the end of the 1880s, some thirty years after the original English publication."[38]

Darwin was introduced into Norway by Asbjørnsen's article on his theories in *Budstikken* (February–March 1861), though the article neglects to explain the mechanism of natural selection by which transmutation of species occurs and sets up the idea of increasing perfection.[39] Asbjørnsen emphasized Darwin's idea of descent while "hardly mentioning his explanatory mechanism for the development of species through natural selection. There is no description given of the principles properly defined as Darwinian."[40]

How, then, did Ibsen become Darwinian? The answer may lie in the broader implications, rather than the scientific particulars, of Darwin's theory. Asbjørnsen stressed the cultural sweep of his influence, affecting every genre and area of human inquiry, with the potential to shed light on human origins, deep geological time, and comparisons of previous and current organic forms. Evolution will help us understand "culture's creations."[41] To some extent, this anticipates the idea of memetics. Mrs. Alving worries about the "ghosts" of old ideas and habits that are passed on through generations and are so hard to shake: "It's not just what we've inherited from our parents that continues in us. It's all kinds of dead opinions and dead beliefs and the like. It isn't alive in us; but it remains nonetheless and we can't get rid of it."[42] This echoes Herbert Spencer's view of human culture as part of cosmic evolution; more important, it shows Ibsen articulating, a century before Dawkins, the idea that "we have the power to defy the . . . selfish memes of our indoctrination."[43]

Ibsen knew Asbjørnsen personally from their membership in the "Learned Dutchman" circle,[44] which he frequented from at least 1859. Asbjørnsen was specifically interested in "questions of evolutionary history and the theory of breeding, together with the possible connection between apes and man. These themes also emerge in several of his popular science articles."[45] These became key questions for Ibsen as well. Ibsen was still living in Norway in 1861; he might have seen the article by Asbjørnsen, followed by likely exposure (especially at a more mature stage of life, during a period of active interest in modern thought, stimulated by Brandes) to Jacobsen's article of 1872 when he was in Dresden. Ibsen likely encountered Jacobsen's translation of Darwin while he was writing *Emperor and Galilean* in the mid-1870s,[46] arguably the most revolutionary, fertile years in Ibsen's intellectual life, when further exposure to Darwin (building on the Asbjørnsen piece of 1861), especially in the atmosphere of "the modern breakthrough" generated by Brandes, is a key part of his development and his turn from verse to prose and from nationalist romanticism to modern realism. The year 1872 was when Brandes's work had its greatest impact on Ibsen; he writes Brandes to tell him as much.

Given his influence on Ibsen, it is important to consider how Brandes interpreted Darwin. In *Det 19nde Aarhundrede*, he mentions Darwin in the social Darwinist context of his chapter on Spencer. For Brandes, Darwin was an agent of transformation: "By means of the concepts of evolution and progress, [he] transformed the dead into a living universe."[47] Brandes regarded Darwin as "the corner stone upon which the intellectual edifice of the nineteenth century rested. So axiomatic had this become to him that, in his *Goethe*, he follows up all the connections which lead from the poet to Darwin. . . . It is the natural law of evolution which provides the basis for his monumental work *Main Currents*," which so deeply influenced contemporary writers like Ibsen.[48] In fact, Brandes cast this work as a drama: the "six acts of a great play" whose hero is the psychology of Europe as expressed through its literatures (France, Germany, and England) and that culminates in the victory of liberal thought in Young Germany. This "drama" shows that the "fittest have survived" and a "new species" can take root.[49] Certainly, Brandes's close affinity for, and appreciation of, evolutionary theory cannot be overlooked as another likely influence on Ibsen.

Significantly, just as Asbjørnsen had done in his mediation of Darwin, Jacobsen in *Nyt dansk Maanedsskrift* looks beyond the scientific impact of Darwin's theory to hail its implications for "human thought in general."[50] In light of Darwin's theories, he says, the "'poetry of miracles' must cease to be"—a striking analogy to Ibsen's declaration around this same time that he must write in prose because verse drama is dead. But Jacobsen immediately goes on to say that these new teachings suggest another kind of poetry: "the poetry of personal emotions humbly yielding to satisfy the severely beautiful law of higher spiritual necessities."[51] Thus, Jacobsen's interpretation of Darwin is decidedly positive, and it appears repeatedly in subsequent articles in the journal. Jacobsen sought a way for Darwin's work to serve as "the basis upon which to build a positive general philosophy of life, an ethics, even a religion."[52] It was to be a natural, not a supernatural, religion. Yet, like Darwin and so many others embracing evolution yet not quite stomaching the possibility of reversion (evolution going backward), Jacobsen talks of "perfectibility" and the "constant growth in human intelligence."[53] The controversy aroused in Denmark by Jacobsen's articles on Darwin was, Gustafson claims, as great as that in England, with a direct parallel to the Samuel Wilberforce–Thomas H. Huxley debate as well. Yet Kjærgaard and coauthors question the "myth" of the importance of Jacobsen's translations of Darwin.[54] They argue that in fact it was Haeckel's interpretation of Darwin that Jacobsen was conveying, downplaying natural selection as the sole mechanism of evolution.

It is interesting to speculate on how far Haeckel, rather than Darwin directly, could have influenced Ibsen. Haeckel's recapitulation theory was immensely influential and extended well beyond the sciences; Freud saw in "childhood behavior a recapitulation of primitive, tribal behavior,"[55] and the idea haunts several of Ibsen's plays, as in Ellida's longing for her evolutionary past.[56] Johnston claims that in a note to *Emperor and Galilean* (1873), Ibsen states: "The individual must go through the evolutionary processes of the race." As Johnston puts it, "Ontogeny recapitulates phylogeny, culturally as well as biologically."[57] This artistic interpretation of the recapitulation theory has more in common with Pierre Bourdieu and with Richard Dawkins's concept of the meme, though, than with its original biological context. Later in his career, Ibsen might well have encountered (perhaps with the aid of his son, Sigurd) Haeckel's monism, which gripped the imaginations of so many playwrights, artists, and writers. Ibsen lived in Germany for many years and would have read German papers and periodicals giving accounts of Haeckel's work, if not (though we cannot know this for certain) reading the works themselves. He also would have heard discussion of it at the Scandinavian Club. Indeed, Ferguson argues that there is more of Haeckel in Ibsen than of Darwin, and Tjønneland notes a connection with Haeckel's progressivism: in his rendering of Darwin, Jacobsen channeled this progressive interpretation (for example, quoting the final paragraph of Haeckel's *Natural History of Creation*), which then makes it into Ibsen's speech in 1887 showing such enthusiasm for evolution.[58] Downing Cless suggests a fascinating link between Haeckel and Ibsen, first in Haeckel's idea of the family unit as representative of how all the earth's organisms interact (so Ibsen's plays, usually featuring a single family and how its members live intimately together in mutual cooperation as well as conflict, are likewise a metonym for a larger-scale ecological system) and second in their shared belief in a "pantheistic spirituality" that developed throughout their careers.[59]

The German naturalist playwright Gerhart Hauptmann drew his inspirations from two main sources: Ibsen and "the conceptions of modern science."[60] His plays of the 1890s are almost entirely derivative of Ibsen; for instance, *Lonely People* echoes nearly verbatim the final scene of *A Doll's House* in such exchanges as "You can't deny that you owe a certain duty to your family. . . . And you can't deny that I owe a certain duty to myself."[61] He makes much more explicit and overt allusions to science than Ibsen does, extending beyond text to the staging when, for example, he requires "engravings of modern men of science . . . among them Haeckel and Darwin" to hang on the walls in *Lonely People*, while the conversation (at times rather clumsily) revolves around them ("that old Haeckel and that stupid Darwin, they do nothing but make you unhappy").[62] The central speech in this play is uttered "warmly, passionately" by John, the devotee of the new science, who yearns for "a nobler state of fellowship between man and woman," a new era in which "the human will preponderate over the animal tie. Animal will no longer be united to animal, but one human being to another."[63] Hauptmann is one of the most devoted dramatic exponents of Haeckel's monism; this further underlines Haeckel's strong appeal to dramatists of the time, particularly those like Ibsen who had German connections.

~

Let us take stock, then, of what we know about Ibsen's exposure to evolutionary thought. It begins with, but moves beyond, Darwin; Ibsen engages with the full range of evolutionary thinkers at the time. He knew about Darwin not only from one source in Norway but also from another one in Denmark, a decade apart, both known to him personally. In addition, during the period of living in Germany, it would have been hard to avoid Haeckel's ideas, and both Jacobsen and Asbjørnsen conveyed a Haeckelian version of Darwin that downplayed natural selection. This would be consistent with what Robert Brustein calls the "strong mystical overtones" of Ibsen's Darwinism.[64] Brustein maintains that in his later plays from 1890 onward, Ibsen abandons Darwin in his turn to symbolism: "*The Master Builder* is free from all considerations of biology, determinism, and Darwinism (even the humanistic doctor is now assigned a secondary role)."[65] This is not quite true; in a draft of the play, Solness asks, "How have you become what you are, Hilda?" as she goads him to climbing the tower despite knowing about his severe vertigo, echoing the overarching question raised by evolutionary theory: how have we become what we are?[66] Perhaps a better way of putting it is that rather than "abandoning" Darwin in the 1890s, Ibsen's engagement with evolution develops a clear openness to the Haeckelian emphasis on the influence of the environment on the organism.

But the influence of Darwin and Haeckel varies from play to play, and certain works stand out as particularly deep and idiosyncratic in their engagement with evolution. Tjønneland is right to say that *The Lady from the Sea* (1888), more than any other Ibsen play, shows Darwin's "direct influence."[67] The play brings evolutionary concerns center stage: adaptation, environment, human evolution, origins, hybridity. Although the last might seem to be its main theme, Ibsen ultimately refutes hybridity. For example, although "the lady from the sea" is often taken to mean "mermaid," the play is not called *The Mermaid*, a term that would have signaled a cross-species hybrid of pure fantasy. The lady from the sea—or, more accurately but less elegantly, "the wife from the sea"—indicates instead a focus on the relationship between a mature woman and her environment, while "wife" directs attention to the institution of marriage and women's roles within it.[68] In the course of the play, the term *mermaid* does come up, but Ellida's strange affinity with the sea (which Wolff points out is thalassophilia) never renders her a freak of nature.[69] In the draft, Thora (Ellida) is suffering from some psychological condition that is "inexplicable to the understanding of our time. . . . To the science of our time."[70] She is "the woman from the free, open ocean," yet a liminal creature that "stands on the border-line, hesitating and doubting."[71] Wangel says that she "belongs to the sea-folk," "the people who live out by the open sea and are *like a race apart*. . . . And they never bear transplantation."[72]

Successful transplantation depends on how well the organism adapts to its new environment, and this is one of the play's main concerns. Martin Puchner objects to the dominant narrative of the development of modern drama that has dramatists after Darwin endlessly depicting humankind struggling against a hostile environment as an "old story" because it omits philosophy.[73] It may be an old story to us, but in Ibsen's time, it was excitingly fresh and no one—not even Darwin—knew how it would end. Ibsen in *The Lady from the Sea* takes this idea of struggling with one's environment to a new dimension, exploring the organism's changing relationship to a varying (not static) environment and showing that change is indeed possible. Ellida transforms in the course of the play from ill-adapted to her current surroundings—the air is clammy and oppressive, the water of the fjord too warm, she is unable to relate to her stepdaughters—to a successfully acclimatized specimen. Yet the initial critical response to her showed a general lack of understanding for this "curious creature," this "unnatural" woman.[74]

One of the motifs of the play is a peculiar stutter on the word *acclimatize* every time it is said by Ballested, drawing attention to it at each utterance. This happens repeatedly, and the fact that the word appears frequently, right from the opening of the play to the end, shows its centrality, at the same time highlighting its defamiliarization through the difficulty in articulating the term. In addition, the third-person usage in the draft ("human beings really *can* acclam—acclimatise themselves") becomes first person in the final version: "But I have accla—acclimatized myself."[75] In chapter 5 of *Origin of Species*, Darwin has a section, "Acclimatisation," in which he explores the question of how far it is possible for organisms to adapt when introduced into significantly different climates. It is not so much the more exotic beings that interest him as those closer to home. He mentions some inconclusive examples of exotic plants transplanted to colder climates and then notes the "extraordinary capacity" of domestic animals of "not only withstanding the most different climates but of being perfectly fertile (a far severer test) under them." Darwin concludes that "adaptation to any special climate" is "a quality readily grafted on an innate wide flexibility of constitution, which is common to most animals."[76] He muses on how far habit, use and disuse, contrasted with the mechanism of natural selection, play a role in acclimatization. Whether Ibsen read this section of *Origin* or heard about it through discussions going on around him and in the newspapers and journals he constantly pored over, the same question clearly interested him enough to make it a central element of *The Lady from the Sea* and to pose it in human terms. Ellida may seem exotic and foreign, and she may complain about the physically as well as emotionally stifling, inhospitable climate of the fjord, but in the end, she becomes successfully acclimatized through a conscious process.

Indeed, along with his remarkable notes to *Ghosts*, Ibsen's notes to *The Lady from the Sea* indicate a direct engagement with evolution during this phase of his career in the early 1880s. Take, for instance, his use of the sea. Ellida's thalassophilia is tied directly to larger questions of human origins and descent. Ibsen muses on "the sea's power of attraction" and our "longing for the sea" and says that "human beings [are] akin to the sea. Bound by the sea. Dependent on the sea. Compelled to return to it. A fish species forms a primitive link in the chain of evolution. Are rudiments thereof still present in the human mind? In the minds of certain individuals?"[77] An element of this can already be found in *A Doll's House*; Nora knows her children will have a surrogate mother in their nurse, who was Nora's wet-nurse, and she leaves them because it is the only means of breaking the hereditary cycle whereby children inherit their parents' (and their culture's) old ideas—the ghosts Mrs. Alving talks about or the "rudiments" of our evolutionary past. Our social institutions are the most destructive of these "rudiments," especially marriage; Ibsen's plays are one long, continuous assault on it. In these notes Ibsen further alludes to "pictures of the teeming life of the sea and of that which is 'lost for ever.' The sea possesses a power over one's moods that has the effect of a will. The sea can hypnotise. Nature in general can do so. The great mystery is the dependence of the human will on that which is 'will-less.'"[78] This conceptualization of the will is often seen in relation to Arthur Schopenhauer and Friedrich Nietzsche. But could Ibsen also be engaging with Lamarck here?

Lamarckism was still popular, and for Ibsen to refute it, to contest the notion that there could be an element of will in evolution and to embrace instead the Darwinian sense of a "will-less" transmutation, would have been utterly characteristic of a dramatist whose approach to the main intellectual currents of his age was questioning and contrarian. "Tværtimot!" ("To the contrary!") is what Ibsen is supposed to have uttered when the nurse attending him on his deathbed told those gathered around that he had taken a turn for the worse. The watershed speech of 1887 quoted earlier exemplifies his frequent use of the phrase *on the contrary*. Ibsen was a welter of contradictions; Ferguson claims he wore a mask of liberal humanism that "obscures several paradoxes" because his position changed so that he was "a sometime republican" who was also "an abject monarchist," "distinctly a feminist, yet strongly anti-democratic," and opposed to party politics.[79] This contrariness marks his entire career, his whole approach to playwriting: going against the grain of received opinion, being unafraid to question, being skeptical and hostile. Ibsen could even seem to go against his previous contrariness, flouting whoever was attempting to label him something or other, as in his notorious "I am not a feminist" speech in 1898. Contrarianism also marks his response to evolutionary thought. For example, where Darwin hails domestication as positive because it yields greater variety in species, Ibsen equates domestication with degeneration and as therefore negative. This "creative misprision"—the misunderstanding of domestication as weakening the organism—becomes a brilliant dramatic stroke in plays like *The Wild Duck, The Lady from the Sea,* and *When We Dead Awaken* (1899), whose very title goes against "science" in the first place.[80] Yet, as in most other things, Ibsen is not predictable or dogmatic; as Tjønneland points out, he reverses this idea in *An Enemy of the People* in the comparison Stockmann makes between poodles and mutts.[81] He simply cannot be pinned down to any single dogmatic idea or program.

This contrarian tendency colors Ibsen's overall sense of how evolution works—whether it is progressive or degenerative, whether (as he puts it in his notes to *Ghosts*) "the whole of mankind [has] gone astray."[82] Ibsen does not seem to see human evolution as progressive. Not only are we going "astray," but we are on a downward evolutionary trajectory, possibly heading for extinction. This is the common equation of evolution with degeneration and regression—a conflation that runs through so much literature of the period and culminates in Max Nordau's attacks on Émile Zola, Ibsen, and their contemporaries.[83] As late as 1897, Ibsen writes: "The development of the human race took the wrong turn from the start. Our dear fellow-men ought to have evolved themselves into maritime creatures," an idea Downs suggests Ibsen got from Haeckel's demonstration that fish "stand in the direct evolutionary line instead of leading down to man."[84] Over a decade earlier, in the draft notes to *The Lady from the Sea*, in almost exactly the same language, Ibsen muses: "Has the line of human development gone astray? Why have we come to belong to the dry land? Why not to the air? Why not to the sea? . . . We ought to possess ourselves of the sea."[85] In the final version of the play this becomes Ellida's theory—stated in act 3, right in the middle of the play—that human beings do not belong on the land, and that if only we had learned to live our lives on the sea, "or perhaps even *in* the sea," our development would have been markedly different, "both better and happier." Arnholm's response is half joking: "Well, what's done is done. We've once and for all ended up on the wrong track and have become land creatures instead of sea creatures."[86]

Closely allied to the concern with evolution having gone "astray" is the idea of racial senility or worn-out genes. The paleontologist Alpheus Hyatt's work (1866, 1884, 1889) with fossil cephalopods, such as ammonites, led him to suggest the possibility of "the senility and death of whole invertebrate groups," the larger implications of which included the individual's loss of growth energy and the decline toward old age and death—the idea that "the group will eventually use up all of its evolutionary energy . . . declining in parallel through stages of increasing simplicity toward racial senility and extinction."[87] This is not the same thing as degeneration, however. In a draft of *Little Eyolf*, Allmers reveals his theory that "every family—that has breeding, be it observed—has its ascending series of generations: it rises from father to son, until it reaches the highest point the family is capable of attaining. And then it goes down again."[88] A few moments later, he refers to his weakling son as "the summit and crown of the Skioldheim stock."[89]

Two further examples illustrate this concern running throughout Ibsen's works and show that it is frequently tied to the idea of extinction. In *John Gabriel Borkman*, the terminally ill Ella is obsessed with the fact that her name will die out: "This thought strangles me. To be erased from existence—even down to the name."[90] As in François de Curel's *Les Fossiles*, she wants to ensure the perpetuation of the name by transferring it to a surrogate child. Wolff notes that several critics have pointed out how "children in Ibsen's drama . . . often appear to be physically marked, or even cursed, by the actions of the previous generation," and she attributes this exclusively to Ibsen's " interest in heredity."[91] It is not only that they are maimed or cursed; often, they die or they simply do not appear at all, which is not related to heredity as much as to the idea of extinction. In *Hedda Gabler* (1890), Hedda tells Brack: "I am the child of an old man—and a worn-out man too—or past his prime at any rate—Perhaps that has left its mark."[92] This may also account for Ibsen's emphasis on hair in *Hedda Gabler* (and elsewhere).[93] His contrast between Hedda's thin hair and Thea's "luxuriant" mane was not thought of until later, in subsequent or final drafts, as William Archer points out.[94] It indicates that Ibsen is focused on not only heredity but also extinction; Hedda is the last of her line, and she seems determined to kill it off. She commits suicide despite (or because of) probably being pregnant, thus deliberately truncating her family tree.[95] Perhaps the "thin hair" symbolizes this sense she has of her genes having become exhausted. But it also relates to her state of mind. Darwin sees women's hair as a marker of their sanity, saying in *The Expression of the Emotions in Man and Animals* that those with "destructive impulses" exhibit "bristling" hair.[96] He also reports several cases of women's hair signaling their emotional state. In fact, Ibsen links hair directly to three aspects of evolution: sexual selection (Thea's ability to win Løvborg as her partner is signified through her fuller head of hair), heredity (Hedda's sense that she is a child of worn-out stock is signified through her thinner head of hair), and our close evolutionary kinship with animals as demonstrated by our shared emotional expressions.

Several of Ibsen's plays show a particular interest in animals, and these shed further light on his engagement with evolution. The rapidly expanding field of animal studies is yielding new insights in relation to Ibsen scholarship, most prominently with regard to *The Wild Duck*, which H. A. E. Zwart claims is the first instance in world literature of an animal becoming a subject of experimentation. But, apart from this play, relatively little has been said about the presence of animals in his works generally and what this might mean in the broader context of ethology and evolution. Unlike naturalists like Zola, who conflate the human and the animal to make a negative comment on the human condition, Ibsen keeps them decidedly separate. For example, *Little Eyolf* features a little black dog throughout one scene. The draft of the play shows that Ibsen was already at this early stage planning to have a live dog on the stage, something that is retained in the final version of the script. "A little dog with a broad black snout pokes its head out of the bag" (belonging to the Rat Wife), and it at first terrifies but then mysteriously attracts Eyolf.[97] The dog figures a lot in this scene and—as with babies in so many of James A. Herne's plays—it could not have been fake as that would have completely undermined the realistic tone. If anything, the final version of the play draws even more attention to the dog as something mysterious, threatening yet attractive to Eyolf, a symbol of death in its little black bag.

Let us look more closely at what Ibsen is asking for theatrically here. Once the dog, with its disarmingly sweet name of Mopsemand, has made its appearance by poking its nose out of the bag, things really become interesting, as the Rat Wife beckons Eyolf to come nearer, saying the dog will not bite. Eyolf cowers next to Asta—notice it is not his mother he seeks protection and comfort from—but the Rat Wife persists, asking: "Doesn't the young man think the dog has a mild and lovable face?" Again, it is hard to see how a fake dog could convincingly be used at such a key moment when we are all being invited, along with Eyolf, to study the dog's face. Eyolf is at first surprised but then intrigued; he stares at the dog and says: "I think he has the most awful—face I have seen." The Rat Wife closes the bag and says, "Oh, it will surely come" while Eyolf continues to be drawn to the dog; he "*lightly passes his hands over the bag*" and says "Lovely—he's lovely after all." The Rat Wife warns him that the dog is quite tired out now from his arduous work as her helper in luring rats to their death. He does not bite them to death—they just lead the rats down to the water, the rats follow the boat and Mopsemand swims behind it, leading them out to the deep where they drown. This little black dog with its innocuous name is not just a symbol but an active agent of death. Its presence on stage dramatizes several associations with the workings of nature: death, breeding practices, and domestication, all key means of the human manipulation of evolution.

Although we are not invited to study the animal or its face, *The Wild Duck* similarly foregrounds animality: it not only takes an animal as its title, but also shows an extraordinary theatrical innovation in having a bifurcated stage with one half the ordinary domestic family interior of realistic drama and the other half a mysterious, darkened loft inhabited by a menagerie straight out of Darwin's *Variation of Animals and Plants under Domestication.* Darwin gives several examples of wild duck behavior in *The Descent of Man*. However, Meyer suggests that Ibsen got the image of the wild duck from a poem by Johan S. Welhaven called "The

Sea-Bird," which tells about a wild duck being wounded and diving down to die on the seabed. Whatever the approach, it is common to interpret the wild duck as a symbol of Ibsen's own anxiety at this stage in his career about getting too domesticated, too far from the wild: "One who has forgotten what it means to live wild, and has grown plump and tame and content with his basket, as unlike the author of *Brand* as the duck is unlike the hawk of the earlier play."[98] At the end of his career, Ibsen is still concerned with this concept of wildness as necessary, a key condition of artistic integrity and creativity: in a speech to the Swedish Society of Authors in Stockholm in 1898, he says he eschews membership of any organization because writers have to "go their own wild ways,—yes, as wild as they could possibly wish, if they are to fulfil their life's work."[99] Ibsen's plays consistently suggest that the artist seems to evolve differently from other humans and has difficulty adapting to his environment, perhaps possessing more of animality than humanity.

The theme of wildness versus domestication underpins many of Ibsen's plays and has usually been looked at as a metaphor for human behavior. Torvald's "pet names" for Nora (his little lark and his squirrel) refer to animals that share their environments with humans and whose wildness is constantly threatened by human intervention. What is most striking is that, with regard to animals, Ibsen often had his facts intriguingly wrong, whether on the issue of wildness versus domesticity (Ibsen mistakenly thought that the animal and thus the entire species deteriorated under domestication) or on the issue of extinction (his suggestion that some dodos still existed), further instances of his characteristic contrarianism.

Hjalmar and Old Ekdal are experimenting on the duck to test the concept of adaptation.[100] The loft is a kind of laboratory: the animals and their artificially constructed environment constitute

> Ibsen's ecological concoction of fantastic and real wilderness, a displaced place for both animals and humans. It is a bizarre yet functional "complex web" in which only the severely wounded wild duck thrives, while the rest of the menagerie is fed and reproduces only to be shot for food by Old Ekdal, the once-great hunter who killed nine bears. As such, it is his new ecological niche. . . . Yet, it is ironically or tragically an unsustainable niche, a web that is doomed to unravel and collapse.[101]

Both Una Chaudhuri and Cless see the loft as an evolutionary microcosm or niche environment, a carefully controlled experiment and a significant moment in theatre ecology in which Ibsen gives us "a '*symptomatic* space' and a 'prescient model of the paradox of a man-made nature.'" Cless calls it "astonishing" that "the playwright who more than any other brought the drawing room and domestic situations to prominence on the modern stage dared to give nature vital agency *within* one of those interior, familial dramas."[102]

Ibsen stages nature not only as backdrop but also as agent—as both character and action. Nature becomes completely integral to the themes of the plays, such as rebirth, adaptation, and the destruction of self. This is consistent with Darwin's metaphor of nature as presenting acts in one long geological drama (referred to in the introduction). Ibsen constantly invokes, as well as depicts, the awesome power of nature; in the draft to *Little Eyolf*, for example, there is a reference to the "spellbound fascination" of "nature among the glaciers and the wide open spaces" that holds a kind of "magic." Further in this draft, the "wide open spaces" are invoked again.[103] If we accept Brustein's theory that Ibsen stopped being Darwinist with *The Master Builder*, we need to pay particular attention to *Little Eyolf*; this play contains a fascinating insight into how Ibsen's ideas about evolution were developing and finding expression in his staging requirements as well as in his themes. *Borkman* and *When We Dead Awaken* (and earlier epic dramas like *Peer Gynt* [1867] and *Brand*) require an avalanche on stage, a woman engulfed by the snow and literally snowballing before our eyes, and at least the suggestion of vistas of snow to traverse on skis. His extreme landscapes and the demands they place on directors and designers are a continuation and modification of the nineteenth-century melodramatic spectacle that was discussed in chapter 1, now placed within a specifically Darwinian context that theatricalizes the organism's situation within its dynamic environment.

In fact, Cless notes that the scenic requirements of these final plays place exceptional challenges and burdens on the director or producer and designer. The elaborate staging Ibsen asks for in *When We Dead Awaken*—with its fjord "stretching right out to sea," its "small islets" visible in the distance, its "vast treeless plateau" leading to "a long mountain lake," and the towering mountain range with snow-filled crevices—recalls not only the landscapes of *Peer Gynt* but also the settings of Victorian melodrama, spectacle, and equestrian drama (the breathtaking scale of a piece like *The Cataract of the Ganges*). Critics do not seem to have drawn this parallel, perhaps because we are so set in thinking about Ibsen as a pioneer who moved drama *away* from melodrama and spectacle. But these spectacular natural landscapes are a consistent feature of Ibsen's plays: *Ghosts* with its endlessly rainy fjord setting, *Brand* with its avalanche, and *Peer Gynt* with its desert sands, storm at sea, and mountainous terrain. To see the play as it is sometimes seen as a closet drama, not meant for the stage, is to miss entirely its organic connection to the theatrical conventions of its time. In modern revivals, such extensive scenic requirements are usually scaled back or ignored due to budget constraints and perhaps also our undervaluing their literalness, reading these settings as reflecting something psychological rather than ecological.

One of the most interesting tensions or contradictions within Ibsen's work is the geographical binary between the subterranean and the mountaintop. In an early poem, Ibsen figured himself as "bergmanden," the miner (literally, "mountain man") chipping away at the underground rock face—the subconscious, dark tunnels of our souls. Hammering at a rock face is a common motif for nineteenth-century writers; for example, Nietzsche's *Twilight of the Idols* was provisionally referred to as "philosophizing with a hammer."[104] The metaphor figures writing as hard physical labor, struggling to uncover buried truths that are difficult to confront, to bring into the light. Ralph O'Connor's study *The Earth on Show* illuminates the cultural resonance of the "geological hammer" and its associations with a new kind of scientific heroism. In this context, it is significant that Ibsen continued to identify with this motif throughout his career (even his tombstone has a hammer engraved on it), indeed building it into his penultimate play *John Gabriel Borkman*.[105]

A complementary metaphor to the hammering of art is one of nature as a series of wedges, each representing a species, and each wedge subject to displacement by the hammer of natural selection. Darwin's early notes on species change use the metaphor of the wedge to describe what he would eventually call natural selection. Jessica H. Whiteside writes that "in an 1856 manuscript, Darwin compared nature 'to a surface covered with ten thousand sharp wedges . . . representing different species, all packed closely together and all driven in by incessant blows.'

Sometimes a wedge, a new species, driven deeply into this imaginary surface, would force out others, affecting species across 'many lines of direction.'"[106] In the first edition of *Origin*, Darwin describes "the face of Nature" as "a yielding surface, with ten thousand sharp wedges packed close together and driven inwards by incessant blows, sometimes one wedge being struck, and then another with greater force."[107] Stephen Jay Gould especially liked this metaphor of the wedge, which Darwin removed from all subsequent editions of *Origin*; consequently, most people are unaware of it. Whiteside calls this wedge metaphor "prescient" for its ability to depict "both the interconnectivity of species and the importance of external impacts."[108] She notes the "subtle point embedded in Darwin's abstract metaphor of the wedge: that life evolves not with a monolithic tendency toward perfection but continuously, adventitiously, but without direction, and in intimate interconnectedness to its fellows."[109] In giving up this wedge metaphor, Darwin succumbed to the pressure to see evolution as optimistic and directed toward improvement. Although there is no overt link between Ibsen's hammer and Darwin's wedge, Ibsen's plays show a tension between his call to hammer beneath the earth, in the subterranean darkness, and his longing for the "face of Nature" and the light of day, the wide open spaces and high mountaintops. Both writers conceptualize the workings of art and nature through powerful forces.

So far, we have seen in Ibsen's and Darwin's writing similar, often awe-struck, conceptions of nature and the natural. They share a sense of wonder at natural forces, the vastness of time and space, and natural beauty, but also nature's random, blind, and ultimately death-bringing mechanisms. But Ibsen does not share Darwin's positive view of the effects of experimentation with breeding. What Darwin wrote in the beginning of *Origin* regarding variation under domestication (and leaving aside for the moment the question of whether Ibsen might have actually read Darwin's *Variation of Animals and Plants under Domestication*, or at least known about it, and that this rather than a perusal of *Origin* sparked him to put these issues into dramatic form in *The Wild Duck*) is as follows: he aims to correct a misconception put about by some naturalists that "our domestic varieties, when run wild, gradually but certainly revert in character to their aboriginal stocks. Hence it has been argued that no deductions can be drawn from domestic races to species in a state of nature." Darwin puzzles over why this assertion is made when "we may safely conclude that very many of the most strongly-marked domestic varieties could not possibly live in a wild state. . . . If it could be shown that our domestic varieties manifested a strong tendency to reversion,—that is, to lose their acquired characters, whilst kept under unchanged conditions," then he might concede—but "there is not a shadow of evidence in favour of this view."[110] We can breed plants and animals infinitely.

In his remarkable notes to *Ghosts*, Ibsen dwells on the idea of artificial breeding and the fear of the deterioration of the human race: "The complete, finished person is no longer a product of nature, he is artificially produced like grain, and fruit-trees, and the Creole race and thoroughbred horses and dogs, the grapevine, etc."[111] This theme will preoccupy him in the next few plays, developing ideas that have much in common with Francis Galton's work *Hereditary Genius* (1869); in fact, dogs and horses feature in the very first lines of Galton's book:

> Man's natural abilities are derived by inheritance, under exactly the same limitations as are the form and physical features of the whole organic world. Consequently, as it is easy, notwithstanding those limitations, to obtain by careful selection a permanent breed of dogs or horses gifted with peculiar powers of running, or of doing anything else, so it would be quite practicable to produce a highly-gifted race of men by judicious marriages during several consecutive generations.[112]

Quite apart from the dogs and horses, what is significant is the shared idea of selective breeding applied to humans. Galton had been publishing his views for many years, though he first used his term *eugenics* in 1883. People were shocked at Ibsen's portrayal of syphilis and near-incest in *Ghosts*, but Downs sees this in the context of discussions of breeding at the time; the "calm manner in which Mrs. Alving . . . contemplated the possibility, even the desirability, of an incestuous union between Oswald and Regina—which no one would have condemned in the case of domestic animals—betrays the distance by which Ibsen had outstripped most of his co-evals in the acceptance of the new science. He not only drew his own conclusions from it, but interested himself in some of its details."[113]

In 1874, Georg Brandes complained of Ibsen having preposterous eugenic views: "Fancy—he seriously believes in a time when 'the intelligent minority' in these countries [Norway and Denmark] 'will be forced to enlist the aid of chemistry and medicine in poisoning the proletariat' to save themselves from being politically overwhelmed by the majority. And this universal poisoning is what he wants."[114]

Wolff identifies two main ways in which eugenics figures in Ibsen's dramatic vision: "the potentially utopian question of breeding for an improved human being, a possibility demonstrated by horticulturalists and stock breeders through experiments in biological restructuring, and the question of responsibility for unborn children."[115] Eugenics is already part of Ibsen's dramatic vision as early as *A Doll's House* and as late as *Little Eyolf*, with its killing of the deformed child by a mysterious, symbolic force (the Rat Wife).

In one draft of *A Doll's House*, Dr. Rank refers to suffering for another's sin. "Where's the justice of it?" he asks. "And yet you can trace in every family an inexorable retribution. It is my father's wild oats that my poor spine must do penance for."[116] This was a popular notion in Europe in the mid- to late nineteenth century, due in large part to the work of French scientist Benedict A. Morel and his theory of degeneracy presented in 1857, which included the idea that physical weaknesses or degenerate traits could be produced by a debauched lifestyle and carried on genetically.[117] In the full draft of the play, Dr. Rank has much more to say about degeneracy, referring to the feeble-minded, "unfit" specimens of humanity who are drunk or dishonest.[118] Rank has no sympathy with such people, who he says merely exploit the good will of others; when Mrs. Linde points out that in fact they are the ones who most need sympathy, he replies: "But we don't need the depleted examples of the race; we can do without them." He explains: "The stronger tree draws life from the weaker ones and directs them toward its own use. The same thing happens among the animals; the bad individuals in the group must yield to the better ones. And in this way nature progresses. It is just we people who with violence and power prevent progress by looking after the bad individuals."[119] Ibsen links evolution with progress here, and Tjønneland identifies this as marked social Darwinism and sees it again in *Hedda Gabler*.[120] In fact, in the first act of the first draft of *Hedda Gabler*, Hedda mocks Tesman's "survival-of-the-fittest" mentality: "You are so fond of saying that the strongest always wins," and the act ends with his uneasy echo of this as he begins to doubt this idea: "The strongest, yes—."[121]

Ibsen cleverly undermines Dr. Rank's eugenic speech, though, by having the doctor suddenly remember a critically ill patient he is supposed to be seeing (a miner whom he refers to as "that beast"), at which Fru Linde says: "Is that also a bad example of humankind, Doctor?" The doctor says that the man, in a drunken episode, has managed to shoot his right hand, so he will be useless if he survives, and Fru Linde says ironically: "But then it's surely best to get him exterminated." The doctor—voicing sentiments we will hear again in Dr. Stockmann—agrees, saying that this is a thought doctors often have, especially when tending to the poor, but who is going to take on this responsibility? "No, Madam, *we haven't evolved far enough yet*."[122] Quite apart from the direct reference to evolution, this is a fascinating insight into how Ibsen's engagement with contemporary thought on heredity was developing and could take on so many modes, from strident eugenicism to satire. Ibsen's notes to *Ghosts* continue this borderline eugenic concern with heredity and breeding: what is natural and what is artificially produced, and why do "we allow lepers to marry; but their offspring—? The unborn?—."[123]

Eugenics permeates *An Enemy of the People*, full of Spencerian ideas about survival of the fittest, lower and higher orders, and so on. According to Zwart, in this play Ibsen "staged the emergence of the modern scientific outlook."[124] I would argue that the play expresses this through its eugenics as well as its environmentalism. Doctor Stockmann looks forward to the day when there will be an aristocracy of the liberated; he talks about the "vermin" and the "curs" up north (i.e., the uneducated and impoverished) whom he was forced to treat as a doctor but whom he would be perfectly happy to see eliminated; he invokes the "minority" who is always right, the man who stands alone versus the might of the unenlightened masses. It certainly seems that Stockmann bears greater affinities with Galton's ideas than with Nietzsche's due to the specific allusions to breeding and heredity. Wolff sees Stockmann as shaping raw materials, and this is evident especially in the ending of the play, when he will nurture the "mongrels" in his new school, sowing the seeds of new thinking for new generations—an important about-face (similar to the undercutting of Dr. Rank by Mrs. Linde mentioned previously) after the eugenically tinged speeches of act 4.[125] This ending also connects with that of his next play, *Little Eyolf*, in demonstrating the need to recognize "human responsibility" (the subject of Allmers's planned book) and suggesting that human evolution proceeds by cooperation, not competition.

Nevertheless, it is difficult to be sanguine about the eugenic overtones of Stockmann's central speeches. I would like to probe the connection with Brandes a little more here. We know that Ibsen was greatly influenced by Brandes. But Brandes began to think twice about his fiery radicalism of the 1870s that had made such an impact on Ibsen and on Scandinavian literature generally. In 1884, he denied being a democrat in his political beliefs; in 1886, he discovered Nietzsche, and the conception of the Superman became for him "the final end of cultural values."[126] Brandes focused on Nietzsche in a lecture series published in 1889 as *Aristocratic Radicalism*, in which he "rejects categorically most of the ideas characteristic of his earlier criticism," including as he himself put it "certain theories of heredity, with a little Darwinism, a little emancipation of women . . . a little free thought."[127] One critic argues that Brandes "never had been a democrat in spirit."[128]

Just as with William Shakespeare, we want to have it both ways: Ibsen must be both unique—the lone genius, the exception—and deeply of his age.[129] We have danced gingerly about his eugenic views for too long, uncomfortable with them yet reluctant to confront them. It may well be that he hopped off the eugenic bandwagon after a few years. But we need to accept his temporary embrace of eugenics as part of his response to the intellectual package of evolution as it was then understood, to accept that he was in this respect a man of his age, just as Darwin disappointingly rehearses the dominant Victorian views on the inferiority of women in the final section of *The Descent of Man*. In both cases, these views are thoroughly undermined by the inherent radicalism of the works themselves.

A more salutary aspect of the "modern scientific outlook" of *Enemy* is its prescient environmentalism—specifically its exposure of eco-hubris. Cless, Zwart, David S. Caudill, and Greg Garrard have explored this aspect of the play, which, astonishingly, has often been dismissed as merely allegorical—the pollution of the baths as simply a metaphor for the corruption of the spirit. But what has polluted the baths in the first place? Just as the "infusoria" Stockmann has discovered in the baths are invisible, so are the physical sources of them kept from the audience's sight—the tanneries and polluted swamp referred to in the play. The scenes of the play are set indoors, allowing both the characters and the audience too readily to "shut out awareness of the sources of the problem, thereby losing track of the environmental stakes as the play reaches its peak in the town meeting."[130]

Not only do we not see these sources, but the intensity of the human conflicts in the play (Stockmann versus his brother, then the newspaper editors, then the entire town) further obscures them from our observation. Yet ultimately the play enacts one of Ibsen's abiding concerns: how to live harmoniously with nature even while exploiting it. As he puts it in a draft of *The Lady from the Sea*, we must "learn to harness the storms and the weather. Some such felicity will come."[131] In the draft to *Rosmersholm*, Hetman (later called Brendel) proposes to liaise with some local businessmen-entrepreneurs to manufacture some as-yet-undiscovered product that will require

> *all the oxygen that is contained in or brought to the atmosphere of the county—or will require all the carbon in the air. We—I and the other two or three capitalists might be using it to make diamonds of. But in both cases the air of the whole county would be unserviceable for men and other animals and for everything organic. Everyone of them would have to buy his portion of vital air from us—perhaps at an exorbitant price.*[132]

Is this just a ham-fisted swipe at capitalism, or is it a satirical comment on our use of precious natural resources? Brendel/Hetman goes on to say that while this scenario of course sounds preposterous, "I have only been trying to emphasise the fact that we all agree that the air and water of our planet are common property to everybody. But when the solid earth is in question—the ground under our feet, that no one can do without, well, *das ist was Anderes*! Nobody breathes a word against the solid earth of the globe being in the hands of a comparatively small band of robbers, who have made use of it for centuries, who are making use of it to-day, and who propose to make use of it for all futurity."[133]

Henry Arthur Jones hailed Ibsen's influence in a lecture at Harvard in 1906 in terms of natural history, using vast, catastrophic imagery from nature: Ibsen "looms darkly through a blizzard, in a wilderness made still more bleak and desolate by the gray lava streams of corrosive irony that have poured from his crater. Yet by this very fact he becomes all the more representative of his age."[134] It is striking how often writers characterize Ibsen's impact in terms of vast geographical events or movements (such as Rainer Maria Rilke's idea of Ibsen as "subterranean").[135] It is as if his impact on drama is on an evolutionary scale, transforming the very landscape, foundations, and elements of drama, but especially important is that this impact is not Spencerian (allied with progress and improvement) but Darwinian. By the 1890s, Ibsen is a declared pessimist, rather like Thomas Hardy. He has turned from the consolations of teleology, redemption, and progress. He seems to embrace the Darwinian concepts of blindness, randomness, chance, and even ultimate obliteration.

This is perhaps one of the ways in which he seems "so evolved," in Henry James's words. James once wrote of the contrast between Ibsen's form, "so difficult to have reached, so civilized, so 'evolved,'—and the bareness and bleakness of his little northern democracy."[136] He puts quotation marks around the term *evolved*, drawing attention to it, using it self-consciously. The very qualities critics like James and Shaw praise in Ibsen's plays, such as their "cold fixed light" and "hard frugal charm" (terms in this same *Borkman* review), evoke Darwinian nature with its seeming indifference to suffering. In the same review, James begins his discussion of the play by saying that, for many people in London, Ibsen is "a kind of pictorial monster, a grotesque on the sign of a side-show," because of what he does with form.[137] James calls Ibsen "an extraordinary curiosity."[138] This is deft as well as thoroughly evocative of contemporary evolutionary language: on the one hand seeming to concede that Ibsen seems an oddity, a human freak, yet on the other hand praising him as highly "evolved"—the quintessence, perhaps, of contrarianism.

Ibsen's imaginative and transformative encounters with evolutionary thought became a rich source for other dramatists to mine, and he sets a key precedent of contrarianism. This is especially true of his treatment of women's own contradictory situation—biologically destined to be mothers, yet doomed to exclusion from the very society they have peopled with their offspring.

4

"Ugly . . . but Irresistible"

Maternal Instinct on Stage

> *My personal revolt was feminist rather than suffragist. . . . What I rebelled at chiefly was the dependence implied in the idea of "destined" marriage, "destined" motherhood—the identification of success with marriage, of failure with spinsterhood, the artificial concentration of the hopes of girlhood on sexual attraction and maternity.*
>
> Cicely Hamilton, *Life Errant*

This chapter explores how plays in the 1890s and the first few decades of the twentieth century engaged with an aspect of evolutionary theory that had become particularly vexed. This was the idea of gender essentialism: whether motherhood was the true calling for women, whether the bond with the infant was inevitable and instinctive, whether woman's evolutionary role was to select the superior mate for the continued improvement of the species. Henrik Ibsen made such questions about women's changing roles central to his plays, and other playwrights followed suit in a range of plays that address the issue of motherhood and maternal instinct in ways that, I argue, self-consciously relate to, and often directly challenge, the scientific discourse on this subject. One of the threads connecting the plays in this chapter is what Sally Shuttleworth identifies as the "potential violence" of all the stages of a woman's reproductive life: from "the cannibalish [*sic*] longings of pregnancy" to the "onset of puerperal insanity" and the constant threat of hysteria women were under simply by virtue of having a uterus.[1] Evolutionary discourse focused particular attention on the burden that biology placed on women.

Sociobiologists have posited that motherly behavior is not ingrained, innate, or deterministic. Women are not hardwired for it. In rats, notes Sarah Blaffer Hrdy, "the act of caring for pups leads to reorganization of neural pathways in a mother's brain, making her more likely to respond more quickly to pups the next time. This is one reason why experienced females tend to be more responsive to pups than first-time mothers," and in fact confronted with newborn pups a virgin rat may simply eat them. "This is especially true in primates, where learning is critical for competent caretaking."[2] There is a postpartum scene in the play *Alan's Wife* (which culminates in the act of infanticide) when Jean's mother is badgering her to show an interest in the baby and encouraging her to pick it up, showing a process of trying to teach a behavior that she—and the audience—assumed Jean would have innately. The scene also shows that allomaternal assistance is on offer, as Jean's mother is clearly ready to take on the rearing of the baby should Jean fail to do so. But what is a standard practice in nature seems unnatural to humans who so doggedly insist on the mother's dominant role in caring for offspring.

Isadora Duncan was one performer who believed that "women's emancipation was not possible unless it began with women's bodies."[3] She wrote in her autobiography that her thoughts about "dance as an art of liberation" and "the right of woman to love and bear children as she pleased" were "considerably in advance of the Woman's Movement of the present day."[4] Yet such seemingly natural, healthy ideas were often misrepresented in the discourse of the late nineteenth century, which typically linked women's emancipation with "distorted sexuality."[5] August Strindberg is only one extreme example of those with a tendency to regard the emancipated woman as half-male, half-female, pathologized by medical science in works like Richard von Krafft-Ebing's *Psychopathia Sexualis* (1889). For Herbert Spencer, as for Patrick Geddes and J. A. Thomson, the female was weaker because she had to conserve her energy for procreation and child rearing. Many believed that the gender difference was based on actual cell differentiation, with male cells being the source of variety and female cells having continuity and stability, and that this was not conjecture but a "biologically ineluctable" fact.[6] Havelock Ellis likewise framed as scientific truth his theory of men as energetic and creative and women as supportive and nurturing.[7] At the crux of the analysis here is the fact that "biology cut both ways; embracing development, it also enabled an argument *against* change and, expressly, social change, which would protest that excessive education would damage the reproductive health of women."[8] Theatre proved to be at the center of this tension and indeed became in important platform for the public discourse around it.

Essentialism on Trial

The sanctity of motherhood was deeply engrained in late-nineteenth- and early-twentieth-century discourse.[9] This was a widely held, cross-cultural assumption about women's biological destiny; Andrew Sinclair suggests that "the very popularity of [Charles] Darwin and [Thomas H.] Huxley in conservative American circles was due to their emphasis on motherhood as woman's vital role."[10] Grant Allen "dreamed of a society in which women would seek no goal beyond motherhood, which he figured as their natural and only true function,"[11] a sentiment shared by several New Woman writers, including Sarah Grand and George Egerton.

The emphasis on the maternal role as inherently at odds with education and work links to long-held assumptions about the essential biological differences between men and women, with scientists giving biological explanations for the assumed inferiority of women's intelligence. Darwin struggles with the conventional wisdom and the scientific evidence on this toward the end of *The Descent of Man*, noting several times "the present inequality between the sexes" and suggesting that early education of girls and boys might be the key to rectifying this, at the same time agreeing with Francis Galton that "the average of mental power in man must be above that of woman."[12] It was a chicken-or-egg situation. Drawing on Jean-Baptiste Lamarck's and Gustave Le Bon's writings, Emile Durkheim wrote that the division of labor that had increasingly separated men from women in modern societies "had the effect of enlarging men's brains (and intelligence) while diminishing women's," a phenomenon one could actually see by measuring brain sizes in men and women. The advance of civilization explains the "progressive gap" between the sexes, causing "the considerable development of the male skull and . . . a cessation, even a regression in the growth of the female skull."[13] The popular press is full of such essentialist assumptions; *The Spectator* in 1894, for instance, mentions "that placid content of women with their unalterable destiny."[14] The mainstream theatre generally adopted this line, producing such affecting dramas as *Theodora* (1866) by Watts Phillips—whose depiction of an awakened maternal instinct was praised—and *Queen's Evidence* (Grecian Theatre, London, 1876) by George Conquest and Henry Pettitt, in which a blind woman is led by maternal instinct to recognize her supposedly dead young son. The popular melodrama *East Lynne* likewise revolved around the motifs of anguished mother and dying son, immortalized in the lines "Dead! And never called me mother!" Such productions affirmed the sanctity of a woman's sexual purity, her natural role as mother, and the inescapability of the maternal bond.

There were those like John Stuart Mill, Mona Caird, and the philosopher David G. Ritchie who argued that "womanhood was largely a social construct," but they were countering culturally embedded biases shared by even the best-educated and most scientific of men. Darwin and Huxley favored education for women, but believed that women's biological disadvantages made them less able to compete with men outside their natural domestic sphere: in creating women, nature had found a way of "offsetting, to an extent at least, the ravages of the struggle for existence, of counteracting the seemingly harsh regimen to which all life was subjected."[15] Caird was quick to look beyond Darwin's own conclusions about women's supposed inferiority in *The Descent of Man* and see how aptly suited Darwin's work was for feminism by freeing it from the notion of gender essentialism to focus instead on the constant process of change and flux at the heart of evolution.[16] Caird begins her celebrated essay on marriage by attacking the "careless use" of terms like *human nature* and *woman's nature*, arguing that far from being immutable, these have "an apparently limitless adaptability" that rigid social norms have failed to accommodate.[17] In the realm of fiction, George Egerton's *Keynotes* and *Discords* and Charlotte Perkins Gilman's works are just some of many texts showing ambivalence about motherhood as woman's true calling.[18]

In short, essentialism was coming under increased fire. If women were programmed to be mothers and sexual selectors, how did these two roles play out in reality? Despite the many examples he gives of female choice in the animal world, Darwin ignores it in his consideration of human evolution in *The Descent of Man*; he does not make the imaginative leap. This did not escape women writers like Eliza Burt

Gamble, who, in *The Evolution of Woman* (1894)—published the same year as the first edition of Ellis's popular *Man and Woman* but now virtually forgotten—emphasized female choice as a key feature of sexual selection. Gamble thought that women were "the intelligent factor" in human evolution; she also believed in the role of the woman's will.[19] She "made novel use of sexual selection for the purpose of underlining the injustice of female subordination. For her, nature provided a model for the organisation of social relations and sexual selection based upon female mate selection should be once again allowed full expression."[20] Likewise, feminist writer Antoinette Brown Blackwell embraced natural selection but criticized Darwin's *Descent of Man* for not recognizing sufficiently the distinct evolution of gender lines: "With great wealth of detail, he has illustrated his theory of how the male has probably acquired additional masculine characters; but he seems never to have thought of looking to see whether or not the females had developed equivalent feminine characters."[21] This is exactly the criticism Bryony Lavery's recent play *Origin of the Species* levels at evolutionary theorists more generally—that their ideas relate solely to males, and "evolution" is taken to be universal when its findings are based on only one sex.

Hrdy calls this "the road not taken" by evolutionary biology, with the popularity of Social Darwinism drowning out Blackwell's voice of dissent. But the issue did not go away, least of all for women writers, feminist campaigners, and—as this chapter will explore—theatre practitioners and playwrights. Gilman addressed this issue in *Man-Made World or Our Androcentric Culture* (1911), arguing that "the human species" was "an aberration in the natural order" because it was "the only one where the male selected females and where the female was dependent for her livelihood on the male."[22] Though Gilman believed that women's "natural work" was motherhood, she also saw men's natural work as fatherhood. Helen Hamilton Gardener likewise lectured that the natural order did not set a "line of demarcation between the sexes on moral grounds. The male and the female differ in qualities, but neither is 'better,' 'purer' nor 'wiser' than the other." It is "artificially built up conditions" that have caused the inequality between the sexes, that have "weakened woman and brutalized man."[23] She states that nowhere in nature, "from the protozoan to the highest beast or bird," is there any distinction between male and female "of right, or opportunity or privilege as to the occupation, life, liberty or the pursuit of happiness . . . until we reach the one species of animal where one sex has been subordinated to the other by artificial industrial conditions."[24]

These were by no means the only writers to put women on equal footing with men. Ellis in *Man and Woman* (1894, running to eight editions by 1934) dismissed the belief that women were inferior just because of cranial and brain size, but echoed John Ruskin's "separate but complementary spheres" idea, arguing the male sphere was the cultivation of industry, exploration, and the arts while women's was "the bearing and raising of children and domestic activities."[25] These separate domains were universal. But Ellis undermined his own efforts at redressing the gender imbalance because his "complementary spheres" model still defined women in terms of their maternal and domestic roles. Far more radical was the belief that inequality was not natural but was socially determined. Johann Bachofen's *The Law of the Mother* (1861), which argued that the earliest humans lived in a harmonious matriarchal social arrangement, enjoyed a certain currency.[26] "These were powerful arguments, and the portrait of an original sexual equality becoming subverted by social processes would assume paradigmatic status in socialist and feminist circles."[27] In the United States, sociologists such as Lester Frank Ward and William Isaac Thomas went even further, producing an "equally paradigmatic version in which females were initially *superior* to males."[28]

The discourse surrounding women's roles that began in the Victorian period came to a head in the Edwardian preoccupation with motherhood and its relation to growing eugenic concerns, and theatre became a constant and visible site for such debates. The decades on either side of 1900 saw intense struggle for suffrage, and women were being scrutinized as never before and from all angles. This directly shaped how they were represented on the stage. As Sos Eltis notes, the increasingly problematic relationship between feminism and motherhood played out on the stages of many countries.[29] Ibsen's dramas helped launch this scrutiny, as they gave unprecedented space and depth to women's issues in the theatre. But many other playwrights and, in particular, actresses took an interest in creating plays and performances that explored fundamental questions about motherhood, marriage, and the kinds of roles women could take on in society. This period also saw the "eclipse" of Darwinism and a greater openness to alternatives to natural selection. It is this convergence that interests me in this chapter: how the changing discourse on women's roles maps onto the changing discourse on evolution, how they mutually inform each other, and how and why this is reflected and explored in the theatre in particular.

Theatrical treatments of maternal instinct are often controversial and provocative and directly related to the ways in which gender roles were being radically reconceptualized in the light of evolutionary theory. But it was not only on the actual stage that this took place; theatre could also provide a useful model for journalists debating these issues in print. In a short piece for *Reynold's* in May 1894, Walter M. Gallichan captures the key debate then raging over women's roles through the form of a dramatic dialogue between Helen, "a married doctor of medicine"; Sybil, a middle-class younger woman; and Ruth, a private school teacher. All are gathered at the house of Minerva, "a middle-aged, unmarried lady, with extreme views upon the rights of her sex."[30] Because there are no men present, the women can "speak without reserve," and they do so in the language of Darwinism. Helen declares that "there is no such thing as absolute sexual equality and there never can be" because "this a hard world for female animals of every species, and for female humans in particular. I should be very glad to alter certain natural fundamental laws, but I can't. What we women have to determine is our adaptation." She says women nowadays are "in a condition of wholesome discontent," and things are slowly beginning to change for the better; but we cannot "defy the implacable mistress, Nature."[31] Minerva says that "we *can*, and *do*, defy Nature. The human brain is continually outwitting the arbitrary and irrational in what is termed 'natural law.' Our crowning achievement will be the struggle *against* the struggle for existence amongst women."

Helen agrees, and says she is certainly not "one of those who believe that a woman's 'highest duty' is maternity. The burden of reproduction falls upon us, and not upon men, hence the view that child-bearing is our chief social obligation. Men are not told that it is their 'highest duty' to fulfil the function of paternity." There is clearly a double standard. (At this point, Sybil, the young, timid, unmarried woman, nervously says to Ruth "Do you think I had better go away?" confessing she does not understand politics and her father knows best.) They discuss Spencer's theory of women's incapacity for mental labor. Helen is ready to accept this inferiority. "In a few generations of marriages," she maintains, "you can change nerve tissue, but you cannot alter sex functions."

Minerva cites numerous examples of societies in which women do play an active political role: "Researches in human evolution prove that the women of primitive tribes were the chief founders of civilization." She compares the situation for women in Britain now with that of the slaves in America. But, in a surprise twist at the end, it is Helen, not Minerva, who comes out as the truly progressive one: "There has never been any question of obedience in my marriage. Mine is neither the patriarchal nor maternal family; *the partnership is equal*. . . . I have no more desire to subjugate my husband than he has to rule over me."[32]

This short dramatic dialogue offers a glimpse of how the question of women's roles and rights was being contextualized in terms drawn directly from scientific writings on evolution and the emerging discourse around it. The use of the dramatic framing directly acknowledges the popularity of theatre as a site of discussion of such issues (only a few years later, Elizabeth Baker's *Partnership* dramatizes the same idea of a new model of marriage based on complete equality), and it also employs female "types" that were instantly recognizable to audiences: the young, submissive woman; the older suffragette; the woman torn between these two extremes. Theatregoers, in other words, were well versed in the evolutionary implications of feminism, and it registers this developing discourse on gender in astonishingly diverse and innovative ways.

Expressing Maternal Love

In his investigations into the expression of the emotions, Darwin noticed something paradoxical about motherly love. The love of the mother for her baby is "one of the strongest [feelings] of which the mind is capable"; indeed, "no emotion is stronger than maternal love."[33] Yet it is "inactive," lacking a "peculiar means of expression"; it does not register in an extremely marked or noticeable way, apart from the usual manner in which our bodies register affection more generally ("a gentle smile and some brightening of the eyes").[34] Painting and sculpture lend themselves better to the expression of this essentially static emotion than do the dynamics of the stage. Henry Arthur Jones's *The Physician* captures this paradox when the Reverend Peregrine Hinde says, "I never saw a mother's love, but I'm sure it's about the realest thing on this side of the grave."[35] How, then, do you stage motherly love?

Breast-feeding is one active and iconic expression of maternal love, but although it is a common theme in the visual arts, especially in religious art, it rarely makes an appearance on the stage. This chapter explores the single instance of breast-feeding on stage that I have been able to discover, in James A. Herne's play *Margaret Fleming* (1890). There are plays that refer to this act without actually staging it, such as Eugène Brieux's *Les Remplaçantes* (1901), which exposes the corrupt wet-nursing system in France. Breast-feeding on stage needs to be set in the context of the rise of interest in infant feeding as a public issue as well as a medical one. Recent studies have richly documented the decline of the wet-nursing systems in Europe alongside the medicalization of infant care and the takeover of that domain by the male-dominated profession of physician.[36] By the turn of the century, what had been an exclusively female role had become yet another male-regulated province.

Victorian scientists like Spencer saw breast-feeding as what distinguished mammals from the lower animals, and it was a powerful symbol of maternal instinct (synonymous with it), leading to the assumption that if you produced milk for your young, you loved them and wanted to nurture them and were biologically programmed to do so. The biologist and novelist Grant Allen firmly embraced this view, thinking of women as "not even half the race at present, but rather a part of it told specially off for the continuance of the species. . . . She is the sex sacrificed to reproductive necessities."[37] But there are plenty of exceptions to this model in the animal world, examples of mothers neglecting their young, refusing to feed, or even killing them, as *The Descent of Man* revealed. For the sake of brevity, here are just a few of the instances given by Darwin describing bird behavior. In the quail-like *Turnix* species, the female is much bigger than the male: "The females after laying their eggs associate in flocks, and leave the males to sit on them."[38] The male ostrich is smaller than the female, and it is he "alone who sits on the eggs and takes care of the young." In breeding season, the female is highly "pugnacious."[39] Likewise in one type of emu, the female is much bigger and more powerful than the male, displays ornate plumage when angry or excited, is usually "more courageous and pugilistic," and makes a loud booming noise, while males are more slender and "more docile, with no voice beyond a suppressed hiss when angry, or a croak." The male does all the incubating (as is also true of the African ostrich male, despite being larger than the female), and he also defends the offspring from their mother:

> "For as soon as she catches sight of her progeny she becomes violently agitated, and notwithstanding the resistance of the father appears to use her utmost endeavours to destroy them. For months afterwards it is unsafe to put the parents together, violent quarrels being the inevitable result, in which the female generally comes off the conqueror." So that with this emu we have a complete reversal not only of the parental and incubating instincts, but of the usual moral qualities of the two sexes; the females being savage, quarrelsome, and noisy, the males gentle and good.[40]

Thus, there are many bird species in which the males are smaller and weaker and have "not only *acquired the maternal instinct* of incubation, but are less pugnacious and vociferous than the females. . . . Thus an *almost complete transposition of the instincts*, habits, disposition, colour, size and of some points of structure, has been effected between the two sexes."[41] Darwin's seemingly oxymoronic suggestion that maternal instinct can be "acquired" questions the gender essentialism that anchored Victorian discourse on women's roles. And, if allegedly gender-specific instincts are fixed and innate, how can they be transposed from one gender to the other?

This issue is important to address and to set in its original context because it is still ongoing through such controversial works as Elisabeth Badinter's *The Myth of Motherhood*, which challenged the assumed innateness of motherliness and posited instead that this quality was socially conditioned and varied according to social norms.[42] The fact that we are still debating whether all women are born with maternal "instincts" and whether that is the same as maternal love makes it all the more relevant and useful to recapture the historical context from which the discourse on this issue arose and with which theatre directly engaged, often in pioneering ways.

In this period, as Rae Beth Gordon points out, "instinct" generally suggested atavism, savagery, and regression to the Victorians because of the association with animal behavior and with the unconscious.[43] How does the maternal "instinct" fit into that tendency? How can it be both "savage" and saintly? This paradox is staged in Herne's *Margaret Fleming.*

Margaret Fleming: The Play About the Breast

As noted in chapter 2, James A. Herne was once characterized as "the American Ibsen," and *Margaret Fleming* was once studied as the first attempt to establish a Théâtre Libre in America, initiating the little theatre movement there.[44] When the play premiered in 1890, American drama had just been declared nonexistent by the hugely popular playwright Dion Boucicault.[45] Although as one critic puts it, the Ibsenism that Herne attempted to usher into American drama was "a sadly emasculated affair" that melded social problems and romantic melodrama, Herne was an important force in helping to develop a distinctive native drama throughout the 1890s that drew on European influences and spectacle transposed into a more realistic, regional key.[46] Although the play was successfully staged in 2007 in New York off Broadway, revivals have been extremely rare. Until the 1950s, scholarly articles were still being written about *Margaret Fleming,* but then it gradually dropped off the critical radar, despite its recognized status as "a significant milestone in American dramaturgy."[47]

The original production of *Margaret Fleming* largely came about because of the dogged support of the writer Hamlin Garland, one of the leading exponents of American realism, who admired Herne's attempts to write realistic drama and who shared his interest in science generally and in Darwin specifically, mainly through the works of Spencer. Garland and Herne premiered the play at the Lynn Theatre in Lynn, Massachusetts (close to Boston), on July 4, 1890, for a few performances; then, rebuffed by a succession of theatre managers in New York and Boston, they decided to put the play on themselves and rented Chickering Hall in Boston, a "small auditorium" with a capacity of five hundred that they "converted" into a theatre for their production of the play there on May 4, 1891.[48] Just as with many of Europe's independent theatres, here the audience was select, the intellectual elite, the cream of the literati in Boston society.[49] The play ran for three weeks. In a memoir of the production published in 1914, Garland recalled that "we all looked forward to the experiment with intense eagerness, for the reason that in this small hall Herne had determined to produce *effects of intimate realism hitherto unknown on our stage.*"[50]

According to one biography, Herne "could not have chosen a worse time for his bold venture into advanced realism" because the plays that were on in Boston at the time were sentimental melodramas like *East Lynne, The Octoroon,* and *Ten Nights in a Bar Room* or "romantic adventure dramas" like *Beau Brummel, Shenandoah*, and *Don Juan*; Richard Mansfield was starring that year (1890) in *Dr. Jekyll and Mr. Hyde*, and another hugely popular play was *The Soudan*.[51] Ibsen's *A Doll's House* was also on Boston's stages at this time: in a production with Beatrice Cameron (Mrs. Richard Mansfield) on October 30, 1889, and in the groundbreaking English production by Janet Achurch and Charles Charrington that played in New York, Boston, and other cities during the 1889/1890 season. The final scenes of *A Doll's House* and *Margaret Fleming* revolve around the same ideas, suggesting that Herne most likely had seen at least one of the these productions.[52]

What is often overlooked entirely about *Margaret Fleming* is that the definitive text no longer exists: the script used for the original production was lost in a fire in 1909, and the printed version now in circulation is a much later reconstruction from memory by Katharine Herne, the wife of the playwright and the actress who played Margaret (and largely created the role). Herne was apparently not only extremely motherly and loved babies but also deeply religious.[53] These qualities clearly shaped her reconstructed text of *Margaret Fleming*, which was first published in 1929, decades after the production, when Arthur Hobson Quinn helped to rescue the play from oblivion by publishing Katharine Herne's version in an edition of his widely read anthology *Representative American Plays*.[54]

Many theatre historians seem unaware of this fact, either taking Katharine Herne's reconstructed text in Quinn's anthology to be James Herne's original as played in the 1890s or seeming not to consider this textual instability to be significant.[55] Yet the two versions differ significantly, and the assumption of what is meant by "the play" in discussions of *Margaret Fleming* is thus far more problematic than has been acknowledged. Indeed, precisely because we have to rely on secondhand information about performance and reception to gain any insight into the nature of the original text, *Margaret Fleming* is intriguing not only in itself (whatever that elusive self is) but also in providing a reversal of the usual procedure in theatre history whereby the text becomes the main artifact of theatrical production.

According to written accounts of the original version,[56] Margaret is the young wife of Philip, who owns a mill. Together, they have a lovely baby girl and seem very happy. But already in the first scene, the local doctor tells Philip that he knows Philip is the father of a baby who has just been born to a young woman named Lena, who used to work for the Flemings and who is now dying. He appeals to Philip's conscience and gets him to go to Lena. Lena's sister, Maria, works for Margaret looking after her baby, and is of course distraught at Lena's grave illness but does not yet know who the father is. In sympathy for Maria's distress, and also pitying and identifying with the new mother, Margaret decides to visit the dying Lena. There, she finds out that Philip is the father, and she immediately loses her sight due to a hereditary weakness in her eyes that means any sudden emotional shock can trigger blindness. Lena dies, the starving baby cries piteously, and Margaret takes him to her bosom.

To this point, the reconstruction and the original are nearly identical. But descriptions of it at the time suggest a fairly preposterous and contrived ending to the original: an interval of five years elapses between the last two acts, during which Margaret seems to have gone temporarily mad as well as blind. Her own daughter, Lucy, has been stolen by Maria in an act of revenge and is now helping out in a tavern run by Maria and her no-good partner Joe, who also used to work for Philip; little Lucy is sipping the foam off the beer as she serves the customers when her parents appear. The play ends in a brawl in a police station, with Philip and Margaret parting after she refuses to forgive him: "The wife-heart has gone out of me. Only the mother-heart remains."[57] The illegitimate baby seems to have been entirely forgotten, and we can only infer that he has died.

Bizarrely, this original version was said to have a plot that "moved forward in a quiet and thoughtful manner with simplicity of effect."[58] However contrived it might seem, one must not overlook that this original ending was far more radical in two ways: by not reconciling husband and wife and by Margaret accepting the bastard baby. John Perry calls this a "searing resolution [that] . . . diluted the work's original strength."[59] By contrast, the reconstructed version seems dignified and unsensational, clearly the product of a different period of American drama when the psychological realism of Susan Glaspell and Eugene O'Neill had supplanted Victorian melodrama. In the reconstructed version, only a week elapses between the two last acts, and it is Philip who has disappeared out of shame and weakness. He has attempted to drown himself but ended up in the hospital, and he now returns to find Margaret, still blind but happily looking after both babies. Margaret sticks up for the bastard son, insisting that Philip accept him as Lucy's equal; the curtain falls on what looks like an idyllic nuclear family, though without a definite reconciliation. "Motherhood is a divine thing," says Margaret: "Remember that, Philip."[60]

This is the text most scholars are referring to when they speak of *Margaret Fleming*, effacing key distinctions between the two versions. In the reconstructed version, motherhood is a religious calling, a "divine thing" whose biological function becomes enshrined within the sacred ideal of the eternal maternal, echoing Comte's notion of the Divine Woman. The pure and virtuous Margaret is made to embody this ideal through the play's most significant act: the depiction of breast-feeding. According to one biography, Katharine herself suggested this breast-feeding scene to Herne.[61] This may be partly why she retained it when she reconstructed the play much later, despite watering down the rest of the piece.

The play may not quite bare the breast in the style of Janet Jackson, but Margaret is shown to be "openly nursing the baby on the stage."[62] When Margaret is alone on stage with the starving, wailing baby, she has just gone blind and is groping her way around the room, reaching out to comfort the baby. Her blindness emphasizes even more the inexorable pull of natural instinct as she first "feels for the child and gently pats it," then picks it up and goes over to a low chair, "then *scarcely conscious of what she is doing*, suddenly with an impatient, swift movement she unbuttons her dress to give nourishment to the child, when the picture fades away into darkness."[63] The stage directions indicate this is pure instinct, shown theatrically by the sudden action and the immediate dimming of the lights; it is obvious what is happening, but the audience only glimpses it.[64] It is clear from contemporary accounts that breast-feeding was very nearly depicted, as Katharine got as far as "unbuttoning the front of her dress" before the curtain "drops on her display of underclothing."[65] An audience member described in his journal how Margaret, "in sight of the audience, prepares to suckle" the baby.[66] There were references to "nursing a baby on the stage," and one critic was convinced that Margaret really bared her bosom.[67] Julie Herne later explained that "the curtain fell just as Margaret started to nurse the child," but this hardly helps to clarify what the audience actually saw.[68]

There are many possible literary inspirations for depicting breast-feeding, and in fact Herne was accused of plagiarizing parts of Wilkie Collins's *Hide and Seek* (1854) by his fiercest critic, William Winter.[69] In Collins's novel, Mrs. Peckover takes a starving baby to her own ample breast when she finds her crying beside her dying mother at the side of a road; she later adopts the baby. "The milk which nourishes little Mary is, literally, the milk of human kindness," notes Catherine Peters.[70] It is significant that Mrs. Peckover narrates this act herself and in so doing indicates how instinctive it was for her to nurse the woman's baby: "I whispered to her again, 'Why don't you suckle it?' And she whispered to me, 'My milk's all dried up.' I couldn't wait to hear no more till I'd got her baby at my own breast."[71] Much further in the novel, the child's uncle tries to show his gratitude for Mrs. Peckover's act by paying her a sum of money, but she refuses on the grounds that "the mother's love transcends the dealings of the marketplace."[72] To put this act on stage, to render these feelings without the aid of narration, was a much more radical gesture, and it did not go unnoticed.

This must have been delicate to depict on stage in 1890. Even today in the United States and the United Kingdom, the issue of breast-feeding one's own child in public is controversial, let alone doing it for someone else's baby, which is practically taboo now in Western culture.[73] In *The Descent of Man*, Darwin notes that "birds sometimes exhibit benevolent feelings; they will feed the deserted young ones even of distinct species, but this perhaps ought to be considered as a mistaken instinct," and though he does not explain why it is mistaken, it must be because this altruistic act interferes with the Malthusian mechanism of nature taking its course.[74] Certain parallels can be drawn here, for while Margaret and Lena are obviously of the same species, socially they are utterly "distinct" from each other.[75] It is significant, then, that both the original and revised versions seem to dwell on this scene—although it was "tempered" in the reconstructed text so that "Margaret's maternal action is merely begun as the curtain goes down."[76] The original version may in fact have been much more forthright about the nursing, as the reviews indicate. According to one outraged critic, Margaret "bares her bosom to the public gaze in order to suckle a famished infant,"[77] and *The Cambridge Guide to American Theatre* notes that "critics who rejected the play were especially offended by the spectacle of Margaret breast-feeding her husband's starving bastard," even though they universally praised Katharine Herne's acting.[78]

This is the moment usually cited as the notorious breast-feeding scene in the play. But there is actually another incidence of nursing, and that is when we *first* see Margaret. The opening stage directions for Margaret's first appearance are explicit about her natural role as mother: "Margaret is seated in a low rocking chair near the fire with the baby in her lap. A large bath towel is spread across her knees. . . . Margaret is putting the last touches to the baby's night toilet. She is laughing and murmuring mother talk to her. A shaded lamp is burning on the table to the right. The effect of the light is subdued . . . making a soft radiance about Margaret and the child."[79] Margaret's first lines are gently spoken to the baby while she "fastens the last two or three buttons of her dress": "No—no—*no*! You little beggar. You've had your supper! No more!"[80] In his review of the play at the time, Garland did not mention this near-nursing moment; he merely called Margaret "happy in her motherhood, with her baby—a most intimate and beautiful home picture," so we do not know how explicit Katharine chose to be about the breast-feeding.[81] But these stage directions are clear about what has just happened, as other critics also have noticed; for instance, Gary Richardson, in his plot summary, describes how "the scene turns to Margaret at home, where she has just finished bathing and breast-feeding her baby."[82] Richardson notes that "the tender stage business between Margaret and her baby not only invokes traditional visions of the purity and warmth of motherhood, but also implicitly contrasts the sexuality sanctioned within marriage with the type of amoral intimacy that surrounds the other mother and child that have claims upon Philip." Yet, echoing male critics' objections at the time, Richardson seems to consider her stance toward the philandering Philip as morally rigid (citing her "sternness" toward him) rather as than a dignified response to his infidelity.[83]

The critical response was predictably prudish. One reviewer complained of Herne "bringing into such frequent evidence the functions of maternity."[84] This sense of the play as having captured the "natural" dominates the initial response to it. William Dean Howells, another leading exponent of American realism, wrote in *Harper's Magazine* that "the naked simplicity of Mrs. Herne's acting clutched the heart.

It was common; it was pitilessly plain; it was ugly; but it was true, and it was irresistible"; even Edward A. Dithmar, one of Herne's most hostile critics, admitted that "the piece is . . . realistic in everything. We see human beings as they are. . . . The author steered clear of all the old conventions of the drama."[85] The term *irresistible* is particularly suggestive as it echoes Matthew Arnold's rousing injunction to his countrymen to "organise the theatre" because "the theatre is irresistible"—acknowledging the power of the stage as a cultural force.[86] It also hints at the powerful pull of the role of Margaret both for the actress and for the audience.

These comments generally suggest a thorough-going naturalism on stage, as Émile Zola had called for decades earlier: "human beings as they are" driven by forces beyond their control, guided by biological instinct. In discussions of theatrical naturalism, rarely is the connection made to the displays of human "naturalness" enjoyed by thousands of spectators at the zoos of Europe (discussed in chapter 1). Photographs of human "primitives" being exhibited show the subjects either completely naked or scantily clad, always with the women's breasts exposed, and the women always with a baby on their backs or a small child nearby. Contemporary reports and the scholarship on these exhibits are silent regarding whether the women publicly breast-fed their babies, but it is highly likely that they did.[87] Whether the act took place or not, it was implicit in the exposed breasts and the milling babies. This provides essential contextualization for Herne's depiction of breast-feeding on stage that is missing from analyses of the play and its reception, as it links the play to a popular performance mode that is usually left out of theatre history, and flags the contradictory associations of breast-feeding (and indeed babies themselves) with both "savagery" and altruism. In nursing her husband's illegitimate baby, Margaret is not demonstrating maternal instinct (how can she be, if the baby is not her own?) but a supremely altruistic act of moral generosity that goes right to the crux of contemporary concerns, expressed most clearly by Huxley, about evolution and ethics.

Herne's bold treatment of breast-feeding in 1890 stands in stark contrast to Brieux's *Les Remplaçantes*, which revolves around this act but never shows it.[88] Brieux attacks the corrupt wet-nursing system that still prevailed in France whereby poor provincial mothers would leave their own babies to go to Paris and earn money as wet nurses for wealthy families, with the result that often their own babies would die and much of the money they earned went into the pockets of provincial brokers, usually mayors or other powerful men. These wet nurses had "to sell to the wealthy what by nature belonged to their own infants."[89] This was a time of great national concern in France over the declining birth rate and high infant mortality, a general problem across Europe, including Britain; in fact, Darwin calls it a paradox that "savage" races breed rampantly, checked only by natural factors, whereas "civilised" races have developed moral strictures that inhibit reproduction, such as "the senseless practice of celibacy," which "has been ranked from a remote period as a virtue" because it requires "self-command."[90] The larger social agenda of Brieux's play was to stimulate population growth by reestablishing maternal bonds and the centrality of the nuclear family; this is, of course, in marked contrast to his play *Maternité*, which protested the use of women as breeding machines without reproductive choice or limit.

Les Remplaçantes insists on a woman nursing her *own* child, not someone else's. The play opens with a happy couple crooning over their baby, and then the mother, Lazarette, goes behind a curtain to breastfeed it, interrupting a conversation about wet nursing with her husband in which she has just vehemently asserted that she could not give her own baby the bottle while giving "my milk—his milk—to a little Parisian." To her baby she says lovingly, "Tiens, justement, le voilà qui s'éveille et qui va réclamer son déjeuner" and then "*she begins to unbutton her blouse and goes behind the curtain to nurse her baby*."[91] This is the closest we come to bare breasts, even though the characters talk incessantly about breast-feeding throughout the entire play.

In his analysis of the legal implications of revealing breasts on stage, Jeffrey D. Mason notes that "American law presents the 'female breast' (the statutory phrase) as a danger zone, a territory that authority must control and restrict, an agent of transgression and obscenity. . . . The breast becomes a synecdoche for the entire body, a focal point of vulnerability that suggests the jeopardy of the whole, whether the performed body or simply the body on public display."[92] Depending on context, the female breast can signify either sexual or maternal functions, and confusing or conflating the two (as arguably happens in *Margaret Fleming*) offends the audience. In fact, such neat taxonomies are problematic: "Even those whose nudity the law condones occupy the margins in various ways: naturists, swingers, nursing mothers, those under ten years old, and models for art classes."[93] The implication seems to be that, banished to the borders of society, the breast-feeding mother sits somewhere between a swinger and an innocent child.

Babies on Stage

Although "babies appeared in every genre of Victorian theatre from pantomime to the play of ideas,"[94] and during the early nineteenth century, "it was standard practice for actresses to appear on stage with their own babies,"[95] this practice ceased in the final decades of the century. By 1879, the New York Society for the Prevention of Cruelty to Children had made it illegal for anyone under the age of seven to work as an actor, and the society was protesting at an actress using her six-month-old baby in a play in which she was acting.[96] So, by the time of *Margaret Fleming*, the public, "still hungry for stage babies, had to be satisfied with rhetoric and parody."[97] Yet, judging from contemporary accounts of *Margaret Fleming*, a real baby was used in the first productions; for example, the *Evening Transcript* "loved the baby," though William Winter fumed that "several babies are introduced."[98] In fact, Herne generally seemed completely unfazed by this legal development; "the Baby" features on the cast lists of most of his plays, leading one critic to declare: "No Herne play is complete without a baby and a good dinner."

Herne's training in melodrama and spectacle had taught him the tremendous show business potential of babies: "A child is a great attraction in a play—a *baby* a power."[99] To guard against the unpredictability of babies' behavior on stage, he gave helpful hints in his stage directions about the need for actors to adapt to the babies as they adapted to their stage environment, to play spontaneously off what the babies did. Yet infants were for Herne far more than mere audience pleasers. He depicted them in ways that go beyond extraneous stage business to become thematically resonant and deeply suggestive of an engagement with evolutionary theory. His sustained interest in showing the care, nurturing, and playfulness of babies on the stage constantly opens up questions of parenting roles: What is fatherhood? How much of the motherly instinct is innate? How stable are such roles? His first major stage success, the heart-wrenching melodrama *Hearts of Oak*, features a baby and foregrounds fatherhood by inverting the formula of successful melodramas like *East Lynne*. Rather than a woman agonizingly reunited with a dying child who does not realize it is her mother, in *Hearts of Oak* it is a man, and the emphasis is on the intense bond between father and child: "For six years I have been longin' for one clasp of my baby's arms."[100] The baby is on stage for a long scene in act 3 called "The Baby" in which she must carry out actions that undoubtedly require a live infant, not to mention a precociously good actor. When Uncle Davy "puts his finger in the baby's mouth, the baby is supposed to bite it," prompting an examination of her mouth to see her new teeth; he then plays patty-cake and peek-a-boo with the baby, sings to her, and holds her up "so she can see and reach his spectacles. . . . The Baby takes his spectacles" and tries to put them on.[101] Notably, within the time frame of the play, the baby can be no more than thirteen months old.

Shore Acres (1893) went even further by having what one reviewer called a "swarm of pretty babies" on stage.[102] One of Herne's greatest hits, the play arguably takes babies a step further than in *Margaret Fleming* by showing a perfectly contented three-month-old baby without her mother, who is hiding in the woodshed while the baby is being happily passed around and cooed over by her male relatives after being "discovered" on the doorstep by Uncle Nat. The signal to the audience that Martin, the baby's grandfather, has suddenly transformed from harsh and ignorant to loving and humble is his repeated insistence to Uncle Nat to "give me that baby!" Nat playfully refuses so that when Martin finally gets the baby in his arms and stands center stage with her, there is no question that this union of proud grandfather with baby is the emotional culmination of the scene and deeply satisfying for the audience. As in other plays, Herne foregrounds the male bond with the child, and this extends to all the men present, even those not related to the baby. Blake, a gruff but kind-hearted bachelor and family friend, "*takes the baby carefully in his arms, and looks lovingly at it*" despite the motherly Ann (Martin's wife) protesting, "Be keerful, you ain't used to handlin' babies, Mr. Blake."[103]

It may well be that Katharine Herne's motherliness influenced Herne's writing, but this has obscured the importance of his own interest in celebrating on stage the paternal bond and suggesting that the role of the father is equally nurturing. It is also worth noting that Helen is described in the stage directions as "the modern girl" in contrast to her mother, "the old-fashioned, submissive wife, awed and frightened" by her husband.[104] Helen's modernity is also explicitly linked to her desire to choose her own mate, one of the first things we learn about her.[105] There is a quietly radical act here as Herne celebrates motherhood without linking it to the sacrifice of female autonomy and also insists on giving fatherhood a central place.

In the reconstructed *Margaret Fleming*, the role Katharine gives the babies is similar: scenes hinge on the powerful pull of the live infant, indeed, *two* infants—Margaret's own baby contentedly playing in her lap after breast-feeding and Philip's bastard baby crying with hunger, then instantly calmed by the breast. The reconstructed text may be strongly narrative in this sense, but there is extraordinary theatrical potential here as well. In these scenes, more than just the visual sense is at stake: we must not only see but also hear the baby, the sounds ranging from contented burbling to fierce crying to finally the sound of peaceful, relieved sucking. Breast-feeding is, after all, a multisensory affair, but this is not theatrically viable; quite apart from the legal issue, of all the logistical nightmares involved in theatrical production, a live baby or two would be the worst. As Anne Varty shows in her study of babies in the Victorian theatre, they almost always behave exactly counter to the emotion of the particular scene they are in, turning serious drama into farce and vice versa.[106]

It is instructive to look at how later productions of the play have handled the breast-feeding moments, for despite technological advances the challenges (audience squeamishness, embarrassment, outrage) remain exactly the same. In the revival at the Metropolitan Playhouse in New York in 2007, an intimate space that makes simulation even harder, a prosthetic baby was used in the form of a wireless speaker. The actress portraying Margaret downplayed the implication of breast-feeding in the opening scene, possibly to the extent that this was indicated through the lines more than through action or costume business. The director was sensitive to what an audience might be likely to tolerate:

> Regarding the opportunity to surprise the audience with a fully convincing breastfeeding—an act still able to provoke excited debate in board-rooms and public parks, let alone on stage!—it probably would not strike me as

> necessary. If the point is to tell the story, we can do so with a fairly conventional modesty. If the point is to surprise the audience with a breach of convention, we have to consider whether that "shock" to bourgeois playgoers' sensibility, which includes an acute awareness of the actors' modesty, is a constructive element in the experience of seeing the play.[107]

At the climactic moment of breast-feeding the bastard baby, the actress did not expose her breast but "went through the motions of unbuttoning her blouse beneath the dress bodice, and then bringing the swaddled child [a wireless speaker] to feed. . . . With the baby already raised, a fairly slight rearrangement of her clothing was sufficient to illustrate her intention to the audience without revealing either her bare body or the fact that she was not truly bared."[108] Metropolitan's production was praised as "a fine and clear-headed revival" of a "fascinating" play.[109]

It may seem odd that directors still need to consider with such sensitivity the possible audience response to breast-feeding on stage, but in fact, by the time *Margaret Fleming* was written, breast-feeding had moved into the technological sphere it occupies today and a discourse on it was springing up in relation to pharmaceutical developments. Sally Mitchell notes that by the mid-nineteenth century, "glass bottles and rubber nipples became available" and that in the 1890s "the importance of sterilizing the milk, water, and bottles was understood"—so "for the first time, conscientious mothers could consider bottle feeding when it was difficult to nurse."[110] Thus, Philip Fleming's reaction to seeing his wife about to breast-feed his baby is typical for the period—he "pauses in horror" and "stands in dumb amazement, watching her"—simultaneously admiring the purity of motherhood and revolted by the primitivism often associated with breast-feeding, especially in the historical context of class division over nursing, with well-to-do mothers employing wet nurses. Ibsen alludes to this in a draft of *A Doll's House* when Nora tells her nurse that she realizes she could not have wet-nursed Nora unless she had had her own baby as well.[111] Though ostensibly horrified at the surrogate nursing of a baby not her own, Philip (and the audience) are really aghast at the display of an instinctual, biologically driven behavior that overwhelms Margaret and shows what Ian Hacking has called "the false dichotomy between acting and thinking."[112] The stage directions pack an entire discourse around the vexed issue of motherhood into a single facial expression.

Maternal Instinct and the Displaced Male

Phillip's response not only signals his guilt, shame, and revulsion but also suggests the problem of the displaced husband, relegated to second place by the needy infant, and the impossible position of women having to live up to the two competing ideals of motherhood and wifehood that permeated Victorian gender ideology.[113] H. M. Harwood's play *The Interlopers* (published as *The Supplanters*) scrutinizes this issue and serves as an extended theatrical discourse on motherhood.[114] "Is motherhood their [women's] highest function?" asks the play, and the answer comes from the disillusioned father, Jack, who feels neglected by the single-minded devotion of his wife, Margaret, to their children. In a startlingly bitter tirade, Jack asks:

> *Do women think that? And for this we have cultivated our senses—for this we have ransacked the whole world to make our lives fuller and more varied—so that at the end of it all we may be told that the highest function of women is the performance of a duty that is done better by* any *savage than by the finest flower of English womanhood, and which is done worse by* any *woman than by the meanest member of the animal kingdom.* (With deep disgust.) *If I thought that was true I'd go out and drown myself in the nearest pond.*[115]

This is met with stunned silence by the ladies in the room, including Margaret, until his sister asks him why on earth he got married. He confesses: "I couldn't help it." He tells his wife that he thought she was different, not "one of these morbid modern women who think about nothing but having children."[116] In an extraordinary inversion, modern women here are defined as morbid by their excessive love of their children, whereas a decade or so earlier it was precisely the rejection of motherhood in favor of other pursuits that got women derided as "morbid."

Notably, it is the childless and stereotypically dry spinster in *The Supplanters*, and not the tender, devoted mother, who utters the cliché that marriage is "a means to the end that they [women] may fulfil their highest function." Jack's backstory about his overly maternal mother explains his bitterness against women who give too much of themselves to their children, alienating their offspring by their excessive attention, killing the joy in their marriages, and effacing their own personalities. He ran away from home because his mother essentially suffocated him. "My mother was a woman with the maternal instinct—like you, Margaret. She was a good mother. She gave to her children all her time, all her energy. It was her boast that we always knew where to find her," but, Jack adds bitterly, "she always knew where to find us." Jack's father "adapted himself" by turning to his books and his gardening, no longer even trying to communicate with Jack's mother. "They might as well have inhabited different planets." It was a "horrible tragedy": "The children, for whom it had all been done, gone, and my father and mother gazing in hopeless silence at each other across the wilderness of the dining-room table; an impassable gulf between them—made by my mother's hands. 'Never marry a woman with a maternal instinct.' Those were almost the last words my father spoke to me."[117]

According to the *Oxford English Dictionary*, *to supplant* means "to dispossess and take the place of, esp. by underhand means," while *interloper* means "an intruder" or "a person who interferes in others' affairs, esp. for profit." Both of these meanings imply a degree of cunning not normally associated with young children, strangely discordant with the play's focus on a woman's all-consuming attention to her young children at the expense of her husband. The titles not only are inaccurate but also direct the blame at the children, who after all did not ask to be born. Harwood's play taps into an ongoing discourse on the correct balance between women's roles that was proving harder and harder to achieve, and "maternal instinct" here becomes an easy target for his social criticism. The copy of the play submitted to the Lord Chamberlain bears the working titles *Paternity* and *Wife or Mother,* suggesting that the play's themes could revolve either around the male or the female. The point is that in the final version of the play, *both* parents are in crisis over the way socially constructed perceptions of their natural roles have affected their relationship.

"Baby Science"

In its foregrounding of infants on stage and of the simultaneously "primitive" and "natural" act of breast-feeding, *Margaret Fleming* raises questions not only about the biological basis of motherhood but also about how babies were actually conceptualized at this time, largely through science. What is remarkable first of all is the very centrality of babies to this play, the way the drama depends on actually seeing babies on stage (as opposed to Ibsen's *Doll's House*, which can easily dispense with child actors). As Varty has shown, children were linked to the primitive instinct, the persona before it had become civilized. Building on Gustav Jahoda's work on the idea of the savage in Western culture, Varty notes that Spencer and many other social anthropologists equated savages and children, based on "the child's capacity for imitative dramatization . . . [casting] the child actor as a species of primitive or savage humanity. It offered an irresistible spectacle for the Darwinist and imperialist Victorian public. The logic of Spencer's argument pointed towards a domestication of the figure of the savage during an era dominated by the politics of imperial expansion. Strange peoples encountered at the borders of distant British territory" not only were being brought home to be put on display, as we have seen, in zoos, but also could be found "at the very heart of every civilized home, in the nursery."[118] The nursery and the zoo were not as far apart as they seemed.

For American audiences, the play's surrogate breast-feeding would also have had the potential to evoke strong racist associations with the tradition of African American wet-nursing in the South, using black slave women to nurse white babies, and the dangerous link to "savagery" that implied.[119] Yet in many ways, *Margaret Fleming* flatly contradicts the standard practices of wet-nursing in America at that time. Margaret is bourgeois, not poor; she is not sacrificing her own baby to save another, but taking both on equally; she is doing it completely altruistically, not to earn a living as a wet nurse; she is not a fallen woman, as indeed her act implicitly redeems the "fallen woman" who has just died; and her act goes against the unspoken eugenic aspect of wet-nursing, as articulated so well in George Moore's novel *Esther Waters*, whereby the poor baby dies so the rich one may live. As Margaret Wiley puts it, wet-nursing may well be "the world's second-oldest profession"—the link between prostitution and wet-nursing has always been close.[120]

Margaret's act of wet-nursing came at a time when the practice was fast disappearing from the United States—and its anachronism made it even more startling.[121] Wealthier women were beginning to seek alternatives and were turning to doctors to help them; the infant and all its needs were fast becoming the province of the sciences and the male-dominated medical profession. In addition, the play makes visible the practice of "women who wet nursed in their own homes," which remained "a relatively invisible sector of the labor force."[122] It also reverses another trend of wet-nursing in the United States, that of using usually "poor, young, institutionalized, foreign-born single mothers":[123] Margaret, the American woman, takes the baby of her foreign-born employee to nurse, rather than the other way around.

Most important, *Margaret Fleming*'s emphasis on the baby within the *theatrical* framework simultaneously confirms and contradicts the status of the baby as scientific object. It was an unsettling, almost defamiliarizing move, seemingly illogical at a time of growing realism and naturalism in the theatre, with less tolerance for fakery. The foregrounding of the baby in *Margaret Fleming* is therefore worth looking at more closely in relation to the rise of "baby science," the study of infant physiology and behavior, which began to attract attention from the British scientific world in the late 1870s. Shuttleworth argues that one reason for its belated elevation to a subject worthy of scientific study was its perceived "feminine" nature; it was "a field of science which undermined disciplinary boundaries by taking place not in the well-equipped, and hence fortified, laboratory, but in that feminizing and infantilizing arena, the nursery."[124] Yet, as Shuttleworth notes, the flip side of this elevation to scientific status for babies was precisely the suggestion that they could be specimens, mere objects of study and experimentation—a new appreciation dangerously bordering on the inhumane. Logically, then, if you are a good mother, you will object to your baby being treated as a specimen. The science of infants grew steadily so that by 1886 the journal *Mind* was recognizing "Baby-lore" as "a separate study"[125] and not just for specialists: by the 1890s, "the scientific field of infant study opened up by *Mind* had become a major area of interest across the periodical spectrum."[126] At the same time, another threat to the nursery and to the mother's sacred domain came from the growing science of infant feeding and the rise of pediatrics, both male-dominated areas as discussed previously. In short, this period witnessed "a battle between the sexes over the cradle."[127] By repeatedly putting babies into his plays Herne not only showed an interest in probing and to some extent destabilizing assumptions about maternal instinct but also directly intervened in this battle over the cradle by having the babies in his plays circulate in and out of men's arms.

Tares: Mothers, Babies, and the "Course of Nature"

The fundamental and unsettling question of what a baby actually is, and who "owns" it, thus permeates culture in the 1890s, and many playwrights of the period were writing plays that explored it with outcomes radically different from Herne's. Herne explicitly links Margaret's maternal qualities to her saintliness and has the highly moralistic doctor say approvingly, "This world needs just such women as you,"[128] but several female playwrights argued forcefully against this idea that in the natural order the female inclines to motherhood. For example, *Tares* (1887) by Aimée Beringer, produced at the Prince of Wales's Theatre and later at the Opera-Comique, argues that "motherhood is a matter of individual temperament, not biology, and there are women unsuited for it."[129] The conservative critic Clement Scott wrote that this certainly was "an original motive" for a plot.[130] In the play, the fallen Rachel Denison returns after seven years to claim the little boy she left as a baby at the house of the selfless Margaret, who has always loved the child like her own and its father, Nigel Chester. Rachel's "heart is softened by the little boy, who works upon her feelings by his artless prattle," and she forgoes her vengeance; after a fatal injury, she conveniently dies and Nigel and Margaret are reunited.[131]

Yet the key motif of this anti-biological-determinism play is the phrase "the course of nature," repeated throughout and reinforced by a subplot that appears to serve this one thematic purpose. As the play opens, a young couple, who will not figure in the rest of the play at all, are flirting. The lad is longing for a kiss; the old gardener Giles is angling to get the girl Rosie to kiss him instead. Who is the more natural for her to choose? The process of sexual selection is enacted before our eyes, and then the little boy the grown-ups are fighting over bounds on stage to remind the audience of nature's consequences. *Tares* also offers a parallel with *Margaret Fleming* in the idea of the surrogate mother of the illegitimate child (in both plays, a boy). Margaret in *Tares* is gripped by the same instinct to nurture a child not her own that overwhelms Margaret Fleming. An onlooker describes how she changed inexplicably when Jack was left as a baby: "Never shall I forget the look in her face—it wur as if she had turned to stone. And yet, she wur more womanly-like, and it might be my fancy, but it seemed to me as if she knew all at once how to manage the child. She took the milk from me, and fed it, and then she hushed it to sleep, just as if it wur her own."[132]

Despite its implausible plot, *Tares* is more subtle and complex than it might seem and more ambitious in its social and scientific criticism. The motif of the "course of nature" is presented in such a way as to destabilize its meaning, to present several angles of interpretation so that in the end the audience is forced to ask, what *is* the course of nature? Giles, the leering old gardener, utters this phrase, using it as justification for men to sow their wild oats; Peggy declares at one point that Giles is "the most unmoral man I ever see!"[133] "Course of nature" is also applied to the rector, Margaret's father, just returned from hunting and gloating over how much fun it is to pursue the fox when she reminds him that it is not so much fun for the fox.[134] Beringer systematically questions assumptions behind this cliché about the course of nature and shows how it can be applied to justify *any* behavior. Even the pure and lovely Margaret wonders if she is unnatural, calling herself "Frankenstein" twice in the play.

In an interesting twist, *Tares* shows that least natural of all is the biological mother: "Ugh—the nasty unnatural creature," sniffs Peggy about the villainess Rachel, Jack's real mother.[135] She returns thinking that she will slide naturally into her motherly role, only to find that it is not part of the course of nature after all. Beringer signals Rachel's unnaturalness by linking her with theatricality; she is associated from the beginning with acting, histrionics, and artifice. When she meets her son, Jack, in a highly emotional scene, she realizes simultaneously her instinctual love and her own failure to be a real mother to him as Margaret is.[136] At least we think she does; but how can we tell what is genuine maternal instinct and what is feigned? The play succeeds in thoroughly destabilizing such certainties. Luke sees her passionately kissing and hugging Jack, asking him not to forget her, and he sneeringly says: "For whose edification are you playing this affecting little scene?" and "Allow me to congratulate you on the success of your first appearance in the *rôle* of a devoted mother. You do it well—and it has all the charm of novelty."[137] Such lines present motherhood as performance rather than innate. Luke sneers: "You're too clever a woman to make a fool of yourself, puling sentiment over a child you wouldn't have known for your own unless you had been told it belonged to you."[138] This is maternal instinct gone wrong, distorted. It relates not only to Darwin's observation about the difficulty of expressing maternal love but also to his emphasis on mimicry and simulation in nature; how can you tell the "real" maternal love from the fake? It also summons the vibrant image of Eleanora Duse a few years later harnessing physiology to "fake" an authentic blush. Is this real feeling or is it acting? Right up to the moment of her death, writhing in agony after her accident, the real mother is seen as an actress feigning misery, despite her desperate insistence that "I am not acting—it is true—real. For mercy's sake get help for me—I am dying—call—someone—for God's sake!"[139]

This questioning of what is natural, particularly with regard to motherhood, runs through play after play in these decades before and after the turn of the century. Oscar Wilde wrote privately that *Lady Windermere's Fan* was based on the idea of "a woman who has had a child, but never known the passion of maternity (there are such women)," whose latent "maternal feeling—the most terrible of all emotions," is suddenly triggered.[140] Yet his plays often problematize the link between maternity and self-sacrifice. Kerry Powell notes that by the 1890s there were "dramatic revisionings of maternity. . . . Plays by Victorian women redefine motherhood for every woman, and even a choice to leave a child, or kill it for that matter, could in some situations be justified or at least imagined without condemnation."[141] In *The Turn of the Wheel* (1901) by Blanche Crackanthorpe, a play refused a license by the Lord Chamberlain, the main female character rejects her newborn baby: "I care nothing for the child," she says in act 2; "I don't even care enough for it to hate it. Let it live, let it die, what does it signify to me? It's not me—it's not part of me, I tell you—I've done with it."[142] The character who says this does have a change of heart in the last act, but just showing such radical views was daring. In Robert Buchanan's *The Charlatan*, Darrell, the character associated overtly with evolution (and made a comic butt of it), says epigrammatically that "Nature herself is, of all things, the most unnatural."

Elizabeth Robins and Florence Bell's *Alan's Wife* (1893), which I discuss in full in chapter 6, puts this issue of what is "unnatural" in a radically different light. There is hardly a play less like *Margaret Fleming* in all respects. Jean breaks the maternal bond while Margaret celebrates it; Jean retreats into silence while Margaret finds her voice (accentuated even more by the loss of her sight); Jean arguably reaches a moral relativism whereas Margaret cleaves to established moral absolutes. *Alan's Wife* is short and sketchy in its few brief scenes compared with the full-length, multiple-act form of *Margaret Fleming*. Yet the two plays raise similar questions about the ontological status of the infant in contemporary culture and the supposed biological basis of motherhood: Innocent babe or savage? Helpless infant or scientific specimen? Object of maternal love or rival for the husband's affections? In posing such stark questions, these plays relied on performance as much as text to trick or unsettle viewers about an issue they thought they knew (what a baby is), for example, witnessing infanticide in *Alan's Wife* without actually seeing a baby on stage. And, while two "real" babies populated *Margaret Fleming*, the act of breast-feeding them tapped into Victorian fears about their closeness to "savagery." There was a cult of childhood as well as a cult of motherhood; this was fairly recent and exerted supreme pressure on women especially because the child now enjoyed a privileged and sacred status unheard of in previous epochs.[143] *Alan's Wife* shows a mother taking sole control of her infant's life and confusing both her role and the baby's—*she* is not being traditionally maternal, while *he* is reduced to a mere specimen, a scientific curiosity, by the withholding of the exact nature of his deformity. *Margaret Fleming* seems to show the exact opposite and thus to adhere much more closely to the natural order. But this is complicated by the fact that Darwin makes it clear in *The Descent of Man* that infanticide *is* part of the "natural order."

Margaret Fleming changes over the course of the play, gaining resilience and maturity through emotional suffering as well as her loss of sight. In the original production, according to Garland, "after having refused reconciliation with her husband . . . Margaret was left standing in tragic isolation in the middle of the stage, and as the lights were turned out one by one, her figure gradually disappeared in the blackness, and the heavy, soft curtains, dropping together noiselessly, shut in the poignant action of the drama and permitted a silent return of the actual world in which we live."[144] In this moment of darkness, the audience briefly experiences her isolating blindness. Even in the seemingly happier ending of Katharine Herne's reconstructed text, Margaret is alone on stage "watching" Philip go into the garden to her babies, and it is not at all clear how she will reconcile herself to him after taking such a high moral tone just a few lines previously. The capacity of women to change in relation to circumstance and the pressures of their immediate environments is linked to the wider context of Darwinism's emphasis on change as fundamental to the struggle for existence.

The reconstructed version of *Margaret Fleming* hints that she does not really want Philip back—and she certainly does not *need* him, an aspect of the play that was sidelined in the fuss over the breast-feeding. *Margaret Fleming* balances two images of femininity, two approaches to gender essentialism: woman as instinctively procreative and nurturing, yet also capable of doing without the male once successful mating has taken place. From an evolutionary standpoint, this makes perfect sense and happens all the time in the animal world, as the numerous examples Darwin cites in *The Descent of Man* indicate. Herne was hardly alone; other plays at the time were beginning to depict women going it alone with their illegitimate children, questioning the need for a father, and standing up to their family's outrage, such as Hermann Sudermann's *Magda* and St. John Hankin's *Last of the De Mullins*.[145]

A breast-feeding woman could signify the sacred and the profane—something Herne captures in his stage directions for Philip quoted previously (the mixture of admiration and horror). And, though a wet nurse might do the job of feeding as well as the biological mother, there persisted the idea of a moral dimension to mother's milk, "those spiritual truths [Levin] had imbibed with his mother's milk" that Leo Tolstoy alludes to at the end of *Anna Karenina*. By this reasoning, breast-feeding nourishes the child both spiritually and physically, and morality is not innate but is transmitted through the mother. Women are thus the source of moral values, passing them on through procreation and nursing.[146] This both burdens women with the sole responsibility for moral improvement of the species and exculpates men. Modern science has shown how false this is, how rarely across species one finds "the innately self-sacrificing mothers envisioned by Spencer and his predecessors, men who were more nearly moralists than naturalists, projecting onto nature what was essentially wishful thinking."[147] Long before sociobiology showed this, theatre challenged and disrupted this idea. Herne gives us a seeming paragon of motherhood, yet one who can do without the mate; Robins and Bell suggest that there is a difficult spiritual truth for some women that trumps biological determinism and is beyond the comprehension of most viewers and readers of the play.

John Perry calls *Margaret Fleming* "a hybrid between humanism and science, an example of evolutionary idealism."[148] The play's central concern is "the influence of environment."[149] Most critics see the play as emerging directly from Herne's admiration for Spencer. But there is another crucial link that has been less explored, and that is with Gardener, the pioneering feminist and great friend of Katharine Herne's whom I discussed previously. Gardener made her treatment of evolutionary themes distinctive through two strategies: first, by framing them within a eugenic discourse, and second, by thinking in what we now recognize as sociobiological terms. She continually remarks on how the "lower animals" behave, contrasting the power of the female in nature with the unnatural position of women: "Many of the lower animals destroy their young if they are born in captivity. They demand that maternity shall be free."[150] She argues that the animal kingdom has total equality between the sexes, but "when it comes to the human animal . . . the male, for the first time, becomes the whole idea." Slavery does exist in the animal kingdom, but it is not usually based on sex; in fact, sometimes it is the males who are the slaves to a queen.[151] "The great fact of maternity (everywhere else in nature absolutely under its own control)" links childbearing to "sex subjection," to financial dependence and subordination generally for "that part of the race which is the producer of the race."[152] This idea that women are enslaved to men to give them sexual pleasure and to bear them children leads her to eugenics. Maternity is "an awful power" and a blind force, a "fearful menace to mankind" because women who are made mothers against their will or in negative circumstances (such as poverty) will produce only the same kinds of mediocre people.[153]

Margaret Fleming reflects this feminist and eugenic ethos as Margaret makes repeated gestures to improve the race. Through her altruistic act of nursing the other baby as well as her own, she is giving both children the benefit of not only her milk but also, implicitly, her superior moral nature. Her refusal of Philip in the end will also have eugenic implications, as if to limit the number of children she has and thereby improve their situations and that of their offspring. Rather than being yet another slave to the "awful power" of maternity, she makes it a power she can wield. As Gardener puts it in one lecture, "not more children, but a better kind of children is what is needed."[154] Mothers owe "a higher duty to their offspring than that of mere nurse"; women have to be liberated from their current enslavement to men before they can produce high-quality children.[155] Echoing Nora in Ibsen's *A Doll's House*, Gardener proclaims: "Until mothers are both educated and rank before the law as human beings, they will never be able to give that kind [of better children] to the world. . . . Women owe it to themselves, and to the world which they populate, not to allow themselves to be made either the unwilling, or the supine, transmitters or creators of a mentally, morally or physically dwarfed or distorted progeny."[156]

Far from being a shocking illustration of regressive, atavistic behavior, then, Margaret's instinctive breast-feeding stemmed from the assumption, shared by so many social Darwinists but also reinforced by Darwin's own thought, that the future of the species lay in the hands of women as selectors of mates and nurturers of offspring. We saw how Ibsen, too, expressed this idea. But it is arguably the direct influence of Gardener, a self-described "student of Anthropology and Heredity," that is most palpable in the case of Margaret's notorious nursing.[157] And it still appears to be the only instance of breast-feeding on stage in the history of the theatre—perhaps not surprising in a cultural climate that can barely tolerate a bit of accidental nipple baring by a pop star.

Coda

It would be an oversimplification to assume that, as time passed, playwrights became more progressive in their depiction of motherhood. St. John Hankin's play *The Last of the De Mullins* exemplifies this. The play has a bearing on at least two aspects of evolution: this question of maternal instinct and, as its title suggests, the issue of extinction. Janet, the main female character, asks rhetorically: "What on earth were women created for, if not to have children?"[158] She is the outcast daughter who has had a child out of wedlock and now returns after having established herself in "trade" as a successful shop owner, and she "ends the play celebrating unmarried motherhood as infinitely preferable to spinsterhood" and suggesting that her sister get herself a lover and child as soon as possible if she wants to stave off premature aging.[159] "All wholesome women" feel the maternal urge, and since no suitable candidate presented himself to Janet for a husband, and her biological clock was ticking, she took matters into her own hands.[160] Far from being ashamed, Janet (like Mrs. Arbuthnot in Wilde's *A Woman of No Importance*) would do it all again and in fact explicitly indicates that she found fulfillment in her affair, realizing her "true womanhood" and saying that the seduction was completely mutual; indeed, Janet took the initiative as she was seven years older than her lover.[161] In her rousing final speech Janet declares:

> *Whatever happens, even if Johnny should come to hate me for what I did, I shall always be glad to have been his mother. At least I shall have lived. These poor women who go through life listless and dull, who have never felt the joys and the pains a mother feels, how they would envy me if they knew! If they knew! To know that a child is your very own, is a part of you. That you have faced sickness and pain and death itself for it. That it is yours and nothing can take it from you because no one can understand its wants as you do. To feel its soft breath on your cheek, to soothe it when it is fretful and still it when it cries, that is motherhood and that is glorious!*[162]

On the one hand, we have a truly liberated woman, free of the requisite guilt a "fallen woman" should feel and in fact briskly establishing herself as a successful businesswoman as well as a good mother; on the other hand, she is a traditionalist who identifies motherhood as woman's greatest achievement. Janet, notes Eltis, "locates woman's fulfilment exclusively in her sexuality."[163] Max Beerbohm reviled Janet's act of sexual selection: "The maternal instinct is strong both consciously and unconsciously, but no normal woman seeks to fulfil it by the rough-and-ready means of selecting, without reference to any effect that he has on her emotions, the first likely man who comes by."[164] He noted the "sociological" cause of the play and Hankin's "admirable" views but could not stomach the "tedious and jarring" Janet, who, he felt, "patronises and tramples on" everyone (except her child) and finally "kills the play."[165]

It is important to recognize that these developments in theatre went beyond the geographical scope of Europe. Sophie Treadwell's little-known play *Constance Darrow* (1908–1909) frankly addresses the question of how innate maternal instinct really is as one male character, Mr. Mathews, asks whether "all women are essentially housekeepers and mothers?" His term *essentially* goes straight to the issue that was at the forefront of the discourse on women—whether motherhood was the essence of a woman or socially constructed. The reply (by another male, Ben) is a prompt "Of course, given their natural environment," but Mr. Mathews protests: "I had begun to think that as fatuous as that all men are born warriors and providers of game!" Ben responds to this questioning of stereotypes or exaggerations of human evolution with his own version of evolutionary adaptation: "That's just it! Men have digressed, differentiated, changed through civilization, but women, well—nature has not allowed women to go very far from the primitive." Mr. Mathews's response to this is, "You mean, men have not allowed women to go very far from the primitive."[166]

The terms of this discussion come directly from the discourse of evolution as it was then developing. The play is noteworthy in part for its treatment of women's choices about reproduction and for the flexibility of Constance's stance on such issues, which develops and changes throughout the piece (despite the "constancy" signaled by her name). "In the end," writes Jerry Dickey, "she comes to reject Ben's arguments about economic and biological necessity as unchanging determinants of women's domestic roles, arguments that she herself inferred earlier in the play."[167] On the maternal issue, *Constance Darrow* is especially interesting. Constance welcomes her pregnancy, in contrast to the Young Woman in Treadwell's best-known play *Machinal* (1928), who is forced to have a child by her repugnant husband and who refuses to breast-feed or even see her newborn baby in the hospital scene titled "Maternal."[168] But both women's reproductive rights are denied by their husbands; Ben pressures Constance to have an abortion (*Machinal* also features a couple discussing a possible abortion), and only at the end of the play does she put up "meek opposition" to him. Both plays thus end with "the forced stifling of a woman's protestations."[169] The refusal to speak to anyone, to breast-feed, or even to see the baby is strikingly like Jean Creyke's postpartum behavior in *Alan's Wife*.

If playwrights were interested in probing the question of maternal instinct, they were equally concerned with the closely related issue of breeding, given new meaning by the rediscovery of Gregor Mendel's work in 1900. The nineteenth-century preoccupation with heredity and degeneration gives way to new questions prompted by the rise of genetics, and playwrights are quick to engage with them. But this needs to be seen against a broader backdrop of evolutionary ideas all being engaged in various ways by a wide range of dramatists. Before getting more deeply into the question of breeding and other reproductive issues, let us look briefly at what other playwrights in Ibsen's time and directly afterward were doing with evolutionary ideas.

5

Edwardians and Eugenicists

The full cross-cultural story of European theatrical engagements with evolution is far too broad for in-depth discussion here, but I look at several playwrights whose works reflect a serious interest in evolutionary themes and signal the potential of theatre to explore them. The first is Swedish playwright August Strindberg, whose interests in science, and in evolution particularly, set a key precedent for later playwrights such as Bernard Shaw, Thornton Wilder, and Susan Glaspell. My purpose here is threefold: to see what playwrights across different cultures were doing with evolutionary ideas (especially in comparison with Ibsen, who influenced so many dramatists), to set Shaw's work in relation to all of these, and to gauge the extent to which ideas about breeding begin to dominate theatrical engagements with evolution at this time.

In contrast to Ibsen, Strindberg had an overt and well-documented interest in evolution, expressed in many of his non-theatrical writings; unlike Ibsen, and very like Shaw, he took great trouble to lay out his views as clearly as possible. He was, like Charles Darwin, a transformist, but he also embraced the mystical side of evolution implicit in Ernst

Haeckel's monism.[1] In 1870, as a student at Uppsala, he would visit a friend who was studying the sciences to "peer through his microscope and learn about Darwin."[2] His scientific studies took in everything from "Darwin to the occult, from Naturalism to Supernaturalism, from physics to metaphysics, from chemistry to alchemy."[3] In one book of essays alone (or, as he called them, "vivisections"), he touched on the "zoology" of women, racial stereotypes, Maurice Maeterlinck, and other topics; then, in the collection *Jardin des Plantes,* he wrote substantially about natural history.

Strindberg's engagement with evolution changed drastically over the course of his career. In his naturalistic phase, in the 1880s, he approaches playwriting as he approaches science, depicting characters "in terms of the survival of the fittest, natural selection, heredity, and environment" and calling his play *Miss Julie* (1888) "a simple scientific demonstration of the survival of the fittest" in which the male is shown to be the fittest.[4] The play shows, in characteristically misogynistic fashion, the catastrophic results of female sexual selection. After naturalism comes Strindberg's "inferno phase" (1892–1898); during this period, he sees science as the highest calling and urgently explores what he perceives as a tension inherent in nature "between chance, coincidence and discontinuity, on the one hand, and order, relationship and coherence, on the other."[5] One essay, "The Death's Head Moth," proclaims "the role of chance in the origin of species!"[6] Another, "Indigo and the Line of Copper," provides a highly technical chemistry discussion followed by the conclusion that "this is the unity of matter made manifest, and the doctrine professed by every modern scientist since Darwin, although a goodly number have recoiled from the consequences."[7] But, by 1907 (when he finds religion again), he denounces evolutionary theory as "unscientific rubbish" along with contemporary science generally and is heavily criticized for it.[8]

There is a tendency to regard Strindberg's "inferno" phase as an aberration, a somewhat embarrassing parenthesis between the two major creative phases of his career (naturalism and expressionism/modernism) in which he loses his sanity and does futile alchemy experiments and nearly kills himself in the process. Yet his essays show how important Strindberg's scientific writings and experiments were for him as a creative process, how central this phase was to his work and how far it informs his thinking more generally.[9] In one typical example,

"To the Heckler" (1896), he asks: "Has not the cyclamen shown you that all botanical systems are arbitrary and vain, and that nature does not create according to a system? . . . Enlighten me and all the other people who believed that the exact sciences do not work with fictions and fantasies."[10] He believes the key is combining imagination (arts) and the purely animal world (sciences). He is, at this stage, a true interdisciplinarian.

Strindberg had several ambitious projects in different genres that show how he conceived of evolution on a vast cultural as well as physiological scale. He planned—and, indeed, wrote the first three plays of—a "cycle in forty-five acts" depicting world history.[11] He also planned a huge book in 1895 that would encompass all of evolution; his letters give a detailed description of its proposed contents, beginning with "searching out the primal elements of the world," then descending into the oceans to observe life emerging from the water, ascending into air to explore the atmosphere, then returning to earth to track the development of life from the moment when plants and animals first parted ways to the arrival of man.[12] These epic projects indicate an interest not only in the mechanisms of evolution but also in its mode of progression; he seemed to see history's course as saltationist and catastrophic rather than gradual, moving "with glaring surprises and violent jumps, proceeding relentlessly and with a certain lack of feeling for mankind's suffering. . . . [It is] in keeping with nature's insensitive method of employing thousands of years to form a geological strata, only to break up and recast what it has made."[13] Here again is the familiar rhetoric of the Victorians that nature is insensitive, indifferent, cruel. These vast projects also look forward to Shaw's and Wilder's use of epic in *Back to Methuselah* and *The Skin of Our Teeth* respectively in order to stage the whole of human evolution.

Along with his born-again faith, what finally turned Strindberg from Darwinian evolution was the fear of reversion or retrogression. He addresses this in his essay "Whence Have We Come?" which was published in *Vivisections II* (1894) and marks a distinct aloofness to Darwin; by 1907, with *A Blue Book* and its three sequels (1908–1912), he is fully anti-Darwinian. "So, we come from heaven, and descend towards the anthropomorphs. Is life then evolution backwards?" The "true end of life" seems therefore to be "an ape . . . so much trouble just to become an ape!"[14] He expresses the same thought in "In the Cemetery" (1896): How could the beautiful child before him, so angelic, be descended from an ape? Yet he could believe that an old man is. "Progress backwards, then, or what? . . . Are we the degenerate offspring of those blessed ones, whom we can never forget?"[15]

Strindberg also had trouble accepting extinction as a concept. In "The Death's Head Moth" (1896), he imagines a child asking where the light goes when a candle is blown out: "In the last century scientists would say that it returns to the primary light from whence it came. Our scientists, who pronounce energy indestructible, nevertheless say that it has ceased to exist! Ceased to exist, to be perceived? But nothing can cease to exist."[16] By 1903, in "The Mysticism of World History" (echoing Haeckel's recently published *The Riddle of the Universe* and Shaw's *Man and Superman*, performed the same year), he is positing a Schopenhauerian "Conscious Will" in history, not just a random and chance set of events but an Ur-pattern.[17] He derides the "Darwin monomania" that he detects everywhere which for him means a cultural obsession with heredity.[18] He embraces telegony, an alternative to biological heredity; the term, meaning "offspring at a distance," was used by August Weismann in 1892 to describe instances of the phenomenon of psychic heredity.[19] As Marvin Carlson has shown, Strindberg wove this idea of the heredity of influence into his plays, as did Ibsen in *The Master Builder*.[20]

On the one hand, Strindberg managed to engage with a wide range of evolutionary themes, beyond the preoccupation with heredity that dominates his contemporaries.[21] On the other hand, his trajectory is that of so many of his contemporaries: moving from a firmly Darwinian stance to a rejection of natural selection in favor of a mystical, monist vision. This makes Anton Chekhov all the more exceptional. He "liked Darwin terribly" and said in 1894 that science was "performing miracles everyday."[22] Chekhov had a "strong but sensible" faith in humankind; he was a doctor, a firm materialist, interested in the earth beneath his feet and the natural order, saying in 1889, for example, that "outside of matter there is no experience or knowledge, and consequently, no truth." He relished many fields of inquiry, including sociology, zoology, gardening, psychiatry, and the philosophy of science.[23] Downing Cless argues that "natural stakes are present in each of Chekhov's four major plays," such as the "nostalgic affection" for the cherry orchard vying with "eco-hubristic" entrepreneurship.[24] He also notes the odd mixture of people in Chekhov's plays, their "unrelated relatedness": "They are put together like random species" whose ability to survive is not easy to predict.[25]

Like Strindberg and Chekhov, the French avant-garde playwright Alfred Jarry was biologically inclined and well informed. Rae Beth Gordon contends that Jarry's interest in biology, natural history, and evolution was the "cornerstone" to his thought and that it is "the most important, but unexplored, key to understanding the *Ubu* cycle."[26] Jarry—who played the Troll King in the French premiere of Ibsen's *Peer Gynt*, in the same year and same theatre as *Ubu*'s premiere—was fascinated by the concept of the biological hybrid, especially by the idea of the missing link (for him the ultimate and elusive hybrid, the bridge between man and animal). Like so many avant-garde theatre practitioners, he was interested in the frontier between human and animal and between human and machine, the idea of what it is to be human in a rapidly changing modern environment. Gordon argues that Ubu is a hybrid that Jarry thought of even before he read Haeckel and began to consider himself a naturalist, and in *Ubu Roi,* Jarry created "a hybrid form of theater, at once popular theater and avant-garde," fusing Shakespeare's *Macbeth*, Jacobean revenge tragedy, Grand Guignol, and abstract art.[27]

Of a completely different dramatic stripe was Jarry's contemporary Eugène Brieux, once "the most popular dramatist in France,"[28] whose fans in the English-speaking world included Shaw, H. L. Mencken, and St. John Hankin. In his sustained theatrical exposé of social problems, Brieux shows particular concern with sexual health and with the roles of women in society and in nature. He deals with sexually transmitted disease, abortion, infant mortality, and other reproductive issues, showing them as the result of ingrained and often-corrupt societal systems that adversely affect women. *Les Trois Filles du M. Dupont* (*The Three Daughters of Mr. Dupont*, 1897) exposes the damage done by the dowry system; *Le Berceau* (*The Cradle*, 1898) deals with divorce; *Les Remplaçantes* (*The Wet-Nurses*, 1901) shows the tragic repercussions of the corrupt wet-nursing system for women and babies and for the nation as a whole, in the grip of a low birth rate. This crisis prompted an urgent national debate for many years, in which artists, doctors, playwrights, and politicians alike participated: Brieux; Clémence Royer, Darwin's first French translator; Henri de Rothschild, doctor and dramatist;

playwrights François de Curel and (in satirical mode) Guillaume Apollinaire. Stepping up his intervention, Brieux's most famous play, *Les Avariés* (*Damaged Goods*, 1901), deals with syphilis acquired by men visiting prostitutes and shows its devastating consequences for the family; *Maternité* (*Maternity*, 1903) pleads the necessity of birth control; *Simone* (1908) "denies the right of a man to kill his wife for adultery," while *Suzette* (1909) "promotes the rights of mother and child."[29] These are just a few of his plays; Brieux was hugely prolific, and Shaw strenuously promoted his work in Britain, yet he has all but dropped off the theatrical radar. His plays are rarely revived or discussed, some are still unavailable in English, and there is no biography of him in English.[30] Both Mencken and Arnold Bennett had reservations about his work that may help to explain this subsequent neglect: "Violent reformers," wrote Bennett, cannot be serious dramatists.[31]

Shaw Re-creates Evolution

Shaw complicates Bennett's easy assertion, producing by far the most thoroughgoing and innovative theatrical engagement with evolution after Ibsen. Ibsen's plays—"steeped in the gloom of mid-nineteenth-century science"—furnished Shaw with dramatic inspiration for characters and situations as well as modeling how to use science as a theatrical foundation.[32] Yet Shaw created a completely different dramatic idiom, full of inventiveness, wit, debate, stylization, and overt metatheatricality. He combined a deep devotion to Ibsen with an equally profound antipathy to evolution by natural selection, epitomizing the contrarian position demonstrated by so many of his theatrical contemporaries (despite William Archer's stinging claim that Shaw's "great intellectual foible is credulity").[33] He lamented that Ibsen's plays show "no trace . . . of any faith in or knowledge of Creative Evolution as a modern scientific fact."[34] In other words, Ibsen's only flaw is that he is not a Shavian.

Shaw rejected the Malthusian harshness of Darwinian natural selection; he had a "natural abhorrence of its sickening inhumanity."[35] He repeatedly objected that natural selection "explained the universe as a senseless chapter of cruel accidents."[36] This reaction was a "characteristic scientific aberration" that Shaw explained "with admirable lucidity and wrongheadedness" in the preface to *Back to Methuselah*.[37] Dismissing the idea that "Darwin invented Evolution," Shaw's plays consistently endorse non-Darwinian evolution.[38] Influenced by Arthur Schopenhauer and Friedrich Nietzsche as well as by Buffon, Lamarck, Samuel Butler, and Herbert Spencer, Shaw believed in the role of the will in human evolution. Creative Evolution is his name for Lamarckian evolution, or functional adaptation. The idea "underlies almost every one of [Shaw's] major plays," and his works have an "underlying unity" through their sustained commitment to it.[39] He gave a trenchant account of his views in a letter of 1907:

> I believe that all evolution has been produced by Will, and that the reason you are Hamon the Anarchist, instead of being a blob of protoplasmic slime in a ditch, is that there was at work in the Universe a Will which required brains & hands to do its work & therefore evolved your brains & your hands. I have the most unspeakable contempt for Determinism, Rationalism, and Darwinian natural selection as explanations of the Universe. They destroy all human courage & human character; & they fail utterly to account for the most obvious facts of life.[40]

Shaw explains in his preface to *Back to Methuselah* (1921) that his idea of the "will" is a synthesis of Schopenhauer and Lamarck: Schopenhauer's *World as Will and Idea* (1819) is "the metaphysical complement to Lamarck's natural history."[41] It is not the same as Ibsen's treatment of will, which emphasizes past events and their impact; for Shaw, this aligns Ibsen too much with Darwinism as it rules out the agency of the individual.[42] Yet in one respect, that of sexual selection, there is some significant resonance with Darwin, as shown by Shaw's depiction of women as the Life Force, a concept based on vitalism and similar to Bergson's *élan vital*.[43] One of the reasons Shaw liked the theory of sexual selection when he loathed everything else about Darwinian evolution was that sexual selection reinstated some kind of intentionality to human evolution after *Origin of Species* had expunged it. We *choose* our mates. That is some comfort in the teeth of natural selection's randomness, blindness, and threat of regression. Sexual selection allowed for "some intelligence driving the motors of evolutionary development after all"[44] and dovetailed well with Shaw's Lamarckian belief in the role of the will in human evolution.

Apparently rejecting any form of determinism, Shaw placed the freedom of the human will at the core of his evolutionary vision.[45] Yet biological determinism drives *Man and Superman*, and it is centered in women. Women's "highest purpose and greatest function" is "to increase, multiply, and replenish the earth," to choose the best mate for the perpetuation and improvement of the species, and "a man is nothing to them but an instrument of that purpose."[46] Women's vitality is therefore "a blind fury of creation"[47] and the social conventions of courtship and marriage a mere sham to hide the real locus of power in sexual selection:

> TANNER: *The will is yours then! The trap was laid from the beginning.*
> ANN [concentrating all her magic]: *From the beginning—from our childhood—for both of us—by the Life Force.*[48]

This determinism ("from the beginning") is complicated by a moral element that Shaw, echoing Huxley, introduces to offset the troubling blindness of the relentless process of "construction and destruction" that constitutes evolution.[49]

Although both men and woman are the "helpless agents" of the "universal creative energy," women have the burden of mate selection because of their childbearing capacity and thus their innate responsibility to ensure steady improvement of the species: life is a "continual effort not only to maintain itself, but to achieve higher and higher organization and completer self-consciousness."[50] Shaw had been developing this idea for some time; for example, in his play *You Never Can Tell* (1897), Valentine envisions nature as "suddenly lifting her great hand to take us . . . by the scruffs of our little necks, and use us, in spite of ourselves, for her own purposes, in her own way."[51] He explains that nature propels men and women together through some "chemical action, chemical affinity, chemical combination: the most irresistible of all natural forces" (or, as Tom Stoppard puts it in his play *Arcadia* [1993], "the attraction that Newton left out").[52] This mystery of attraction riveted Shaw's attention. In *Misalliance*, a few years after *Man and Superman*, he is still musing theatrically on "the great question . . . the question which particular young man some young woman will mate with."[53]

The famous Don Juan in Hell dream sequence (act 3) of *Man and Superman* elaborates Shaw's theory of Creative Evolution and women's roles within it. First, Shaw castigates Man's destructiveness and stupidity. Then the discussion moves on to Woman, "the one supremely interesting subject." Ann's feminine power is on an evolutionary scale; Shaw's stage directions seem to draw metaphorically on Haeckel's recapitulation theory when he describes her as evoking "a mystic memory of the whole life of the race to its beginnings."[54] Don Juan explains:

> *Sexually, Woman is Nature's contrivance for perpetuating its highest achievement. Sexually, Man is woman's contrivance for fulfilling Nature's behest in the most economical way. She knows by instinct that far back in the evolutional process she invented him, differentiated him, created him in order to produce something better than the single-sexed process can produce. . . . [And civilization is] an attempt on Man's part to make himself something more than the mere instrument of Woman's purpose.*[55]

Shaw's women cannot help themselves; yet their apparent taking of the initiative and overturning the shy, submissive maiden image—charmingly portrayed by Lillah McCarthy as Ann in the play's first production (figure 1)—only replaces one rigid and misogynistic ideal of womanhood (the angel in the house) with another (the biologically driven woman ensnaring men in the service of the Life Force). *The Philanderer*'s Julie and Charteris are characterized in the stage directions as "huntress and her prey," but Shaw's women are just as trapped in their roles of predator as are the males in the "helpless" roles of prey.[56]

As Don Juan puts it, women are the channel for an intelligent force that helps Life "in its struggle upward," when it would otherwise blindly waste and scatter itself.[57] Just as we have evolved an eye for seeing the physical world, through this progressive process we will evolve "a mind's eye" that will see the purpose of Life, and we will finally be enlightened enough to work for that purpose instead of "thwarting and baffling it" through our selfish, short-sighted aims.[58] Don Juan sums up Creative Evolution: "Life is a force which has made innumerable experiments in organizing life itself," whether the mammoth, man, mouse, megatherium, flies, fleas, and Church Fathers, all of which are "more or less successful attempts to build up that raw force into higher and higher individuals, the ideal individual being omnipotent,

Figure 1 "Mr. Bernard Shaw burlesqued in his own play": Lillah McCarthy and Harley Granville-Barker (made up to resemble Shaw) in *Man and Superman* at the Court Theatre, 1907. (© Victoria and Albert Museum, London)

omniscient, infallible," and essentially "a god."[59] Shaw becomes even more convinced of this years later in *Back to Methuselah*: evolution, he writes, builds "bodies and minds ever better and better fitted to carry out" the purpose of the Eternal Life; it is "the path to godhead. A man differs from a microbe only in being further on the path."[60]

He also asserts (as if the mere utterance will make it true) that "a reaction against Darwin set in at the beginning of the present century" that is leading "scientific opinion" to embrace Creative Evolution instead.[61] Shaw's quarrel with Darwinism seems straightforward: he cannot accept death, he dislikes gradualism, and he rejects the lack of agency—the aimless drifting, rather than purposeful direction, of natural selection. His views have much in common with those of Strindberg, who posits that world history proceeds not toward Spencerian homogeneity but with a mystical and "secretive . . . unconscious aspiration of mankind which is unaware of the goal but at the service of the conscious will."[62] Both needed to reinject agency into evolutionary theory, in contrast to Ibsen's acceptance of its absence; intelligent design is implicit in their work. We are "tools in someone's hand, someone whose intentions are incomprehensible to [us], but who looks after [our] best interests," and Strindberg calls this being "the invisible legislator . . . the creator, the dissolver and preserver."[63]

In *Misalliance*, death is pronounced as unnatural. "There wasn't any death to start with," explains Tarleton; then we appeared on earth and had to start dying off in order not to become overcrowded. "And so death was introduced by Natural Selection." Even old age is the "invention" of natural selection, a "mask" of wrinkles and gray hair "to disgust young women with me, and give the lads a turn."[64] One of the points Shaw makes in *Back to Methuselah*, and which he had suggested earlier in other plays, is the idea of the invention of birth and death:

> [FRANKLYN]: *Adam and Eve were hung up between two frightful possibilities. One was the extinction of mankind by their accidental death. The other was the prospect of living for ever. They could bear neither. They decided that they would just take a short turn of a thousand years, and meanwhile hand on their work to a new pair. Consequently, they had to invent natural birth and natural death, which are, after all, only modes of perpetuating life without putting on any single creature the terrible burden of immortality.*[65]

Shaw's contrarianism can be understood in the context of other things he rejected, such as the idea of the bestiality of man, of which childbirth and meat eating are two examples he gives. His extreme rationalism led to a "fastidious mistrust" of "life's irreducible physicalness, its messy, unpredictable, irrational uglinesses and beauties."[66] In the first part of *Back to Methuselah*, Cain declares a revolt against "these births that you [Adam] and mother are so proud of. They drag us down to the level of the beasts. If that is to be the last thing as it has been the first, let mankind perish." Like Alfred Russel Wallace, whose article "Evolution and Character" (1908) suggested a progressive, improvement-centered, perfectible human evolution, Shaw wants mankind to be "something higher and nobler."[67]

In addition to rejecting death, which "calls an end to progress, and mocks all human aspiration," Shaw cannot accept Darwinian gradualism. He has no patience for the nineteenth-century insistence on evolution taking "millions of eons" of geological time and not proceeding "by leaps and bounds."[68] Shaw wrote *Man and Superman* during the period of (in Julian Huxley's words) the "eclipse of Darwinism," the phase when natural selection as the mechanism of evolution was challenged by other theories, such as Hugo de Vries's mutationism (1901–1903; translated into English 1910), saltation, neo-Lamarckism, and even Haeckel's monism.[69] Books like Thomas H. Morgan's *Evolution and Adaptation* (1903) were rejecting "both the selection mechanism and the idea that evolution could be driven by the demands of adaptation."[70] In other words, the question was still open (and would remain so until the New Synthesis came about) regarding what drives evolutionary change, and Shaw's plays were unique interventions in that discourse.

But Shaw's antipathy to Darwin is complicated by an embrace of Karl Marx and Continental philosophy, on the one hand, and a vehement antivivisectionismm, on the other, that attacks Darwin and all experimental scientists for chopping the tails off mice and the legs off dogs to prove natural selection. The Life Force has to experiment as it goes along, something Shaw kept depicting in later plays.[71] It is "not so simple as you think. A high-potential current of it will turn a bit of dead tissue into a philosopher's brain," but the reverse can just as easily happen: "Will you believe me when I tell you that, even in man himself, the Life Force used to slip suddenly down from its human level to that of a fungus, so that men found their flesh no longer growing as flesh, but proliferating horribly in a lower form which was called cancer, until the lower form of life killed the higher, and both perished together miserably?"[72]

Shaw's Lamarckian belief in the power of the will to effect physical and mental modifications is taken to outlandish extremes in *Back to Methuselah*, as the Ancients swap stories of extraordinary transformations they have willed into being, such as growing "ten arms and three heads" or becoming "fantastic monsters" with eight eyes and four heads.[73] But these saltations are hardly genetic mutations. Shaw belongs more to the early evolutionists like Buffon than to contemporaries like de Vries. The end of *Back to Methuselah* signals not some mystical perfection of mankind as much as a return to Alfred, Lord Tennyson's idealistic "coming race." That is the fundamental quality of Shaw's engagement with evolution: he refashions old ideas into the seemingly new package of Creative Evolution, often harking back to the pre-Darwinians while insisting on his own innovation.

It is easy with hindsight to dismiss Shaw's Creative Evolution as wrongheaded and logically flawed. He gets the science wrong, and the eugenicist undertones can make his plays not only uncomfortable but also deeply problematic. But, as is so clear from his prefaces, he understands evolutionary theory perfectly well. For all the thousands of words he expends in trying to distinguish his theory from Darwin's, it seems to boil down to a single concept—agency. Are we simply to sit by and contemplate a universe that is "cruel and terrible and wantonly evil" as well as "oppressively astronomical and endless and inconceivable and impossible?" This question still plagues him in late plays like *Too True to Be Good* (1932).[74] If so, we should "go stark raving mad" and stick straws in our hair. No, we must *do* something, must intervene; there must be a purpose, and humans must be its agents; we can improve and perfect and make progress with the raw materials around us. The things that are of no use must be "scrapped" as he puts it in play after play (*Major Barbara, Back to Methuselah, Heartbreak House*), either by the random forces of nature or by human purpose. Between *Man and Superman* and *Back to Methuselah*, this optimistic vision does briefly dim (World War I separates the two), and the latter is "full of ominous warnings that the Life Force may find it necessary to discontinue the human race"[75]—to scrap it, as Undershaft would say. *Heartbreak House*

(1919) illustrates this in graphic fashion through exploding several of the characters. Extinction is all right if something is obsolete or no longer useful because it is in the service of overall improvement, which mandates the permanent killing off of ineffective organisms and (most importantly for Shaw) outlooks and ways of life.

Shaw's interest in evolution has to be seen in relation to other factors besides theories of evolution. It is tied to a wider interest in cultural agency: Shaw the playwright wants to reinvest the theatre with the powerful cultural role it once enjoyed, and in dozens of articles and prefaces throughout the 1890s and early 1900s, he inveighs against the ineffective contemporary theatre, its managers, the censorship, and the playwrights enslaved by the box office. In addition, he is writing at a time when, as Peter J. Bowler notes, evolution itself is being treated differently, assimilated as "part of a more general transformation of biology . . . related to developments in genetics, ecology," and ethology.[76] Given the ongoing "debates over the public and political role of science" (and, one might add, of theatre), it is natural that they should converge in the work of one so given to "polemic, influence, argument, debate."[77] *Back to Methuselah* is one of his most important contributions to this debate, as he attempts to stage evolution as well as stretch the limits of theatre without relinquishing box office concerns. In other words, he wants to have it all: *Back to Methuselah* (subtitled *A Biological Pentateuch*) opens in the Garden of Eden, ends thousands of years in the future, and cumulatively depicts the Life Force that directs evolution toward ultimate perfection by trial and error. Let us look in more detail at this modern megatherium of a drama.[78]

In the introduction, I quoted Henry Arthur Jones's observation about the seemingly impossible challenge for the theatre in engaging evolution: how to represent in a few hours a process that takes millions of years. Shaw cleverly deals with this in *Back to Methuselah* by having variations on his characters in the five plays that make up the cycle, usually played by the same actors. So, for example, the Accountant General looks rather like Conrad Barnabas, and "even the two politicians make a reappearance as a composite in the President Burge-Lubin."[79] Then, in part 5, "As Far as Thought Can Reach," he calls for a character whose face is so ancient that it seems as if "Time had worked over every inch of it incessantly through whole geologic periods."[80] These human variations and the immense amount of time covered in the course of the cycle combine with the tricks of stage make-up to convey the impression of evolution happening before our eyes. Although Shaw saw human beings as the highest achievement of nature, he was profoundly dissatisfied with what collectively had been achieved so far—due, he believed, to too short a life span to be able to accomplish much.[81] Thus, *Back to Methuselah* offers a solution to the earlier despair of *Man and Superman* at how "wretched" our brains are, considering we are "the highest miracle of organization yet attained by life, the most intensely alive thing that exists, the most conscious of all the organisms."[82]

Darwin wrote in his *Autobiography* that "man in the distant future will be a far more perfect creature than he now is," and Shaw also believed this; but would man necessarily be "more perfect" in the way Shaw depicts? Is it not a paradox that the ultimate result of this striving toward perfection is shown to be evaporation, vanishing, the flesh becoming spirit? Lilith's final lines before she vanishes at the end of *Back to Methuselah* are full of monistic rejections of all matter and a yearning instead for a state in which "the whirlpool" becomes "all life and no matter."[83] As at the end of *Man and Superman*, all that remains is "talk," except here it is literally just the human voice, all that materially remains of the now perfectly evolved mind. The only way to achieve this pinnacle is through willing a longer life, an idea Shaw was still hammering at as late as *Geneva* (1938), staging the notion that we must lengthen the human life span to gain enough wisdom to govern ourselves.

The long-lived people in *Back to Methuselah* do eventually die, not of natural causes but through accidents or through "discouragement," a fatal attribute. Apparently "discouragement" came to Shaw through Strindberg, who was convinced that "every meeting of individuals implied a psychic conflict to establish the mastery. This soul-conflict was the spiritual aspect, he thought, of the physical struggle for survival which Darwin had described in *The Descent of Man*."[84] Perhaps Francis Galton's *Inquiries into Human Faculty* (1883) was another source; Galton suggests that such discouragement affects "savages" when they first come into contact with white civilization—for example, Native Americans' "collapse of morale" and "the actual extinction of some races, such as the Caribs and Tasmanians." Louis Crompton notes Edward Bulwer-Lytton's *The Coming Race* as a further influence.[85]

Shaw proposed protracted longevity as a way of offsetting the hugely frustrating fact that at present our "first hundred years" of life are a state of "confusion and immaturity and primitive animalism" because we are dominated by sex.[86] Shaw is obsessed with sex but somehow finds it distasteful; it is as if his view of procreation can be found in Eve's revolted response to the Serpent's whispered explanations. *Back to Methuselah* shows childbirth now replaced by children hatching from eggs: the audience witnesses a full-grown teenage girl emerging from a giant egg center stage. This solves one of the problems of human evolution: that we are altricial—immature at birth, requiring a long period of dependence. In his preface to the play, Shaw pins great hopes on embryologists' "astonishing" discovery of recapitulation to help humans learn how to speed up "into months a process which was once so long and tedious that the mere contemplation of it is unendurable by men whose span of life is three-score-and-ten." He claims there are already examples of this "packing up of centuries into seconds" in the form of child prodigies.[87]

In fact, in Shaw's plays "children rarely make an appearance unless they are of a marriageable age."[88] Thus, paradoxically, there is a gap between woman's compulsion to reproduce and the natural result of this drive; we are shown only the former. Perhaps this is mere theatrical expedient: "Although it would seem logical to conclude that women live for the sake of the children they may have, *this desire is not so easily conveyed in drama*."[89] This relates to Darwin's observation in *Descent of Man* that maternal love is paradoxically difficult to articulate through facial expression. But this explanation ignores Shaw's revolutionary feminism, his championing of women's liberation from the burdens of motherhood, his attack on the institution of marriage for its constraints on women. In his preface to *Getting Married*, Shaw says that the one thing all women are in rebellion against is that having a child necessarily entails being "the servant of a man."[90] In keeping with the developing discourse on women's roles, particularly motherhood and marriage, *Back to Methuselah* enacts the conflict between the reproductive burden on women and their right to independence.

Not surprisingly, there has been much debate over Shaw's depiction of women's roles within his evolutionary vision. A substantial body of critics maintains that Shaw limits women even while he seems to be arguing for their importance and that he only pays lip service to women as biological agents of evolution, as he "ultimately allows the male intellect to dominate in his plays. . . . Both the new woman and the man of genius are dedicated to the male role of contemplation. . . . Man's role as the intellectual agent of evolution is, accordingly, the most important one in his drama."[91] Jill Davis argues that many of Shaw's plays are "profoundly misogynistic," especially *Getting Married* and *Man and Superman,* in which he attempts to "reorganise feminist arguments to masculinist ends . . . [as a] conscious response to feminist threats to male sexuality."[92] Although it may well have been problematic and limited in its suggestion that women's duty is to ensnare a mate to perpetuate (and, crucially, improve) the species, as Sos Eltis points out *Man and Superman* was "one of the first plays where a respectable young woman's sexual drive was both the crux of the plot and a central topic of debate."[93]

It is perhaps a paradox that "a man of Shaw's profound understanding and personal shrewdness gave, for the last thirty-five years of his life, the wrong answers to almost all the questions that have perplexed our age."[94] Note again this idea of wrong-headedness and contrarianism. J. D. Bernal calls him "a violent and reckless supporter of scientific lost causes, like Lamarckian Evolution" and opponent of theories and practices that "in the course of years have proved their truth and usefulness, like the bacterial theory of disease."[95] Theatrically as well as scientifically, Shaw has received criticism. Wilder commented in 1951 that Shaw was a particularly "tiresome" variant of the "bellicose" Irish geniuses who "only discover what they think through the act of repudiating what someone else has thought. This makes for wit and energy as certainly as it makes for overstatement and personal ostentation."[96] This is significant in light of Wilder's own attempt to stage evolution, as explored in chapter 7.

What should be the final verdict, then, on Shaw's engagement with evolution? The fact that we are still so divided over him—feminist or misogynist? theatrically innovative or hopelessly unplayable? backward looking or visionary?—indicates his ongoing vitality and relevance. And, for all his contrarianism, Shaw's evolutionary vision was capacious enough to include a positive nod toward Darwinism. In *Too True to Be Good* (1932), the Elder recalls how at difficult times in his life he sought "consolation and reassurance in our natural history museums, where I could forget all common cares in wondering at the diversity of forms and colors in the birds and fishes and animals, all produced without the agency of any designer by the operation of Natural Selection," rather than "some freakish demon artist, some Zeus-Mephistopheles with paintbox and plasticine."[97] This makes it all the more essential to understand Shaw's plays in their context, to see how he fits into an extant discourse on evolution that may have been shown to be "wrong-headed" much later but nonetheless had a significant following at the time and that poses fundamental questions that are still relevant. He also has his comic side, and this must be taken into consideration; one can take Shaw too seriously and at face value, forgetting that he is a playwright who can write a play that features a microbe on stage ("In shape and size it resembles a human being," he tells us, presumably with a straight face, in the stage directions to *Too True to Be Good*, "but in substance it seems to be made of a luminous jelly with a visible skeleton of short black rods").[98]

Genetics Enters the Scene

That Shavian mixture of paradox and comedy with serious intellectual and moral questions characterizes so many Edwardian playwrights, particularly in their engagements with evolutionary issues. The Italian director Luca Ronconi notes a peculiarly British ability to write plays with "an ironical, not at all pedantic tone," dealing with subjects "in a light way, but without diminishing the seriousness of the topic."[99] In this vein, Shaw's contemporary James M. Barrie, best known for *Peter Pan*, uses the language of evolution and enacts motifs such as adaptation and the struggle for existence in his hugely successful play *The Admirable Crichton* (1902).[100] A now-forgotten contemporary of Shaw and Barrie was Hubert Henry Davies, three of whose plays reflect contemporary developments in genetics and evolutionary theory, sciences that provided him with rich dramatic metaphors. Of these, *Mrs. Gorringe's Necklace* (1903 premiere) is the least innovative; it contains references to heredity and specifically to degeneration and, like *The Admirable Crichton*, suggests the impossibility of permanent change. Davies's comedy (masking a deep tragedy) argues that the "curse of degeneracy" can never be shaken off and will "taint" an addict's children.[101] *Mrs. Gorringe's Necklace* thus sits firmly within a nineteenth-century naturalist tradition. Although some hope is briefly offered that the protagonist, David, can "triumph" over his problem, he despairs that "the curse is always there in my mind and heart. I'm tainted—." He writes a farewell letter to his wife and then takes a revolver out into the garden and kills himself.[102]

Nothing could be further from a play written a few years earlier addressing exactly the same theme, *Man and His Makers* (1899) by Wilson Barrett and Louis N. Parker, which presents degeneracy as not hereditary but all in the mind.[103] No one is fated or doomed by his or her parents' indiscretions. The hero, John Radleigh, is informed by his would-be father-in-law that he comes from a long line of drunkards and opium and morphine addicts. As soon as he hears of it, John promptly begins to succumb to this family weakness. At his nadir, he is penniless in St. James's Park and surrounded by society's other refuse, in a scene of deep pathos that shows a poor woman in such despair that she is taking her baby to the river to drown it, only to discover that it has already died in her arms. John is rescued by his true love, Sylvia, who has been forbidden by her father to marry him, and the play ends ten years later with him with a happy, healthy family and a prosperous job. At the end of act 2, Radleigh proclaims his optimistic stance in the face of all the despair in society, even in the face of science: he alludes to entropy and heat death when he says that men "faint at fresh air; and their only comfort in the sun is that it's gradually going out," which is where he was mentally when he took opium, until he was rescued. Now, "give me fresh air, I say, and Long live the Sun!"[104]—an obvious echo (and transformation) of Oswald's dying invocation of the sun in Ibsen's *Ghosts*. But whereas Oswald has no power against disease, the notion of "flying in the face of science" lies at the core of *Man and His Makers*.[105]

These two directly opposed plays by Davies and by Barrett and Parker show how theatre intervened in an ongoing discourse spread across all cultural forms about the nature of so-called degeneracy and its impact on human evolution. Already by 1899, a certain fatigue had set in, though, and reviewers of *Man and His Makers* found the repeated references to the theory of heredity tedious, even though the play's aims were deemed "admirable."[106] One reviewer simply could not fathom what species of play this might be: "A work of this class hardly comes within the ordinary dramatic categories."[107] The most remarkable review, for its scientific specificity, came from the *Pall Mall Gazette*. The reviewer discusses specific scientists in relation to the play, saying it is hard to tell how far the playwrights are "fellow champions with the 'progressive heredity' and 'the heredity of acquired characters' of Mr. Herbert Spencer, or how far they are attacking the theories of Weissman [*sic*], who would explain the whole of transformism by the 'all-sufficiency of selection.' Probably, like wise men, they thought of neither the one nor the other . . . [but of] whether [the play] were dramatically sound rather than scientifically correct." Quite apart from the feeble punning on "wise men," the critic sets dramatic "soundness" and scientific accuracy in opposition, as if a play cannot be both theatrically viable and scientifically truthful, a debate that is still with us today. The reviewer also notes that the theme of heredity had most recently been dramatized by Brieux in *L'Evasion* at the Comédie Française with the aim of showing that, "however strong heredity, there is always the will to combat it."[108] Finally, the reviewer says that John should have done what any "sensible" man would do on hearing the news that he has a heredity tendency to drink: seek a scientific opinion.

As these contemporaneous plays show, the workings of heredity were not yet clear in the public discourse—what *exactly* could be passed on, what could be combated with will and determination as well as with medical intervention, and how heredity actually worked. Even in the late 1920s, Max Beerbohm criticized Barrie's play *Old Friends* for having an Ibsenic theory of heredity whereby the daughter of an alcoholic will herself become a drunkard. "The child of a drunkard, though its health is in many ways affected by its father's habit, will not, when it grows up, have any special cravings for alcohol," unless it had been brought up to share the father's habit, making it more likely to acquire that condition.[109] He wonders how Barrie can be so ignorant of these basic facts.

Heredity also runs through the play for which Davies was best known, *The Mollusc* (1903).[110] Though now obscure, as late as 1928 *Play Pictorial* was recalling how *The Mollusc* had been "hailed as a triumph of craftsmanship" at its premiere and how "its fame has endured till this day."[111] The leading male role of Tom Mowbray was played by Charles Wyndham, who by then was seventy years old, playing a forty-five-year-old who falls in love with a young woman in her twenties—yet he was praised for his energy and vitality. The "mollusc" of the play's title is his sister, Mrs. Baxter. As Tom explains to her exasperated husband, she is a specimen of "Mollusca, subdivision of the animal kingdom" ("I know that," Mr. Baxter says testily). Davies injects some zoology into the play while pushing the science toward metaphor: "I don't know if the Germans have remarked that mammalia display characteristics commonly assigned to mollusca. I suppose the scientific explanation is that a mollusc once married a mammal and their descendants are the human mollusc."[112] Tom explains that these hybrids, these human molluscs, cling to the rocks while the tide flows over their heads; they have an instinct for molluscry, and it spreads like an infection throughout the household. It stems from women ("mother was quite a famous mollusc"), but it is not the same thing as laziness; it is a kind of resistance: "The lazy flow with the tide. The mollusc uses forces to resist pressure. It's amazing. . . . [It has an] instinct not to advance." There are some people who find routine boring, who "suppose the secret of existence to be mutability, the fact of never knowing, from day to day, what strange adventure shall befall," but Mrs. Baxter every day "looks forward to tomorrow as a replica of to-day."[113] Mrs. Baxter, a "perfectly healthy woman,"[114] is like a mollusc because she clings to her rock while life goes on around her, and she resists pressure, growing a metaphorical carapace. And just as molluscs have long been associated with luxury for humans—pearls, sea silk, Tyrian purple dye—Mrs. Baxter luxuriates in being waited on and living in extreme comfort.

What is interesting is not so much the obvious metaphor but how the biological language enters into the dialogue and how it refers to evolutionary processes—perhaps even counter-evolutionary, if one considers the idea put forth by Tom that molluscs have the ability to resist change and adaptation to their environments and to ignore the environmental pressures on them. This also shows a playwright making assumptions about cultural absorption: throwing certain evolutionary ideas (adaptation and progressiveness) into his plays and assuming that the audience has some prior exposure to a discourse on these themes and issues.

In fact, the play was well received and praised for its successful integration of science, done with a light touch. Percival P. Howe wrote that "nothing could be more delicate than the art with which we are told about molluscs."[115] The *Athenaeum* called it "the lightest of light comedies," portraying "that type of woman who makes everybody around her wait hand and foot upon her, and takes advantage of other folk's good nature to spare herself the least exertion. . . . An indolent, selfish woman who avoids any kind of locomotion."[116] Walpole repeatedly praises the whole of *The Mollusc*, while other Davies plays are only partially successful in his view: "Mrs. Baxter goes gloriously on, far into time, triumphantly conscious that she will get her way so long as there is a man in the world, a way that no votes for women can affect so long as human nature remains unchanged."[117] Beerbohm called it "exquisite" and "truly delightful" and explained that "in the brute creation a mollusc is a shellfish that firmly adheres to a rock, doing nothing but adhere, while the tide ebbs and flows above it. Thus, in the human race, a mollusc is a person who will not lift a little finger to perform any duty" that someone else could do instead.[118]

But Davies's use of this metaphor is, in fact, misleading; in reality, even though they seem not to be doing anything, molluscs are amazing engines of adaptation. Whereas the play suggests that Mrs. Baxter is forcing everyone to adapt around her while she remains static and unchanged, this is the reverse of what happens in nature. Drawing an analogy with performance, Rebecca Stott notes that marine invertebrates played a central role in evolutionary theory because they were "the key to understanding the complicated biological processes of all 'higher' animals."[119] She notes the popularity of the aquarium from the mid-nineteenth century: the first Aquatic Vivarium was set up in the Zoological Gardens in 1853, enabled by Philip Gosse's discovery of how to make artificial seawater that would be viable for marine creatures. The newly opened aquarium became "one of the spectacles of London." These tanks were "theatres of glass in which sea creatures performed epic natural dramas," and soon "aquarium mania began to sweep the country" and people flocked to see such "entertaining and philosophical" creatures.[120] This is precisely the assumption behind *The Mollusc*. The audience—newly trained in encountering natural history through not only such "theatres of glass" but also zoos, geological displays, and museums—is able to detect the underlying drama in a seemingly static show.

Huxley's strenuous work in popularizing Darwin's ideas, including the groundbreaking study of barnacles, might also have played a role. Again in the theatrical vein, Stott notes that Huxley turned the barnacle into something rather comical, "a theatrical zoological sideshow," by his description of it in *Westminster Review*, in which he drew an analogy with Mr. Punch and his hump: "He anthropomorphized the barnacle as a man settling down, gluing himself to a rock."[121] This is exactly how Davies portrays Mrs. Baxter, though turning the condition of being a mollusc into a specifically female malady.[122] This transformation was not lost on Beerbohm. In his response to the play, Beerbohm tries to get past the clichés by drawing on known facts about molluscs. He notes that although the mollusc has been feminized, in reality "the human mollusc is often masculine."[123] But Davies also is on top of his facts; in the play, Tom admits to being a mollusc himself, the difference being that he fights this condition rather than succumbing to it.

Continuing to mine evolution-related topics for theatrical metaphors, Davies in his comedy *Doormats* (1912) reflects the buzz about genetics that William Bateson helped to generate.[124] Davies's highly topical comedy plays with the idea of "dominant" and "recessive" types of people. Working with his pea plants, Gregor Mendel posited that there were dominant and recessive elements; recessives are expressed only when the same element is passed on by both parents, while dominant ones can be expressed even if only one element is present.[125] The play borrows terms from this new science—terms that the playwright evidently felt confident audiences would recognize and understand—and adapts them to suit his purposes, as a metaphor for the two main types of people in the world, masters (boots) and slaves (doormats), represented in the play by Noel, a doormat-like painter who is married to Leila, a boot-like young woman who is self-absorbed, restless, and fickle. By the time Davies has woven this metaphor, the terms have been so modified that they retain almost nothing of their original scientific meaning, for *dominant* and *recessive* are not value-laden terms (i.e., strong and weak) in the way that they become in the play.

Indeed, Aunt Josephine, a wise middle-aged woman who is the play's voice of reason, springs on everyone the surprise revelation that she also is a doormat, and the problem is that doormats "wilfully lay ourselves down to be trampled upon. We *love* being trampled upon. It thrills us to give and it bores us to take. It's of no use *knowing*—with one's brain—how to take if one hasn't got their *instinct*—."[126] This is a surprise because outwardly she betrays no sign of that allegedly negative trait. She has evidently been able successfully to adapt to her condition. It is the same twist that we saw in *The Mollusc* when the highly active and energetic Tom had confessed to being a mollusc himself. Through these characters, Davies makes the point that what might seem to be non-adaptive can in fact be turned to advantage.

Although Davies does not repeat the gendering that occurs in *The Mollusc*, since *Doormats* shows that either gender can be dominant or recessive, he does assume gender essentialism. Josephine says further in the play that some people give and some take, and one cannot change that: "You can't not be whichever kind you are, any more than you can change your sex."[127] The events of the play do question this determinism as Leila, who has left Noel, in the end wants him to take her back.[128] But the final moments of the play show their reconciliation in a highly ambiguous tableau: Noel stands behind a seated Leila, his hands cupping her chin, pulling her head back as the stage directions indicate, while she stares unsmilingly out at the audience, looking like a trapped animal (figure 2). The picture's caption paraphrases the final lines of the play and summarizes its theme: "Noel and Leila find that the 'boots' need their 'doormats' just as much as the 'doormats' need their 'boots.'" This undermines both the suggestion of the ideal pairing of these two opposites and the genetic metaphor at the center of the play.

It is worth looking more closely at how Mendelian dominants and recessives work in this play. Aunt Josephine spells it out (as Tom does in *The Mollusc*):

> *Just as every one is either a man or a woman—not in the same degree of course—but there are* men *and* women—(illustrating with her hands) *at either end, as it were, of a long piece of string; very mannish men at one end and very womanish women at the other. Then—as you go along—men with gentler, what we call feminine qualities—and women with masculine qualities—some with more and some with less—right along—till you come to a lot of funny little people in the middle that it's hard to tell what they are. Just so, it seems to me, is every one a more or less pronounced doormat or boot.*[129]

The metaphor is not all that is of interest here: note the ingenious syntactical construction by which the gender essentialism that has just been established is undermined a few lines further.

The play's use of genetics did not pass unnoticed. According to Howe, "If Professor Bateson of Cambridge University had written [the

Figure 2 "The Reconciliation" between Leila (Marie Löhr) and Noel (Gerald du Maurier), capturing the final moments of *Doormats* by Hubert Henry Davies, from a twenty-two-page feature on the production in *Playgoer and Society Illustrated* (December 1912).

play], no doubt he would have given it a more scientific name. But Mr. Davies' art prefers the more homely analogy of the doormat and the boot." And the play's construction is perfect because the older couple (Josephine and Rufus) mirrors the younger (Noel and Leila): "dominant and recessive . . . only the other way round."[130]

The reference to Bateson is significant. In 1902, Bateson published the first English translation of Mendel's paper. Bateson also coined the term *genetics*, which he first used publicly in a lecture in London in 1906. The leading British geneticist, he helped to found the field, was appointed to the new chair in genetics at Cambridge in 1908, and established the *Journal of Genetics* in 1910; he then became director of the John Innes Institute from 1910 until his death in 1926.[131] As a hard-line Mendelian, he was an anti-Lamarckian who did not believe that environment could influence heredity. In addition, Bateson believed in saltation rather than gradualism: he claimed that "discontinuous variations (saltations) were the real cause of evolution," and this was even before the rediscovery of Mendel's laws.

In the first decade of the twentieth century, as Bateson's Cambridge group helped establish the field of genetics, a deeply divisive and highly public debate developed between Bateson and his followers (the hereditarians, or the Mendelians) and the followers of Karl Pearson (the biometricians). As Daniel J. Kevles points out, "The biometrical–Mendelian debate . . . reached a level of vitriolic intensity" in Britain that was never matched in the United States, Bateson labeling zoologists "nincompoops" full of "ignorance and bigotry" when a paper by one of his collaborators was rejected for publication.[132] By contrast, in America, a vigorous and high-quality level of genetics research obtained, comparatively unencumbered by such bitter disputes. In both cases, though, "the pursuit of genetics was also . . . affected by social forces, notably the eugenics movement."[133] In 1913, Pearson, Edward Nettleship, and Charles H. Usher objected to the general trend to apply Mendelism "wholly prematurely to anthropological and social problems in order to deduce rules as to disease and pathological states which have serious social bearing."[134] In a sense, *Doormats* does just this, although—unusually for a piece so clearly of its time in its direct engagement with genetics—it does *not* play the eugenics card.

It is unlikely that audiences knew all of this or that Davies himself did, but it is safe to assume that genetics was "in the air" in 1912, due to the very public, "vitriolic" debate of Mendelians versus biometricians. Added to this, in 1910, the notion of the inheritance of feeblemindedness was hotly debated as Charles B. Davenport claimed that it was "a simple Mendelian recessive."[135] Davies drops such buzzwords into his play. The central, burning question for many was fitness—physiological, moral, and mental. One of the fascinating side elements of *Doormats* in this context is its use of an artist as the central figure; as Bateson noted, artists served as soft targets for the eugenicists because of their eccentricity and apparent unproductivity: "In the eugenic paradise I hope and believe there will be room for the man who works by fits and starts [such as] Bohemians, artists, musicians, authors, discoverers and inventors. . . . I imagine that by the exercise of continuous eugenic caution the world might have lost Beethoven and Keats, perhaps even Francis Bacon."[136]

Davies's work was, for one reviewer, marred by his "hesitation between comedy and tragedy."[137] The same "lightness" of treatment that earned him praise became in the end his undoing in terms of durability, for even his engagement with science remained for some merely a dabbling: "Davies's mind was essentially conventional, and never more so than when he was taking what he believed to be an unconventional point of view."[138] The author of this appreciation of Davies's work concludes that he is "almost equal to Hankin." The writer is referring to their dramaturgy—tone, form, structure, dialogue—but it is also notable that they share an interest in using biological motifs. Where Davies eschews eugenics, though, Hankin addresses it head on.

Although he is seldom revived and has only a single, slim biography now over three decades old and only a clutch of scholarly articles, Hankin must figure prominently in any consideration of the Edwardian theatre's engagement with evolution. He is closely allied in spirit and dramaturgy to key playwrights of his time, such as Shaw, Davies, Harley Granville-Barker, John Galsworthy, Elizabeth Baker (discussed further in this chapter), Barrie, and Brieux, whom he greatly admired and one of whose works he translated. Hankin incorporated into his works "what was perhaps the most unpopular disposition of his time: an open regard to eugenics and social Darwinism, along with a wry awareness of the impossibility of any tranquil panegyric for society."[139] *The Return of the Prodigal*, his second full-length play, shows Hankin hitting his dramatic stride in "a curious hybrid of comedy of manners and Darwinian philosophy," a mode similar to that of Davies.[140] With its emphasis on the strong versus the weak—and its direct allusion to "survival of the fittest"—the play might seem more Spencerian than Darwinian. But, by not ending the play with marriage between Eustace and Stella, Hankin rejects the Victorian ideal of the "ennobling power of woman and marriage" that is implicit in Spencer's philosophy, just as Shaw does in so many of his plays.[141] Hankin casts life as a struggle between predators and their prey. Eustace, sounding like a mixture of Darwin and Zola, does not believe we have free will; survival is determined by one's environment, heredity, and chance. This explicit espousal of natural selection and adaptation was "rare in English drama and unprecedented in a comedy" in the first decade of the twentieth century.[142]

Deeper consideration of Hankin will be given in the following chapter (on reproductive issues) with regard to his play *The Last of the De Mullins*, but it is worth noting here that his work also encompassed other evolutionary concepts as well, such as extinction, competition, and altruism. In his play *The Charity That Began at Home* (1906), the theme of survival of the fittest is more "muted," foregrounding instead a larger philosophical question about human nature: Are we really meant to help one another, or does that just lead to more misery and suffering?[143] Verreker declares that charity is interfering: "People who try to improve the world have rather an uncomfortable time." Extinction features in *The Last of the De Mullins* (published 1909; produced by the Stage Society in its 1908/1909 season); the child that resulted from Janet's illicit affair is the "last" of Janet's family line, which she dismisses as "your ridiculous De Mullins." As we have seen, familial extinction is a preoccupation of other plays of the period, such as Henry James's *Guy Domville* and de Curel's *Les Fossiles*, even though the seemingly tragic status of being the last of the Domvilles and De Mullins is questionable given the existence of illegitimate offspring.

The theatrical landscape in this period is thus saturated with evolutionary issues and ideas, but it is distinct from its Victorian predecessors. The emphasis on exterior landscape shifts to the internal situation—not only the domestic interior but also the psychological one and, beyond that, the invisible physiological landscape inside each individual that was beginning to be probed and revealed by science. Most prominently, the treatment of heredity is changing: the preoccupation with the inheritance of acquired characters, the pessimistic rendering of the "sins of the fathers" being passed on to the sons, and the generally passive, deterministic depiction of the workings of heredity give way to a sense of identity and genetics as malleable and mutable rather than fixed. When Virginia Woolf famously asserted that human character changed in December 1910, chances are that she was not thinking of the theatre; but considering this statement in light of broader cultural changes at this time gives it new meaning. By this time, genetics has become a well-established field and is rapidly changing the way scientists understand evolutionary mechanisms. Playwrights embrace the possibilities opened up by this new sense of the flexibility of character. It is also the moment when both the suffrage campaign and the eugenics movement gain a tremendous amount of public attention. Theatre becomes a site where new ideas converge, and the spectrum of questions it poses is all-encompassing: What is the nature of identity? Can we liberate ourselves from our culturally conditioned assumptions (suffrage)? Or is our biology our destiny? What happens if we manipulate this biology? What can we do to correct nature's mistakes (eugenics)?

Reviving Freakery

An extraordinary play was produced in 1918 that broached this last question directly to the audience: Arthur Wing Pinero's *The Freaks*. The play speaks to a continued public interest in freakery, but registers a notable change in the nature of that interest: the Victorian fascination with human anomalies as a link to our evolutionary origins is replaced by a curiosity about otherness and disability that was concomitant with the returning war wounded. *The Freaks* is itself something of an anomaly and has received scant attention since its premiere in 1918.[144] Like Edward F. Benson's *The Freaks of Mayfair* (1916), the title foregrounds "freaks" and makes the same argument—that the real freaks are the aristocrats—but goes one step further by bringing actual freaks together with these upper-class specimens.[145] As with much more recent plays like Suzan-Lori Parks's *Venus*, Mary Vingoe's *Living Curiosities*, and Bernard Pomerance's *Elephant Man*, *Freaks* re-creates the Victorian freak show with a view to showing the people beneath the deformities and arousing sympathy for their plight, showing the consequences of their exploitation both by their handlers and by the audiences who paid to see them—including, of course, those watching the play itself.

The list of characters divides the persons of the play between "ordinary mortals" and "extraordinary mortals," the latter all "late of Segantini's World-Renowned Mammoth International Hippodrome and Museum of Living Marvels." The "extraordinary mortals" have been let go by Segantini and are sympathetically described by the well-to-do Mrs. Herrick, who takes them under her wing, as "poor souls . . . Freaks . . . human oddities, doomed to exhibit themselves as a side-show of a circus, or in a booth at a fair."[146] Pinero tests our prejudices by not showing us the freaks themselves until after the ordinary mortals have described them. This forces each audience member to fall back on cultural clichés and stereotypes of freakery that the play will then dismantle.

Pinero's depiction of freakery draws on authentic, real-life examples. Tilney, Pinero's "Skeleton Man," is most likely an echo of Mr. Tipney, the most famous skeleton man of the Victorian period, and the rest of his troupe of human anomalies is also fairly standard: a giant (9 feet tall), a female contortionist, and two "little people." The group was depicted arrestingly on a drop curtain by the well-known artist Claude Shepperson, a frequent contributor to *Punch*, and this was reproduced as a two-page spread in *Tatler*: "A special feature of the production was a 'Wonderful Drop-Scene of the Freaks,' used instead of a curtain between the acts, from a painting by Claude Shepperson portraying the troupe entertaining an audience that was made to appear more freakish than the performers at whom they were gawking" (figure 3).[147]

The production featured prominently in *Tatler*, which published not only Shepperson's painting but various photographs of the actors over several issues. This suggests that the production was very much in the public eye and that its premature closure due to air raids was the reason for its demise as a play, rather than theatrical inviability. Ironically, then, *The Freaks* was a victim of the very circumstances to which it alluded: "This satiric lesson of man's inhumanity to man suffered the effects of two air-raids in its opening days and, despite full houses and an enthusiastic welcome, closed" after a run of fifty-one performances.[148]

Figure 3 "The wonderful drop-scene of *The Freaks*": Claude Shepperson's drop-curtain for *The Freaks*, a play by Arthur Wing Pinero. The curtain was displayed across the stage between the acts and was reproduced in *Tatler* (February 1918), along with photographs of the leading actors. (© Victoria and Albert Museum, London)

It would have been challenging to stage a play that required half the cast to be freaks. How would they be played convincingly, if not by actual freak performers? Pinero alludes privately to the "difficulties of presentation" and theme that might prevent the play being produced in America.[149] His stage directions certainly seem to call for complete realism. Tilney is "emaciated," with "a pale face and sunken cheeks."[150] Eddowes is a giant "between eight and nine feet high," while the two midgets have "lined, wizened features, a tottery gait, and heads which, being too heavy for their necks, have a tendency to lop on one side."[151] And there is no possibility that these figures are just glimpsed briefly; Tilney introduces each one in turn, and he presents them as relics of both an evolutionary past and a theatrical one, expanding in one stage picture the spectrum of the human. Indeed, he refers to them all as "by-products of the animal kingdom."[152] Sheila, the "normal" girl who falls in love with Tilney, probes him on his origins as if getting at a much bigger ontological issue. "Who *are* you?" she asks. "What have you sprung from, Mr. Tilney; where were you born; how did you come to be mixed up with this curious crowd?"[153]

The subtitle of the play (*An Idyll of Suburbia*) directs the focus on the nonfreaks (and most of the theatre audience) who inhabit suburban London and gives the sense that this space represents the status quo that is only briefly jeopardized. In fact, the "normal" and the freak groups turn out to be related: Segantini was Sheila's uncle. The centerpiece of the play is Tilney's "who is a freak" speech, syntactically and thematically echoing Shylock's "what is a Jew" soliloquy in *The Merchant of Venice*:

> TILNEY: *Who is a freak and who is normal in this world? Who shall decide? . . . Are there no Freaks in your list of acquaintances? Are all the women you lip, and all the men you rub palms with, beautiful specimens of the normal—the Christian—type? And yet you sneer at my poor grotesque companions, who, in spite of infirmities of body and temper, have more true love in their hearts, treat 'em kindly, than seventy-five per cent. of the well-formed and the well-endowed.* [Grinding his teeth.] *Freaks, are they! . . . Looking into the faces in front of me at our shows, my hardest task has been to refrain from crying out that we ought to change places*—to change places—*the so-called Freaks upon the rickety platform and the damned sniggering spectators on the tan floor!*[154]

What is so interesting here is the emphasis on changing places, on switching audience and actors and using theatrical practice itself to bring about the very exchange it is suggesting, if not literally—the audience is not actually going to rise up and change places with the actors—then imaginatively. The cultural memory of the Victorian freak show becomes not an awkward anachronism but a potent tool as it haunts the theatre and puts the audience, which probably prided itself on having evolved superior theatrical taste to its Victorian predecessors, in the uncomfortable position of gawking at freaks just as they did. The audience viscerally experiences, not just observes, the blurring of boundaries between freakish and normal that Christine C. Ferguson describes in her study of John Merrick, the "elephant man": "Physical normalcy and irregularity, rather than representing two permanently opposed and inherent states of being, are fluid concepts involved in a recurrent process of dialogue and mutual remaking."[155]

Although the play was praised at the time, it has been virtually neglected; what little attention it has received has focused on the way in which Pinero might have been using his piece to comment on the war.[156] Nadja Durbach, although apparently unaware of Pinero's play, writes of the effect of the war on freak shows in Britain more generally: "The cataclysmic events of the First World War immediately wrought changes to the ways in which British audiences engaged with freakery. This was both because the war triggered changes to the entertainment industry that had long-term effects and because it compelled British society to engage with physical deformity in new ways." During the war, the War Office "took control over several major exhibition venues, including the resort and theme park of Kursaal at Southend-on-Sea and the Crystal Palace in Sydenham. . . . The government encouraged the closing of fairs and shows that detracted from war work, seeking to prioritize the production of munitions and discourage leisure activities. . . . Human and animal acts in particular went into sharp decline, surviving primarily within the context of the circus."[157] *The Freaks* marks this transition.

Furthermore, as Sean O'Casey's *The Silver Tassie* (1928) so eloquently shows, the war created a new generation of "human oddities"—the deformed, mutilated, disabled bodies that were its casualties. This meant that in addition to genetically generated anomalies, there was now a growing number of men permanently disfigured or altered by their environment, and since their deformities were inflicted in the service to their country, they occupied a higher social place than natural freaks. Durbach tracks this change and the resulting category of "disabled," pointing out that in fact the one did not subsume the other: "Indeed, the distinction between disabled veterans and others with bodily deformities not only remained, but in many ways it intensified during the interwar period in Britain."[158] The relationship of the onlooker to the thing exhibited also changed dramatically, no longer safely distant but intimately connected as these war-deformed people were woven back into the fabric of the community and daily life, not objects up on a stage to be stared at and then forgotten.

Despite being closed prematurely, the play enabled Pinero to be rediscovered by critics who had dismissed him as merely an unsuccessful imitator of Ibsen in the 1890s. *The Freaks*, writes the *Saturday Review* critic, "teaches us how Sir Arthur should really be viewed and valued." The characters are distinctive, the story is "ingenious," and the play as a whole shows us "a dramatist in continuous touch with his audience."[159] The critic feels it is a bold move to introduce a group of "nature's freaks" into a suburban household: "Nothing could seem more surely bound in the direction of crude farce. But we are speedily reassured," as Pinero shows us their humanity in all its fullness. The review applauds the play's "sincerity" and "honesty," its compassion for the freaks, its genuine feeling through scenes like the one when the female contortionist begs the vicar to pray for the sick giant lying in bed upstairs—"One of the best [scenes] we have had on our stage for many years." Yet the critic notes that part of the play's honesty lies in its ending: the freaks are not allowed to "marry and live happy ever afterwards." The critic does not spell out this eugenic message, but the assumption here is that the coupling of freaks with nonfreaks would never be natural or socially acceptable.[160]

Pinero's "little play," as he self-deprecatingly called it,[161] is not mentioned in any of the standard works on freakery; one can only assume that it has gone unnoticed by most scholars. It is hardly better known in theatrical circles, though Griffin's study of Jones and Pinero calls it "an extraordinary experimental play."[162] It deserves scholarly recognition and theatrical revival, not least because the play was more radical than even Pinero might have realized. It championed freaks as more human than normal people precisely at the peak of eugenic interest in both the United Kingdom and the United States, when the freak show "became a willing tool of the direst perversions of Darwinism."[163] The subsequent synthesis of Mendelian genetics with natural selection meant that "congenital disability was recast as pathology. Freak shows thus provided a counterdiscourse for the rising popularity of such notions as eugenics and the elimination of 'invalids,' 'defectives,' and 'mutants.'" It was a culture in which "all deviances came to be seen as diseases."[164]

Eugenics on Stage

The Freaks makes a fascinating and important intervention in this discourse on deviation and is one of dozens of examples of how theatre in this period engaged with eugenics, or as its founder Galton put it:

"What Nature does blindly, slowly, and ruthlessly, man may do providentially, quickly, and kindly."[165] Darwin in his *Autobiography* says he is "inclined to agree with Francis Galton in believing that education and environment produce only a small effect on the mind of anyone, and that most of our qualities are innate."[166] A full consideration of eugenics in the theatre lies beyond the scope of this study, and the work of Tamsen Wolff has already significantly expanded our understanding of this area.[167] In addition to the many playwrights she explores in her groundbreaking study *Mendel's Theatre*, there are those like H. M. Harwood, Elizabeth Baker, and Adam Neave whose eugenics-themed work I briefly discuss. Others, in particular Maeterlinck, lie beyond the scope of this study; Maeterlinck's *Betrothal* (1921), for instance, directed by Granville-Barker, combines eugenics, monism, and an epic theatrical vision spanning the whole of human evolution.

It is a striking paradox that although theatre overtly engaged with eugenics, the Eugenics Society lacked substantial membership from "those involved with the media (journalists, freelance writers, *people connected with the stage*)."[168] Geoffrey R. Searle notes that although the Fabians "sometimes employed a rhetoric with a strong eugenical colouration," they were not strong eugenicists: "A certain kind of libertarian progressive in the early twentieth century found eugenics irresistibly attractive because it presented itself to his mind as a refreshing departure from bourgeois conventionality; in this configuration of beliefs eugenics became linked to such things as utopian socialism, vegetarianism, sexual experimentation of one kind or another, proposals for 'the abolition of the family,' etc."[169] This almost exactly describes Shaw's main enthusiasms; after all, "the Life Force is eugenic, not social."[170] In "Treatise on Parents and Children" (1910, his preface to *Misalliance*), he writes that a child is merely "an experiment. A fresh attempt . . . to make humanity divine," and sees reproduction as the mechanism for improvement of the species. Shaw identifies a cruel paradox in the makeup of family life, one that is surely a hindrance to human evolution: that the most intelligent adults are those who will be most fit to raise children but also most annoyed by the interruptions that children constantly make to their work. Yet the responsibility of parenthood "cannot be discharged by persons of feeble character or intelligence."[171] Like Sidney and Beatrice Webb and other fellow Fabians, Shaw's stance is generally that of "positive," "soft," or "weak" eugenics: raising awareness of the need for "better" breeding to obtain improved human specimens, rather than taking measures to prune the "unfit" from society. However, "some Fabians were prepared to flirt with eugenics; both Webb and Shaw sometimes chose to describe themselves publicly as eugenists" and "not *merely* a matter of rhetoric."[172]

Indeed, it is a mark of how pronounced the eugenic overtones of *Man and Superman* were that a decade later Neave could signal his own dramatic warning against this movement by invoking Shaw's play in the title of his own *Woman and Superwoman* (1914). Deeply misogynistic and anti-eugenic, the play both echoes and distorts Shaw in presenting a dystopian vision of the world as it might be in 1963, "in the latter days of European Civilization, when Woman rules and the Millennium is at hand," a choice of time and setting prompted by the passing of a bill in New Jersey in 1911 authorizing sterilization of the feebleminded.[173] Neave explains in his prologue that he wrote the play to enlighten the British public about this alarming development. He depicts the eugenics movement as American led and run by "nasty, noisy, masculine women who boss—that means tyrannise—the country through their feminine majority in and out of Congress, and think they can force their ideals on the rest of humanity, as they've succeeded in doing with their own mankind. They really imagine the Millennium is at hand, and that they are the chosen instruments for bringing it about."[174]

The play lampoons both the New Woman and the eugenics movement, and by linking them together it doubly damages the image of the women's rights campaign, which, if anything, had sought to distance itself from reproductive issues for fear of jeopardizing or distracting from its main purpose of securing the vote for women. Neave swipes at key figures such as Evelyn Sharp and lumps her campaign for the suffrage together with eugenics: at one point, the female doctor, head of the eugenics home, says gleefully that "Miss Sharp of Harvard in her biological treatise entitled *The Passing of the Male*, holds that, as Science advances, some method of artificial generation will be evolved, and the male become unnecessary."[175] Here lies the real concern of the piece: the fear that men will become irrelevant to procreation. Neave makes his hostility to women the iron tongue in his velvet cheek.

A review of the published play in *Bookman* in December 1914 says that it has an "astounding preface describing unheard-of things in the direction of Eugenics," but its central financial idea is actually viable and "should bring success to some Utopia of the future: . . . creating a ministry of borrowing and lending and so abolishing all bankers and money-lenders, and diverting their profits into the pockets of the community." The reviewer argues that the eugenic agenda outlined in the play "with so much humour and force" is actually going on in America; maybe, the reviewer says, some European state will find "the courage to try the plan." The reviewer calls the play "good reading" as Neave's style is "trenchant and his dialogue sparkling and animated," though it is more like "the sledge hammer of Swift" than the "small-sword of Mr. Shaw."[176] There is no evidence that the play has ever been performed.

Harwood's *The Supplanters* (performed under the title *The Interlopers*) engages with eugenics by indirect reference to the ideas of Thomas R. Malthus and Galton, and in several places it directly echoes Hankin's *Return of the Prodigal* (1905). As the play opens, the "barren" spinster Beatrice is trying to persuade Margaret to be president of the society that Beatrice has founded and is secretary of whose purpose is to "teach women to have healthy children and to teach them how to look after them when they have got them, in short, to inculcate in woman a sense of duty to posterity."[177] Peter, Margaret's brother, the *raisonneur* of the play and therefore the voice of authority for the audience, attacks the society for meddling; he believes in "live and let live," letting nature take its Darwinian course rather than try to manipulate it. "England is simply covered with a network of mutually destructive societies for the improvement and reform of other people."[178] His own mother wants him to contribute to her home for misfit children:

> PETER: *It's a home into which she collects all the children she can find who are either dull or degenerate, or rickety, or consumptive, or whose parents are criminal. Thousands of pounds a year are collected for it, and numbers of children are patched up into tolerable specimens of humanity and turned out into the world to produce others like themselves. Now Miss Harbord [Beatrice] has a society—possibly you are a member—a society which would like to spend just as many thousands on preventing these children being born at all. How much cheaper and simpler it would be to persuade people like my mother to cease doing what they conceive to be their duty to others. Then the whole thing would cure itself and there would be no need for Miss Harbord's society.*[179]

Here, Harwood directly echoes *The Return of the Prodigal*, when Eustace deplores the current "humanitarian" approach of our age in which "we coddle the sick and we keep alive the imbecile. We shall soon come to pensioning the idle and the dissolute," a pronouncement that in turn echoes Karl Pearson's sentiment that "£1,500,000 spent in encouraging healthy parentage would do more than the establishment of a sanatorium in every township."[180]

As M. J. S. Hodge points out, Malthus "argued from any species's tendency to exhaust its resources to the conclusion that charity toward paupers will eventually bring even more misery, so that the Poor Law violates the greatest-happiness principle."[181] Peter scorns "people rolling pebbles in the path of nature."[182] Yet Harwood allows the audience to weigh both sides of the issue. Beatrice's assistant, Thorpe (who is "a thin, rather unhealthy-looking person" in the stage directions),[183] says "eagerly": "But we are *assisting* nature, that is our object. We are insisting on the importance of the great natural law—the supreme importance of the child." Peter says: "*Is* that the great natural law?" and Thorpe replies: "Of course it is." Peter says: "Then why interfere?" Thorpe says: "We want to be conscious of our purpose. Women *are* conscious of it."[184]

In addition to his largely unacknowledged tribute to Hankin, Harwood seems to be articulating a distinctly Shavian line in this conception of women as serving an evolutionary purpose and eugenically helping to improve the race. "The prestige of motherhood," writes Brian Harrison, was given a further boost by the eugenicists.[185] This almost divine purpose for women is clearly one of the central preoccupations of the play, and although here Peter just says "Hm," he comes back to it: "The consciousness of purpose that you spoke of—is that what you want men to understand?" If so, he says, that's "dangerous," because "suppose you succeed? Suppose you end by persuading men that they are nothing but the instruments of this purpose of yours—what do you think will happen then?"[186] He is interrupted by the entrance of Jack, Margaret's husband—surely such an instrument if ever there was one. The dire projection is thus left unsaid, for the audience to infer, but embodied in the figure of the superfluous Jack, with whose plight as the supplanted male the audience clearly must sympathize.

So far, most of the playwrights dealing with eugenics that we have looked at have been male. Female playwrights were also interested in this subject, but usually from a radically different standpoint, often connected to sexual selection. Baker's *Bert's Girl* (1927) is full of discussions about breeding in a way that is deeply pessimistic about the male's ability to choose intelligently. One of the first lines in the play is that "men are all alike—beasts!" a sentiment that is echoed several times in this first act.[187] The girl of the title is Stella, whom Bert has just introduced to his bickering and narrow-minded family; they find her inferior because she is not, apparently, beautiful. But Bert's eccentric and, we discover, eugenically inclined Uncle Trent disapproves of Stella marrying such a "toad" as his nephew. "Nature be damned!" he exclaims, and then sets about trying to steer Stella away from Bert for overtly eugenic purposes; as he says, "Is this [Bert] all we can produce after a billion years?"[188]

Baker cleverly constructs the play so that the audience is primed to be prejudiced against Stella, to find her plain compared with the more cultivated Iris and Evelyn, not to mention the vain maid Daisy, who is always checking herself in the mirror. The audience has to face its assumptions about beauty, ugliness, and "fitness," and this issue of fitness drives the subsequent scenes. "Nothing is any excuse for filling the world with ugly people," says Trent.[189] He invites some of his cronies to his flat to see her, and they agree that she is superior to Bert, who is merely representative of "the unfit seeking to rejuvenate itself by means of the fit." As Trent says, "Nature again—damn her. But she shan't do it—if I have my way." Trent's aim is to "stop . . . the rejuvenation of the unfit." In his view, Bert is a bad citizen: "It's his narrow, dirty, slimy little mind that keeps the world so close to the muckheap. . . . A mud-rat only breeds more mud-rats," and the world does not need any more of those. Beautiful Stella should not be used to perpetuate "a race of Berts," says Trent, even though there is the possibility that she might actually "improve the breed." No, that would be too risky: "Let ugliness mate with ugliness and beauty with beauty."[190]

Some discussion also revolves around whether Stella "chose" Bert, and the dailogue is often ambiguous about sexual selection. Bert says that he likes this modern age when "the girls run after the men," even while the stage directions have him put his arm around Stella.[191] In a long central speech, Trent muses on how much choice Stella actually had in the matter, what with having to look after herself (being on her own, no relations, having to work) and being at the mercy of nature, which he figures as theatrical, something "played out" on a stage:

> *Good, kind, loving Mother Nature that you're so devoted to, urging her on to mate almost before she's out of socks. There's Bert—the all-conquering male, also urged on to mate by the same kind maternal spirit. There you have it—the whole comedy, farce, tragedy—as you will. You can see it being played any day. Take away Bert, and within a year there'll possibly be Sidney or Percy or Edwin, and each time the poor fool thinks she's choosing the ideal. Nature won't help her; it's time somebody else had a try.*[192]

Trent bets his friends that he can reverse the course of nature, keeping Stella from marrying Bert, by employing her. But his friends are skeptical: "What about "instinct?" asks Quinton. Puttock likewise thinks "human nature" will trump Trent's manipulations.[193] Trent labels Quinton "an evolutionist. Man wasn't made to a model but is to work to one."[194]

Bert's Girl rivals Shaw's *Man and Superman* in the extent to which it addresses, and hinges on, evolutionary themes. Baker takes risks, exploring what the audience can tolerate in this vein. For example, a stark speech by Trent in a scene set a few weeks later reveals his own origins and why he rejects religion and the idea that God made anyone:

> *Awful responsibility, making people, though you wouldn't think so the way they go about it. Make something hideous, and then instead of being ashamed of it they put the blame on the Almighty.* [Pause.] *My father was a consumptive and my mother an epileptic. But did that stop them doing their bit? Not at all. And the devil alone knows the crowd there'd have been of us if they had had time. But consumptive father coming home drunk and trying to hit epileptic mother, missed her, fell and broke his head and died. So there was a lucky escape for little Fanny and Sarah and George and the rest. And when I fell about on my sticks of legs and grew up ugly and spotted and my eyes bulged and my head cracked, did anyone say I ought to have been smothered at birth? Not they. They all sat round and crooned: "Poor dear Mr. Trent, what a burden the Lord has put upon you, and how beautifully you bear it!"*[195]

As in Harwood's *The Supplanters*, the play features at its center a eugenic diatribe against supporting the unfit. But how far does Baker actually endorse this view? Trent exhorts Quinton to bring one of his students to meet Stella—Quinton teaches disadvantaged boys how to write essays—and Quinton is delighted that Trent has chosen, seemingly randomly, from the pile of essays the boy who is evidently his favorite; he has no physical deformities, just "a little delicate perhaps," so Trent wants to get them together. Quinton is revolted: "I couldn't do it. It's—it's repulsive, this mating. . . . I can't help it. I can't interfere. That beautiful girl—it's treating her like an animal." Trent is "disgusted with him," saying that "so she is—so are you—so's everybody."[196] The play thus has it both ways, contriving agency in the seemingly random process of mate selection.

Trent is delighted when Stella rejects Bert, and marriage generally, the next day. He says some other girl will take her place, and all of the family will live and fight and quarrel "and add to the population and furnish another example of a happy and united family." He beams with pride at Stella's having "snapped your fingers in the face of Mother Nature. She was out to catch you, and you gave her the slip! [*Laughs.*] She'll try again—damn her—but don't worry about that. Perhaps she'll catch you, and perhaps she won't; perhaps you'll sit in a cottage with a cat and a parrot—and perhaps you'll marry a poet or a millionaire. Never mind that. Damn the past, and hold out your arms to the future!"[197] To his friends he announces that "old Mother Nature's lost!" but then Puttock quickly adds: "This time!" [*They laugh.*] [Curtain.][198] We are left to assume that Stella will accept Trent's offer to live with the three old cronies, though in what capacity it is not clear, or that she will end up marrying one of the disadvantaged boys whose essays have been discussed.[199] What is certain is that either of these choices is better for the future of the race than mating with the unfit Bert.

Eugenics brought breeding, choice of mate, and reproductive issues to the fore. I continue this exploration in greater depth, particularly in relation to the emphasis placed on women as selectors and vessels. Theatre immediately registered the impact that this new evolutionary focus on women had on the concept of the family unit, dramatizing the shift in the conception of women from stereotypically submissive angels in the house to powerful arbiters of human evolution.

6

Reproductive Issues

The eugenic program of directing human evolution upward through manipulation of breeding needs to be seen in relation to wider debates that were developing about all aspects of human reproduction, particularly the struggle for women to gain greater control over their reproductive capacities. "Of all woman's rights," declared an unnamed author writing in *Review of Reviews* in 1904, "surely the first and most obvious is the right to say how many times she shall be subjected to the glorious but perilous ordeal of childbirth."[1] This view was expressed repeatedly on the stage, for example by Eugene Brieux in his play *Maternity* (1904), in which a woman goes on trial for having an abortion after her drunken husband has raped her. The play boldly advocates that women not have too many children, as repeated childbearing is cruel to the woman, selfish on the part of the husband, and economically infeasible for any but the wealthy. The archbishop of Canterbury advised the Lord Chamberlain in 1918 that this was not a lesson the English needed to hear in a time of war, and the play was refused a license. As Sos Eltis notes, "Theatre was thus positioned as a valuable tool in deploying women's bodies in the greater interests of the state."[2]

It is fair to say that sex and reproduction are two areas of human life that have always dominated the theatre. We have seen how Henrik Ibsen brought female sexuality center stage: the extreme sexual passion of Beata in *Rosmersholm* (in the draft to the play Rosmer alludes to his dead wife's "unfortunate frenzies of passion, which she expected me to return. Oh, how they terrified me!")[3] and of Rita Allmers in *Little Eyolf*, the sexually predatory Hilde Wangel in *The Master Builder*, and the destructiveness that results from the attempts to suppress them. *Hedda Gabler* celebrates reproduction of a different kind: women's creativity through writing rather than through producing children. For Bernard Shaw, sex and birth clearly belong to a more primitive state that humans at the pinnacle of his progressive evolutionary vision have long surpassed. Both playwrights saw theatre as the natural mode to explore these ideas, defying the obstacles of censorship, critical hostility, audience antipathy, and box office failure. And both register a great shift in cultural attitudes and practices regarding human sexuality, largely prompted by developments in evolutionary theory. Perhaps the most striking change in this regard was the view of the "fallen woman," a mainstay of Victorian theatre. The "woman with a past" remained a theatrical favorite, but "active female sexual desire" became most visible in the new realist drama, "not located simply as sin and temptation, but as an essential human impulse."[4]

One thing that is noticeably absent from the stage is childbirth itself—like breast-feeding a taboo subject for theatrical representation. There is simply no contemporary theatrical equivalent to Mina Loy's graphic poem "Parturition" (1914), although Eugene O'Neill's *First Man* (written 1921; published and produced 1922) does mark the event of childbirth taking place off stage by the screams of the mother dying in labor. Yet how accurate is the assumption that society was not ready to see childbirth enacted on stage? In the Victorian and Edwardian periods, at least in middle- and upper-class families, husbands tended to be present during childbirth.[5] There was a much higher probability of a woman's death in childbirth than we have now, so that the husband truly might not see his wife again; childbirth might be their last moments together. Also, births overwhelmingly took place at home; when in the 1930s it became fashionable to hospitalize women in labor, the attendance of husbands at births dropped.[6] In fact, the uniquely female physiological experiences of menstruation, childbirth, breast-feeding, and menopause rarely figure directly on stage even to this day.[7] Sex as a field of scientific study, of course, was already well established by the turn of the twentieth century, through Havelock Ellis, Edward Carpenter, and many others—and many of these were active in theatrical networks. Some sexologists were already gesturing toward what Alfred Kinsey would later articulate so clearly: "not that nature sanctions any given sexual or social practice, but that it sanctions all practices equally."[8]

Uncoupling Marriage from Maternity

Plays increasingly suggest that although she might be destined for motherhood by virtue of her sex, a woman did not have to be married to fulfill that natural role. In St. John Hankin's *Last of the De Mullins*, the rebellious Janet, a single unmarried mother, tells her dutiful sister Hester (the name evoking her Puritanism): "In your heart you envy me my baby, and you know it." She describes the biological clock ticking: "In a few years you will be too old to have children. . . . Your best years are slipping by and you are growing faded and cross and peevish. . . . You will be an old woman before your time unless you marry and have children."[9] Hankin seems to be voicing an old-fashioned sentiment here about women's fulfillment through their children. But Janet qualifies this: "To do as I did [have sex, have a baby, yet remain single] needs pluck and brains—and five hundred pounds. Everything most women haven't got, poor things. So they must marry or remain childless."[10] Stanley Houghton's *Hindle Wakes* likewise features a woman "falling" by going off to have sex with a man she is attracted to momentarily but not enough to marry. "Love you?" Fanny in *Hindle Wakes* says to Alan, who cannot believe she is rejecting him. "Good heavens, of course not! Why on earth should I love you? You were just someone to have a bit of fun with. You were an amusement—a lark."[11] Houghton and Hankin confront head on the limited choices for women in terms of sex, marriage, and reproduction, separating motherhood from marriage, and they join a host of other theatrical explorations of such issues during this period.

Hankin's feminism sounds convincing, like Shaw's, when he rails against the slavery of marriage for women, but ultimately, he still sees a woman's fulfillment in motherhood, though not, notably, in wifehood. *The Last of the De Mullins* has been called "excessively derivative" of Shaw and Brieux.[12] And, for all the radicalism of Janet's actions, her situation is shown as an exception to the general rule, and her choice not available to most women. Until conditions for women change, the only options for most of them who lack the courage, intelligence, and money (and Janet has had to earn this by herself, through trade—itself a scandalous step for a woman of her social class) is to start a family.

As both a playwright and an actress, Elizabeth Robins constantly questioned this familiar family-centered scenario for women, trying to do something radically new on stage in terms of the treatment of the female body and women's issues and openly grappling with biological determinism. She worked within a context of increasing fetishization of the actress that spills over into a wide range of cultural domains, from women's magazines to erotica.[13] Tracy C. Davis notes the link between theatre and pornography: "The inclusion of actresses and incidents within and around theatres in so much pornography throughout the Victorian period demonstrates the theatre's enduring erotic fascination."[14] How do these limiting views of the female body on stage align with Jane Goodall's claim that theatre played a key role in liberating women from biological essentialism through the increasingly radical performance of the female body? A key to this question lies in the interest of many playwrights in the role of sexual selection in human evolution.

Sexual Selection: Reversing Social Norms

Sexual selection is innately theatrical; it involves "flaunting, display, theatre, extravagance, scents and song."[15] In *The Descent of Man*, Charles Darwin observed that in nature the female of most species selects the mate, favoring males of the greatest beauty and prowess, whereas humans have perverted this pattern so that "money buys beautiful women, and men also seek out women who are rich, and therefore probably favour the outcome of low-fertility families where all the wealth is concentrated in one girl."[16] Alfred Russel Wallace argued for a radical reorientation of the social norms by which men select women, treading a fine line between the hard eugenics of Francis Galton and

Hiram M. Stanley and the "pure sensualism" of Grant Allen.[17] Wallace envisions a time when

> powerful selective agency would rest with the female sex. The idle and the selfish would be almost universally rejected. The diseased or the weak in intellect would also usually remain unmarried; while those who exhibited any tendency to insanity or to hereditary disease, or who possessed any congenital deformity, would in hardly any case find partners. Thus, when we have solved the lesser problem of a rational social organization adapted to secure the equal well-being of all, then we may safely leave the far greater and deeper problem of the improvement of the race to the cultivated minds and pure instincts of the Women of the Future.[18]

In this article, Wallace is unequivocal about the "absolute necessity" of "female choice in marriage." But he had the weight of the scientific establishment against him, as sexual selection—let alone *female* sexual selection—was even less popular among biologists than natural selection.

Why was this idea so threatening? Following it to its logical conclusion, it might ultimately imply "the intellectual superiority of women."[19] Even until fairly recently, the "astonishingly brilliant" idea that human sexual selection was female driven and that women thus held the key to evolution was still finding resistance.[20] The debates are still ongoing regarding the relationship of feminism to sexual selection, with Fiona Erskine on the one hand asserting that "sexual selection is intrinsically anti-feminist" and on the other hand Elizabeth Grosz arguing for its inherent feminism.[21] And with regard to the equally contested area of sociobiology, many of the plays of this period seem decades ahead of the findings of Sarah Blaffer Hrdy and others confirming that the animal kingdom is full of females with sexual appetites that have nothing to do with reproduction. The larger context for their engagements with reproductive issues is what Charles Webster calls the dialectic between the biological and social sciences, "indicating the transference of ideas about human society to the animal order, and then their return to provide a biological rationale for behavior within the human family."[22]

Dorothy Brandon's *Wild Heather* (1917) not only suggests some of the ideas of evolution, such as sexual selection, heredity, and adaptation, but also integrates them into the fabric of the drama, even though the word *evolution* is used only at the end of the play, when a well-bred young woman scandalizes her family by showing her passion for her virile lower-class lover. But her father, a professor, is not so sure John is "beneath" his daughter: "It seems to me—in evolution—the young man's a little further from the animals and a little nearer to the gods than most." Heather rapturously calls her father a "darling," and her mother says, "Oh James what *does* evolution matter?" Heather and John hug and "kiss hungrily and openly" in front of everyone—and, far from discreetly leaving them alone, the professor "turns on the electric lamp to see them better." They face the "staring semicircle of astonished and protesting faces," and as Heather says "I can't help it!" the curtain descends on a happy ending that is clearly showing the power, and perhaps necessity, of sexual selection to transcend social boundaries.[23]

What is striking here is not so much the use of the word *evolution*—though that is, as we have seen, relatively rare in dramatic dialogue—but the fact that Heather is irresistibly attracted to the manly man, a theme that pervades theatrical treatments of evolution as evidenced by Robert Buchanan's *Charlatan* (1894), discussed in chapter 1, and Robins and Florence Bell's *Alan's Wife*, in which Jean has rejected a weedy vicar in favor of a virile and masterful mill worker. Shaw's *Misalliance*, by contrast, is much preoccupied with what makes a good male partner, and manliness ranks well below brains; Hypatia dismisses Jerry as "a splendid animal" but a "fool," whereas Bentley may be "a little squit of a thing," but his brains make him "the best of the bunch."[24] This would be a banal enough judgment if it were not tied so cleverly to a complex and radical discussion of the evolutionary motifs of sexual selection, breeding, and instinct.

As these plays indicate, not only are women on stage becoming choosier in selecting their mates, but also the masculine spectrum—just like the feminine one—has clearly expanded, with two increasingly separate and polar extremes: at one end the "primitive" and "red-blooded" savage, who may not be intelligent but has the brawn to "hold" the female, and at the other end the brainy but effete modern male.

An interesting backdrop to this spectrum is the growing field of ethology. *The Descent of Man* catalogues dozens of instances of insects, birds, and other animals having developed the organ necessary to hold, grasp, or secure the female during the sexual act.[25] The book also enumerates exceptions to the assumed norm of the passive female being wooed by the aggressive male. How does this relate to the conventional notion of Victorian women as ignorant of and uninterested in sex, of passively having to "suffer and be still"?[26] Theatrical explorations of human sexuality and reproduction map directly onto the changing discourse on evolution.

The potential of the theatre to explore and change people's minds about women's issues was certainly recognized at the time. The birth control campaigner Marie Stopes wrote in 1927 that she put her ideas about sex, contraception, and human reproduction into plays because "the clash between *real* womanhood & conventional manhood seems to me the most dramatic thing in the world this century."[27] Not only the innate drama of such a clash but also the live, public nature of theatre seemed ideal conditions for a campaigner like Stopes or Robins seeking rapid change. This was recognized by the anarchist, feminist, birth control campaigner Emma Goldman in a lecture titled *The Social Significance of the Modern Drama* (published 1914). The Edwardian feminist theatre critic Marjorie Strachey argued that modern drama was, for women, a "magnificent and untouched field" as it held enormous potential to represent women at this transitional point in their history—but only if women changed the subject matter of plays. Rather than the old-fashioned, "tedious" theme of love, playwrights should focus on themes such as work ("hundreds of the professions occupied by women") and "women's friendships with women."[28]

One might assume that the suffragettes would heed this call most of all. Suffragette playwright Cicely Hamilton agitated for contraception and abortion rights for women, "the right of men and women (but especially of women) to save themselves suffering, to spare themselves poverty, by limiting the number of their children."[29] But Hamilton was an exception, as suffrage workers tended to distance themselves from contentious reproductive questions to keep the focus on winning the vote, rather than alienate potential allies in this fight.[30] Little connection exists between the suffrage cause and the staging of controversial reproductive issues. In fact, many of the strongest advocates for reproductive rights and for greater frankness about such issues had the least success when they tried to enlist the theatre to their cause—either the Lord Chamberlain or the box office simply could not stand it.

Stopes's "banned" play *Vectia* is an example of this. The author of such popular works as *Married Love* (a pioneering sex manual) and *Radiant Motherhood*, Stopes wrote *Vectia* to show the devastating consequences of women's sexual ignorance.[31] Vectia does not understand why after three years of marriage she and her husband have no children; she seems ignorant about how babies are made. William, her husband, is impotent, and it is suggested that he "self-abused" too much as an adolescent and has therefore perverted himself. So naïve is Vectia that she does not realize how odd it is that she and her husband do not have sexual intercourse. The play's central (and controversial) scene has the shy Vectia getting belated sex education through her faithful friend and confidant Heron, a lawyer who senses her dreadful situation. In order to find out whether Vectia's marriage has actually been consummated, he asks her to draw diagrams since she is far too embarrassed to explain verbally. They sit back to back and silently pass the drawings back and forth along with Heron's written questions. As Vectia slowly reads them, "a rather startled and puzzled look comes into her face as though ideas quite novel to her were throwing light on a difficult subject."[32] Vectia, Heron, and the audience all discover simultaneously that Vectia is still a virgin. This was an attempt to get past the censor by avoiding any actual dialogic references to sex.[33] Shaw uses a similar device in *Back to Methuselah*, part 1, when the Serpent whispers to Eve the secret of reproduction and she reacts with visible disgust; the audience is likewise "tempted" to imagine what naughty things are being said. It was a way of staging the unstageable by implication and suggestion. In addition to this device of nonverbally explaining conception, Stopes uses a particular metatheatrical trick (employed by Shaw in *The Doctor's Dilemma*) of incorporating references to herself in a scene in which William denounces Stopes's books. It is a strange moment for the audience, when real life intrudes onto a staged representation of it.

Although on the surface it is just the story of a decent girl wanting to have a baby by her husband, the plot turns on his sexual impotence and, less explicitly, his possible homosexuality. The play ends with a dramatic moment of sexual selection: Vectia realizes that Heron loves her as she wants to be loved, but that William needs her emotionally as a kind of nurse for his soul. "You two," she says to them, "the weak and the maimed and the strong and the joyous. Which of you needs me most?" Her choice is heavily framed by biological purpose. "Isn't it my *duty* to stay if he's so much in need of me?" she asks Heron. "Your duty is to life and the fullness of life," Heron proclaims.[34] She chooses him, leaving William alone on the stage with a revolver that he uses to bash all her sculptures as the curtain descends and, we infer, eventually to commit suicide. The Darwinian message is clear in her rejection of the feeble, aberrant, and sexually spent William in favor of the healthy Heron, just as Shaw's Candida (the Life Force) chooses the virulent Morell over the effete and decadent poet.

Vectia (under the title of her popular book *Married Love*) was submitted to the Lord Chamberlain along with Stopes's play about birth control, *Our Ostriches*, in October 1923. It was refused a license and thus became "one of the few plays banned by the Lord Chamberlain in the 1920s" when so many other bans were finally lifted (such as those on Shaw's *Mrs. Warren's Profession* and Harley Granville-Barker's *Waste*).[35] The play was never granted a license and has never been performed.

A New Model of Marriage

Plenty of other plays were scrutinizing the institution of marriage, despite its persistence as a social norm and as the main means of channeling sexual reproduction. While plays continued, of course, to affirm marriage, a growing body also challenged it, particularly with regard to reproduction, and repeatedly mooted the idea of a "partnership" instead, as in Dorothy Leighton's *Thyrza Fleming* (1894) and Elizabeth Baker's *Partnership* (1921). Ibsen constantly floats the concept of a "true marriage" having nothing to do with religiously sanctioned, socially enforced union; *A Doll's House*'s lines about "true marriage" in the final scene reappear almost exactly in a draft of *The Wild Duck*, while *The Lady from Sea* makes the question of "true marriage" the central problem of the play. Baker's play boldly presents Kate, an independent career woman, teaming up with an unconventional man named Fawcett in what will clearly be a sexual union. She has just rejected a proposal for a very different kind of partnership from a much older business rival, who presents marriage to her as an opportunity to merge their shops and increase their profit. When Kate sees Fawcett, female sexual selection kicks in: "Their eyes meet . . . her gaze following him as if attracted at once," and she plays the pursuing role while he is the shy, diffident one.[36] The play suggests replacing conventional marriage with a true partnership founded on love, sexual attraction, equality, and independence in which a woman can be a businessperson and still be fully female.

Plays like this reflect a tendency to attack marriage as an outdated institution that has little relevance to humans' biological and emotional needs, especially in light of the idea that once viable offspring have been produced, the father might not be needed. H. M. Harwood's *Supplanters* seems to suggest that the single and celibate lifestyle is the preferable one, echoing the vehement attack on the institution of marriage in Shaw's *Man and Superman*. Jack objects to the idea that he is merely a tool in evolution. "I wanted to be your lover," he tells his wife Margaret, "you made me only the father of your children."[37] This echoes the conclusion of *A Doll's House* but completely inverts it: Nora's reason for leaving her children was that Torvald had made her "only" the mother of his children, fulfilling a natural function dictated to her by society more than by nature, and never gaining broader experience as a woman. She argues that she cannot be a good mother to them until she has educated herself. In Harwood's play, parenthood is also sidelined but for a different reasons; he shows the problem of the male being sidelined by the woman's all-consuming attention to her children primarily from the male perspective. It is instructive to contrast this with Shaw's ability to adopt the female perspective on this issue in *Getting Married* (1908), when Lesbia declares: "If I am to be a mother, I really cannot have a man bothering me to be a wife at the same time."[38] Both Shaw and Harwood take it for granted that these two roles are incompatible. Jack says he'll grant her a divorce but refuses to return to her in "the soul-destroying capacity of father of a family."[39] The play ends with Jack coming back after both husband and wife have made adjustments: he has discovered that he is actually fonder of the children than he thought, while she has learned that she has to care about him as well as them. Compromise is shown to be key, but this invariably involves greater female sacrifice. The play includes two trenchant remarks on a woman's lot: "Life isn't very fair to women" and "a woman can't be everything."[40]

Critical responses to *The Supplanters* were mixed. The *Saturday Review* dwelt on the husband's inordinate sex drive: the problem is that husband and wife have "a serious difference of opinion as to how much love goes to the square meal of an ordinarily healthy man." He should simply face the fact that "from his wife's point of view as the mother of a family he was considerably oversexed—that he must grin and bear it like a man, finding compensation in the responsibilities of a father," instead of "irrelevantly cursing the interloping babies." But the reviewer felt that the play was too much a "farrago of derivations" of Oscar Wilde, Hankin, and John Galsworthy.[41] The *Academy*, however, praised the play as "a serious sociological matter written with admirable lightness and cleverness."[42] It also noted the subplot, a "clever picture of the young girl of our day [Isobel] who does not mind facing the facts of life and telling us so in fresh and racy phrases." While the main plot of *The Supplanters* is busy questioning the roles of mother and father, wife and husband, the subplot between Margaret's sister Isobel and her suitor is questioning the relevance of marriage itself; Isobel scandalizes her parents by indicating that she just wants to live with him, not get married, and that she already knows about sex.[43]

Marriage was remarkably stable despite the many attacks on it that were launched throughout the first few decades of the twentieth century.[44] What really changed was attitudes to sexual behavior before and outside marriage, with a pronounced shift between 1904 and the period 1924 to 1934.[45] Certainly, World War I helped to spur this change, as women entered into jobs normally reserved for men, gained greater social and economic freedom, and began to challenge the taboos around sex outside marriage. Alongside this change, not only in people's sexual behavior but also in general attitudes to sex, came the concomitant "perception of sexuality as a crucial element in the nature of the individual."[46]

That marriage might no longer be a necessary, let alone natural and fitting, way to contain sexuality and reproduction was already being suggested by analogies drawn from the animal kingdom. In *The Descent of Man*, Darwin notes in passing that there is some tension around the meaning of the term *marriage*, that some observers have been too harsh and restrictive in their definition of it: "I use the term in the same sense as when naturalists speak of animals as monogamous, meaning thereby that the male is accepted by or chooses a single female, and lives with her either during the breeding-season or for the whole year, keeping possession of her by the law of might. . . . This kind of marriage is all that concerns us here, as it suffices for the work of sexual selection." He addresses the issue of "communal marriage" but then says that "the subject is too large and complex for even an abstract to be here given, and I will confine myself to a few remarks."[47] Readers of *Origin* would have recognized this as a deft Darwinian evasion, as when he writes vaguely in that book that on the subject of human origins and evolution, "light will be shed" in the future. As far as sexual selection is concerned, writes Darwin in *Descent*, "all that is required is that choice should be exerted before the parents unite, and it signifies little whether the unions last for life or only for a season."[48] He is talking about human relationships here, not just animals. This is a startling and bold suggestion: marriage can be dispensed with, and it is just one of a number of possible mating configurations ranging from monogamy to polygamy to rampant promiscuousness.

The Victorian conception of marriage as an institution of high social, cultural, and religious relevance and a deliberate and rational set of rules thus gives way to a Darwinian sense of the inherent randomness of marriage as a construct and the vast gap between theory and practice in terms of the actual workings of marriage as an institution. This in turn would threaten its survival—if the beast does not adapt, it will not thrive. So marriage is no longer an inflexible, rigid institution but an organic being that is subject to the same laws of nature as every other living thing. There is a concomitant shift in perceptions of children's roles, not only within marriage but also within human evolution more generally. It is not surprising to find plays like Elizabeth Baker's *Chains* depicting bitterness at marriage as inhibiting rather than satisfying because pregnancy acts like a chain on the male or the young mother in Githa Sowerby's *Rutherford and Son* complaining that the "struggle for life" is all there is once the couple has children. Harwood takes this hint much further in his depiction of children as interlopers in marriage, undermining the foundations of family life by usurping the male's status.

Where Are the Children?

Shaw's *Misalliance* and Harwood's *Supplanters* appeared within a few years of each other (1910 and 1913), part of the growing discourse on childhood, on relationships between parents and children and between mothers and fathers, and on the proper place of children within families. There were notable precedents; George Gissing's novel *New Grub Street* (1891) depicts the displacement of a father by his children, and Ibsen's bold play *Little Eyolf* (1895) puts this issue center stage and ties it explicitly to sexual frustration—except in this case, it is the woman and not the man suffering from the moribund sex life brought about by having children. In the draft version of *Little Eyolf*, the mother vehemently rejects the father's newfound dedication to his son, fearing that she will be supplanted. Her husband declares that in a good marriage, "New situations are formed. . . . New duties assert themselves. The children, too, claim their rights. They *have* the first claim."[49] She reacts "almost wildly," asking if he has ceased to love her and insisting, "I want you entirely to myself."[50] She refuses to take second place to her son. After his death, she admits: "I was always trembling at the prospect of his taking you away from me."[51] Ibsen gives us a scenario in which it is the mother, not the father, who sees offspring as interlopers in a marriage. Eyolf's mother goes so far as to say he was "a little stranger boy"[52] and proposes that they try to forget him. She completely rejects maternity; it is just sex that interests her. Allmers realizes this all too well: "You were never a real mother to him."[53] In fact, she cannot wait to resume their sexual life despite the tragedy: "I am a warm-blooded being. I have not fishes' blood in my veins."[54] The final version of the play retains much of this vehemence and indeed pushes Eyolf even further to the margins of his parents' lives.

There are strong links here with Elizabeth Robins, who thought deeply about this question of how far children are central to a woman's identity. *The Silver Lotus* (1895–1896) is an unpublished play about the devastation caused by the tragic death of children. The play deserves publication, scholarly recognition, and analysis, particularly in light of the extraordinary scholarly interest already shown in Robins's other theatrical writing, such as *Alan's Wife* (1893) and *Votes for Women!* (1907). In *The Silver Lotus*, Eleanor's three children have died two years ago, and she is wracked with guilt, because despite her husband's advice, she took the children on vacation to an area that had reported cases of diphtheria: "First the baby, then the two girls; all within a week."[55] She has taken to alcohol to numb her pain, and the play treats female alcoholism "with sophistication and complexity."[56] At the center of the play, then, is a mother deprived of her calling, which at first seems to suggest that it is of the *East Lynne* variety (endorsing motherhood as woman's true calling). But the play turns out to be about much more than that.

Abetting Eleanor in her alcoholism is her devoted but strange servant Dwyer, who helps her obtain the key to the liquor cabinet, which Eleanor's husband, Gervais, has had locked. Eleanor says that Dwyer is "victim to the most incurable of female diseases";[57] while all that is meant is jealousy, this is a brilliant stroke in establishing this idea of "female disease" so early in the play, before we know what is wrong with Eleanor. The first scene also raises the "Woman Question," referred to by Eleanor with a harsh laugh and the comment that "we are an obscure sect—we women."[58] Sexual difference is rendered as irreconcilable apartness. In act 2, Eleanor's friend Camilla—who is secretly in love with Eleanor's husband—sarcastically rebuffs the advances of Grantham, a friend of the family, in terms that make fun of male sexual selection as passé: "Really, men are incredible. Your views about women came over with the Conqueror. If my affections were not already engaged I *couldn't* resist any man who was willing to rescue me from spinsterhood. That's the line of argument."[59]

Eleanor becomes hysterical when her mother-in-law, the sympathetic, clear-eyed Mrs. Onslow, soothingly suggests that everything will be fine once she has another child, once Eleanor's "arms are [no] longer empty." Eleanor reacts wildly to this suggestion that all will be well once "we'll begin to forget our dead."[60] Mrs. Onslow thinks that both Eleanor and Gervais are suffering from depression due to the children's deaths.[61] The pathological landscape of the play thus keeps expanding, and the audience does not even know yet about the alcoholism, which is dramatically revealed in silent tableau: Mrs. Onslow and Gervais watch as Eleanor, thinking she is unseen, sneaks toward the liquor cabinet, unlocks it with a key they thought they had confiscated, and takes a decanter, which she presses to her bosom as she leaves the room. Mrs. Onslow and the audience thus discover the alcoholism at the same time.

We learn the seriousness of the addiction when the doctor tells Gervais baldly: "This is not the case of a strong woman gradually drinking herself to death (*a sharp contraction of the muscles of Gervais's face*) but of a delicate woman whose constitution is undermined, *shattered*—."[62] Gervais bursts out "passionately": "If it were a mere physical disease! If it weren't bound up with questions of will, of liberty, of personal dignity! God! if it weren't so hard to talk about, it wouldn't be so hard to face."[63] We never see Eleanor really inebriated, but we hear about one awful episode when she got completely drunk and was discovered by Camilla, whom we now realize is her bitter rival for Gervais's love: Gervais tells how he found Camilla "standing horror-struck by the bed. Eleanor flung across it unconscious, but talking—talking in that horrible incoherent way. (*he shudders*) A glass half full of raw brandy on the small table—the room reeking."[64]

Gervais's sensible and compassionate mother asks a perfectly logical question of her son: "I can't help wondering why under the circumstances, you have wine on your table here."[65] The extent of the husband's masochism is gradually revealed through such touches: wine at the dinner table of an alcoholic wife; cold, stern, and unaffectionate in the face of her debilitating grief and depression; and—most devastating all for Eleanor—totally unresponsive to her sexual needs.

The true revelation of the play is its suggestion that Eleanor's tragedy is not only the loss of the children (that has already occurred before the curtain rises) but also the loss of her sexual life. *The Silver Lotus* is unflinching in its depiction of a woman's desperation when her natural sexual desire is thwarted. In a powerful scene of reconciliation, something of the urgency of Robins's intimate thoughts in her diaries and letters comes through when she has Eleanor plead with Gervais not to deny her physical love: "We've lived in hope long enough. I've *starved* on it. Let us have certainty now. (*she clings to him passionately*). . . . Love me—love me." He says, "I do—you know I do," and her response is: "But as you used I mean." She wants him to have sex with her, but disengaging himself, he gently says that "things can't be as—as they were till we are *sure*." He says they need to be sure "that you are able to bear the responsibility that might come to you."[66] If she were to get pregnant, she would endanger both her life and the baby's unless cured of her addiction. This is similar to Val and Ethan's precaution in Robins's novel *The Open Question* (1898): a eugenic concern for subsequent generations must be the overriding principle in any sexual relationship. The consumptive Val in that tragic novel is also sexually passionate and another instance of Robins explicitly representing female sexual selection.

The Silver Lotus thus reveals layers of complex emotions and motives that show a deep awareness of the way a piece of theatrical storytelling works on an audience. We are pulled further and further into Eleanor's tragedy and keep having to revise our understanding of its causes as new bits of information are gradually revealed; it is almost a textbook emulation of Ibsen's technique of retrospective arrangement. The audience begins by thinking the children's deaths have caused Eleanor's retreat into lotus blossom-like alcoholism; then, we find out it is that loss combined with her husband's refusal to have sex with her and thus offer any physical and emotional comfort to her; and finally, we realize that the presence of a beautiful female rival masquerading as a friend is sending Eleanor over the edge. The only source of comfort to Eleanor is her faithful servant, yet even she turns out to be more of an agent of death than a savior—a self-appointed mercy killer. In a strange parallel with Robins's treatment of euthanasia in *The Mirkwater*, Dwyer—the (possibly Aboriginal, the text suggests) nurse Eleanor brought with her from Australia, who was Eleanor's nurse as a child (a direct parallel with Nora Helmer's situation)—aids in Eleanor's addiction and eventual death by regularly supplying her with small amounts of alcohol, which she argues is healthier than Eleanor getting at the liquor herself. She thinks she is managing Eleanor's addiction, but she is really just helping her die slowly. "I wouldn't have dared to go and leave her without anything [to drink]. She wouldn't have slept. The nights were awful. I couldn't let her suffer like you did," Dwyer stoutly tells Gervais, who of course does not know of her suffering nights as they sleep apart.[67]

Joanne E. Gates comments that the play effectively probes "the pain in a relationship in which the central memory is the death of children," but it goes much further than that.[68] This is the situation of *Little Eyolf*, which Robins was most likely reading hot off the press while composing *The Silver Lotus* and was deeply involved with as an actress, playing the role of the sexually voracious Rita Allmers in the 1896 London premiere. On the surface, these two works seem to handle the representation of women's grief at the loss of children quite differently. Where Ibsen shockingly and unflinchingly gives us a woman whose focus on her sexual relationship with her husband virtually wipes out her maternal grief, Robins gives us a woman slowly dying from that grief. But, as the play progresses, shades of Rita Allmers encroach, as we see Eleanor's mourning bound up in the loss of her physical relationship with her husband as well. Ibsen's *The Master Builder* may also have been an influence, in the tragic figure of Mrs. Solness, whose children have died; Robins played the interloping Hilde Wangel to great acclaim in the London premiere in 1893. There is also a strong echo of the Rat Wife in the strange death figure of Dwyer.

But *The Silver Lotus* is not merely derivative. The play shows that Robins had an "ear for crisp dialogue" and a "command of dramatic form," as Gates puts it.[69] Its symbolism echoes Ibsen's and Anton Chekhov's, from the title to the way it informs and deepens the play's themes, but it is given a specifically female slant. The silver lotus of the title is a necklace that Gervais gave his wife in happier times and that she cherishes. The lotus is of course also a symbol of the sleep of forgetting that Eleanor finds through alcohol. Eleanor is one of Robins's few female victims; in general, Robins's stage women are adaptive and are more like Shaw's than Brieux's victims of circumstance. But Robins shows us a woman at a biological dead end: she has already spent her "life force" and is barred from sex for fear of spreading her "disease" to the next generation. Even while Robins paints this portrait with great empathy, it is chillingly clear that Eleanor must be sacrificed to the greater good of the species.

One can speculate on possible inspirations for this play about female addiction. There is the life of one of Robins's colleagues, the actress Janet Achurch, who had a stillborn child while in Cairo as part of an extensive tour of *A Doll's House* (1889–1891), continuing to perform only by taking morphia, on top of already having a tendency to drink too much. Sharing the same social sphere, profession, and devotion to Ibsen, Robins would most likely have known about Achurch's problems and addiction. They worked closely together, starring in the London premiere of *Little Eyolf*. Her memoirs recall the momentous occasion of seeing Achurch as Nora in the 1889 premiere of *A Doll's House* at the Novelty Theatre, London—another possible theatrical precedent (a mother who "loses" her three children). Robins might also have drawn on the tradition of the temperance play, a nineteenth-century theatrical staple typified by James A. Herne's *Drifting Apart* (1888).[70] Robins's innovation is to depict the alcoholic as female and to suggest unflinchingly the sexual and reproductive issues relating to her tragic condition.

Infanticide

Robins went to the other end of the maternal spectrum in the play she wrote with Bell called *Alan's Wife* (1893). Grieving for her husband, who has died in a mill accident, new mother Jean kills her deformed baby by smothering it in its crib and then retreats into silence, refusing to defend or explain herself to either her mother or the court. Her life is at stake since under the British legal system, baby killers received the death penalty.[71] Just as *The Silver Lotus* redefines an existing tradition of stage treatments of alcoholism, *Alan's Wife* sits in relation to a surprisingly long history of infanticide dramas and to an expanding spectrum of maternal behavior in the context of the contemporary discourse on human evolution.[72]

Ibsen had already opened up the stage to the extreme end of the maternal spectrum in the climactic endings to plays like *A Doll's House* (Nora leaves her three children to find herself), *Ghosts* (Mrs. Alving will surely kill Oswald—the euthanasia Robins and Bell suggest Jean commits in *Alan's Wife*), and *Hedda Gabler.* Before she shoots herself and her unborn child, Hedda "kills" the "baby" engendered by Thea and Løvborg when she throws their book manuscript into the fire, whispering to herself, "Now I am burning your baby, Thea. . . . Yours and Ejlert Løvborg's baby. Now I'm burning—now I'm burning the baby."[73] In the draft to *Hedda Gabler*, reference is specifically made to infanticide. When she learns (falsely) that Løvborg has destroyed his own work, Thea tells him that she will always think of his deed "as though you had killed a little child." He says she is right: "It is a sort of child-murder."[74] The list of killed-off children in Ibsen goes on. Hedwig in *The Wild Duck* shoots herself, Eyolf drowns, and Irene in the draft to *When We Dead Awaken* says she has had "many children" but "I killed them. . . . Killed them [murdered them pitilessly] as soon as they came into the world. [Long, long before.] One after the other."[75] She is bitter at having been the unacknowledged medium for Rubek's success as an artist—the model he sculpted so successfully, yet in the process erasing the person. What made the baby killing of *Alan's Wife* much more outrageous than all these child deaths put together was its directness (the audience witnesses it), the lack of an understandable motive or justification for it, and the fact that an English mother does it to her own baby. Performing child murder on the English hearth, so to speak, was a step too far.

Josephine McDonagh and others have shown that infanticide permeated the public imagination in the nineteenth century. Novels like Sir Walter Scott's *Heart of Midlothian* and sensational reports in the periodical press fueled this interest; the stage also was an important medium. William T. Moncrieff's *Cataract of the Ganges* (1823), a popular equestrian melodrama that was revived in the 1850s (in both Great Britain and the United States) and as late as 1873 (Drury Lane),[76] protested the practice of female infanticide in parts of India. As with Darwin and other natural scientists' discourse on the subject, stress was laid on "the systematic (and thus foreign) nature of the practice."[77] Infanticide as linked to both savagery and orientalism forms the major theme of *Cataract.* McDonagh points out that knowledge of this practice had been "gradually trickling into Britain since its first documentation in 1789," and in 1823—the same year as Moncrieff's play—Parliament "ordered a compilation of correspondence regarding 'Hindoo Infanticide,'" leading to the publication in 1824 of a detailed record of "official reports, surveys, and communications between administrators and local élites."[78] This in turn generated an outpouring of essays, stories, anecdotes, and articles shot through with "rumour and hearsay" attesting to the ongoing practice of female infanticide in India and blurring regional and spiritual distinctions so that, in the end, "all seem to merge into one under a hazy cloud of exotic names, as infanticide comes to operate as a generalised sign of Indian degeneracy."[79] This discourse emerged strongly in the 1850s, the period of frequent revivals of Moncrieff's play.

How was this "Hindoo" female infanticide usually carried out? The deed was done by applying opium to the nipple so that the breast-feeding baby "'drank in death with its mother's milk,'" as explained one account in 1856 in *Blackwood's Edinburgh Magazine.*[80] This paradox of the life-giving act of nursing causing death brings a new and threatening meaning to breast-feeding and contrasts with images of violent means of death (burial alive, suffocation, strangulation). Above all, the discourse around female infanticide in India drew uncomfortable connections with homegrown baby killing, specifically at this time the child murder epidemic of the 1850s and 1860s in Britain, from which "women workers emerged . . . in a particularly sinister light, as mid-century incarnations of Malthus's prototypical professional woman, Dame Nature: wet-nurses, baby farmers, and even the careless lower-class women who abandoned their own children to their care, all became killers in a massacre of the innocents that endangered the very basis of civilised society."[81] So, for all the exotic setting and costumes of Moncrieff's play, the problem of infanticide was uncomfortably close to home. Only a month prior to the opening of *Cataract of the Ganges*, a melodrama called *Infanticide; or, the Bohemian Mother* had also called attention to this issue, though on a domestic level.[82]

Thus, between *Cataract of the Ganges* and *Alan's Wife*, there were numerous plays on both British and American stages dealing with infanticide, though none sparked the controversy of *Alan's Wife*. In *Alexandra* (1893), translated from the German of Richard Voss and staged by the Charringtons at the Royalty,[83] a mother has intentionally killed her baby, but not in full view of the audience. Nowadays, "the crime of infanticide . . . has become a minor offence," writes the critic of the infanticide-themed play *Jeanie Deans*, a stage adaptation of Scott's novel *The Heart of Midlothian* (1818).[84] Most infanticide dramas revolved around a climactic courtroom scene. In *Jeanie Deans*, audience sympathy is directed toward the woman accused and convicted of infanticide because the audience knows (while she does not) that the child is alive and well. This is exactly the same pattern as Frank Harvey's *Mother*: the audience knows that the mother is wrongly accused of infanticide, having seen the villain do the deed himself; then, it receives a wonderful surprise when it is revealed that the child is alive and thriving. Finally, in the sentimental melodrama *The Scarlet Dye* (1887) by Julia M. Masters, a woman is falsely accused of infanticide but somehow manages to lose her child, which only seems to corroborate the tale. She is arrested for murder even though the child is in the safekeeping of a gypsy, who falsely testifies that he "saw the deed committed and the body burnt." But all turns out well: mother and child are reunited and justice is served, and the affecting songs are encored for the enthusiastic audience.[85]

Despite the long history of infanticide dramas, the influential critic A. B. Walkley criticized *Alan's Wife* for treating an act that was "outside the region of art."[86] But these theatrical precedents suggest that it was not the fact that infanticide is staged in *Alan's Wife* that was controversial; rather, it was the displacement of its context and the thwarting of audience expectations of a pattern that sees the woman falsely accused and then exculpated and finally a happy reunion with the child everyone thought was dead. *Alan's Wife* denies all of this: mother definitely kills child, hence no chance of the kind of turnaround audiences would be used to with the last-minute reprieve of the mother and the baby miraculously produced.

In *The Descent of Man*, Darwin cites many instances of the "fearfully common practice of infanticide," well established as one of the main "checks" on population.[87] Again and again, he uses the word *prevail* with regard to this practice. "Infanticide, especially of female infants, and the habit of procuring abortion are practices that . . . now prevail in many quarters of the world," with infanticide formerly having prevailed "on a still more extensive scale."[88] In a subsequent chapter, he writes that "the murder of infants has prevailed on the largest scale throughout the world, and has met with no reproach; but infanticide, especially of females, has been thought to be good for the tribe, or at least not injurious."[89] Until now, he has not yet indicated *who* kills the infants. Only toward the end of the book does Darwin suddenly and repeatedly identify the child killers as women. Among the New Zealand Maoris are "women who have destroyed four, six, and even seven children, mostly females, though this practise seems to be disappearing."[90] Again: "In the Polynesian Islands women have been known to kill from four or five, to even ten of their children; and we could not find a single woman who had not killed at least one." The beneficial implications are clear: "Wherever infanticide prevails the struggle for existence will be in so far less severe, and all the members of the tribe will have an almost equally good chance of rearing their few surviving children."[91] One way of seeing this high rate of infanticide, then, is in the Malthusian light of the greater good of the tribe.

Modern scientific studies have confirmed what Darwin wrote. Far from being an "abnormal and maladaptive behavior," infanticide is "a normal and individually adaptive activity" encompassing "an ever expanding list of behaviors" that are "not necessarily pathological."[92] Not surprisingly, few critics of *Alan's Wife* adopted this view, though some did attempt to understand Jean's character and motives. The critical response to the play was characterized by Jacob T. Grein as a raging "war."[93] William Archer (close friend, adviser, and probably lover of Robins) in his forty-three-page introduction to the play maintained that Jean Creyke is not insane but is simply "a terribly afflicted woman . . . who acts as, somewhere or other in the world, some similarly tortured creature is doubtless acting at the very moment I write these words."[94] Yet Archer, even while ostensibly supporting the play, publicly made it known that had it been his work, he would have treated the subject matter differently, for example "developing dialogue around the ethical issue rather than presenting emotional drama."[95] He thus undermined the play even as he introduced it, by suggesting that it becomes bogged down in emotion rather than appealing to the audience's intellect as a piece of polemical drama arguing for a specific cause, in this case euthanasia. Archer also states in his introduction that his adaptation would have staged the infanticide more obliquely: after a brief and "incoherent" soliloquy, "*in an inner room, seen but vaguely by the audience*, she was to have done the deed."[96] Robins and Bell were much bolder, bringing the curtain down just as the mother smothers the baby in full view of the audience. Though they needed his influence, they were wary of Archer's tendency "to suppress their creative efforts"; they kept the ending of *Alan's Wife* a secret from him until well into the writing of it, aware that he might "quash it" (Bell's words), and they found his lengthy introduction "overwhelming."[97]

At least one critic did note the play's importance in raising awareness of euthanasia and showing the need for a new critical vocabulary to address such issues; as McDonagh observes, this new kind of tragedy is "a drama of impossible choice" that thus "offers a heroic role for the woman, implicitly turning Medea into Agamemnon. . . . The play presents the [baby] killing as an act of bravery in the context of tragedy."[98] Such an act is briefly discussed in Sudermann's *Magda* (one of Duse's greatest acting successes) when Magda, who has had an illegitimate baby, confesses that she was once so depressed and desperate about her situation that she considered killing both herself and the child. For her also it would have been an act of bravery within a tragic context. But, where Magda becomes almost too verbose and expressive, Jean lapses into muteness: throughout the entire final scene, until the very last seconds of the play, she is silent, even though she knows that not attempting to defend herself and explain her actions will result in the death penalty.

From the early 1890s onward, Robins had been an outspoken critic of marriage and motherhood.[99] Through the device of Jean's silence, the play registers "the impossibility of modern motherhood," according to Katherine E. Kelly, which requires on the one hand a liberated, fully developed female consciousness and on the other an assumption of the role of fully invested, full-time carer.[100] Although Robins and Bell are hardly advocating infanticide as a way out of the predicament for modern women (any more than Ibsen is advocating that all women follow

Nora Helmer and desert their families), McDonagh notes that feminist writers often used the "motif of infanticide as a means of female emancipation," and that the motif is "at its most raw in *Alan's Wife*."[101] Robins's unpublished letters show a woman often agonizing and tormented about her choice to resist what she deemed the "natural" female behavior of motherhood. Significantly, she does not deem *her* choices to be natural. Robins and Bell suggest that maternal behavior covers a broad spectrum—one that can encompass infanticide as well as its extreme opposite, the life-giving nurture depicted by Herne.

The picture that emerges from Robins's unpublished plays and correspondence in relation to her published work is of a woman writer and actress whose position on feminism is far more complicated and nuanced than the label New Woman suggests. Sheila Stowell has stated that Robins was a gender essentialist; she did not "'seek to dissolve gender distinction' altogether, but, rather, 'to intrude [her] own version of "womanliness" into a male-dominated social and political system.'"[102] Robins's feminism, her championing of Ibsen's plays, and especially her own dramas with their resistance to accepted moral and biological roles for women all connect her to a newly opened discourse on the female body that stemmed directly from late-nineteenth-century biology. Yet this new theatrical emphasis on women's experience and relationships was by no means the sole interest for Robins. One of her unfinished plays, probably written in 1911/1912, called *Discretion*, contains a character named Eve who refers to "larks and squirrels"—Helmer's pet names for Nora in *A Doll's House* (echoed by John Osborne in *Look Back in Anger*, a play that puts women firmly back into a narrow domestic sphere). On one page, Robins appears to criticize her play as she notes in the margins, "Error / You make character and exigency / all circle about women / *Consider the man's need*, and exigency." She goes on to say: "3rd scene: let man dominate."[103] This is interesting in light of Robins's relationship with male writers like Henry James and Shaw, whose advice she frequently sought and usually followed. *Theatre and Friendship* shows how James advised Robins to let the male dominate in her work; Shaw admired *Votes for Women!* and encouraged her writing of it, helping to get it produced at the Court; yet his nickname for it was "the Stonor play," which turns the focus of the play to the male character, Geoffrey Stonor, when in fact it is Vida Levering and her protégée Jean. Are the notes in *Discretion* about emphasizing "the man's need" and letting "man dominate" simply Robins's capitulation to the likes of James and Shaw and to audience expectations? Or is she less committed than we thought to feminist concerns?

In her biography of Robins, Angela V. John emphasizes the constant reinvention of the self: Robins cannot be "pigeonholed" and her "love of experimentation and wide-ranging life-style preclude narrow compartmentalization."[104] She resists straightforward taxonomy. Even her friendship with Bell renders her a more complex figure in terms of her feminism, for Bell in 1890 had written a play called *A Woman of Culture*, still unproduced, that "contains one of the most notable portraits of the decade's New Woman" in the character of Diana Chester yet comes down in favor of the male protagonist's insistence on raising Evelyn, for whom they are joint guardians, to pursue the traditional route for women (marriage) rather than political activism.[105] In fact, Chester is shown to be deficient in femininity *because* she is politically active.

Grein called Jean "a fanatic who dies for her cause."[106] But what *is* her cause? Jean's refusal to supply explanations for her act left it wide open—as it still is—for interpretation.[107] Both McDonagh and Julie Holledge see the motive for Jean's infanticide as eugenic: Jean's killing of the baby not only is her rejection of the role of mother, but also is "intricately connected with her decision to marry Alan, whom she describes as if he were a member of a master race."[108] If that superior being cannot live, then his inferior child should not either. Further evidence of their eugenic purposes can be seen in the contrivance of two models of maleness that the play holds up to audience scrutiny: the robust manual laborer Alan versus the delicate intellectual James Warren, Jean's childhood playmate, now the local vicar, and a complete invention of Robins and Bell.[109] Jean's mother wishes her daughter had married James, an idea of "marital selection on the grounds of improving class and species health" that had been around long before Darwin and the rise of eugenic interpreters of his work.[110] Archer uses Darwinian terms to describe Jean and Alan: "She selects as her mate the handsomest, most capable man of her class that comes in her way."[111]

The question of how far Robins and Bell go in this eugenic direction is still unresolved. The baby's disability is never made clear, so we do not know if it could ever mature and reproduce; presumably, it is so maimed that its mother feels compelled to put it out of its misery, rendering the act one of euthanasia but with no eugenic motive. Kelly maintains that Robins took a dim view of eugenics, but her evidence is based on a statement Robins made much later in life (1926) dismissing eugenics as "childish." Given that Robins and Archer shared a deep intellectual interest in the ethics of euthanasia and suicide, and that she treated these issues in *The Mirkwater*, an unpublished play written in the mid-1890s, and her novel *The Open Question* (1898), which argues that suicide is ethically defensible to prevent the spread of hereditary disease, there is a strong case for seeing Robins at this stage of her life favoring eugenics, at least on this individual, case-by-case level.

In the published text of *Alan's Wife*, Jean baptizes the child before killing it and in the moment of suffocation expresses anguish, showing that it is not done in cold blood—a crucial distinction from Darwin's dispassionate reports of women killing their babies. Yet this scene was not shown on stage, only put in the published text, so audiences would not have seen Jean's emotional torment. Also, Jean's child is a boy, and as Darwin points out, infanticide is usually practiced on girls, with an implicit sense of identification between mother and daughter if the mother is committing a mercy killing (saving her daughter from a life of poverty and prostitution or slavery, for example). While Jean kills her baby to end his suffering, "theirs is not a shared affliction. Moreover, the act becomes associated specifically with her emancipation, rather than his: it is her 'one act of courage' through which she reaches a state of transcendence, even a kind of self-fulfilment."[112] In short, Robins and Bell remove any ameliorating factors so that it is impossible to excuse Jean's act—the audience must come to terms with *her* justification for it, not apply any existing legal, moral, or cultural paradigm.[113]

Alan's Wife does not call for actual, or at least highly realistic, infants to be used on stage. It is entirely possible simply to have a baby crib on stage and leave the rest to the audience's imagination. But the moment when Jean does the deed, reaching into the crib with the pillow and suffocating the baby, is surely no less powerful for being left to the spectator's own mental staging, just as Edward Bond's *Saved* (1968) does not require us to see the baby in the pram as it is being stoned to death (or to *see* the baby, who is crying inconsolably throughout an entire previous scene in which all the adults in the room, including its mother, ignore it, an excruciating experience for the audience).

In both cases, the distress is acute, and a prosthetic baby would only diminish this effect since its artificiality would be obvious. The reliance on each member of the audience to imagine the baby simultaneously assumes and questions a collective picture of infanthood already being undermined by the emergence of "baby science" and the anthropological findings on the universality of infanticide.

Alan's Wife has often been seen as a New Woman play, which is odd given that it shows a woman so dependent on her husband that she cannot live without him—a female character so defined by her role as wife that motherhood cannot enter into it. Her mother complains, "Yes, it's always Alan's dinner, or Alan's tea, or Alan's supper, or Alan's pipe. There isn't another man in the North gets waited on as he does." Jean merrily replies, "Is anything too good for him? Is anything good enough?" She revels in having "a husband who is brave and strong, a man who is my master as well as other folks'." She delights in keeping her little cottage "bright and shining" and in getting her husband's dinner ready: "Isn't he the best husband a girl ever had? And the handsomest, and the strongest?" Most of Jean's statements early in this scene—forming our first impression of her—are phrased as questions, giving her a tentative quality hardly in keeping with the assertive aggressiveness of the stereotypical New Woman.[114] The portrayal of infanticide has completely overshadowed this tentative quality and the other aspects of the play that undermine her as a New Woman.

Although vastly different in ideological orientation, the plays discussed in this chapter foreground what Angelique Richardson identifies as the "increasingly biologized" maternal aspect of femininity.[115] Quite apart from the fascinating intersections with scientific discourse, these plays are also worth looking at for how they problematize the New Woman label, whose critical currency is in danger of being eroded by becoming too broad and indiscriminate, automatically applied now to any female character in post-1890 drama that does anything even faintly rebellious, thus absorbing practically any play of this period that deals with issues of gender, motherhood, and marriage.[116] These works provide compelling examples of plays grappling with these ideas that don't fit easily into the New Woman–play paradigm.

The critical literature on Robins and the image that she self-consciously cultivated and projected have helped to build a picture of a woman dedicated unquestioningly and exclusively to her many careers (actor, playwright, novelist, journalist, suffrage campaigner), who after her husband's suicide started a whole new life and never looked again toward marriage or children. Yet in one of her letters to Bell, written in 1892 and sounding a lot like the character of Hedda whom Robins had just so successfully played, Robins seems deeply conflicted about motherhood and what is natural for a woman. She calls herself a "coward, a slave to convention," who hates being "loved" yet sometimes yearns desperately for it; she says, "Why am I so afraid to be natural? . . . In my heart of hearts I don't think those women stronger better nobler who resist successfully as I the Mighty Mother's Call, than those who have the courage to obey her.—Ah it's probably a mistake this bondage women are born under and grow so accustomed to they *refuse* freedom as I do." This long and anguished letter probes this question—so forcefully addressed by Mona Caird and others at the time—of what is natural, what it means to be a "hot-blooded woman" whose bed is a "furnace" as she is tortured by her resistance to love, sex, and motherhood.[117]

Contraception

Situating these plays within the cultural discourses on evolutionary theory that they both reflect and challenge complicates a progressivist narrative of women and theatre and allows a new perspective on the female body in performance in this period. Women taking control of their own bodies through contraception was a key issue in debates about reproduction throughout the first decades of the twentieth century, and the stage becomes a prominent site of such debates, in an astonishingly wide range of theatrical modes. Evolution relies on unions that result in viable offspring, but humans have evolved ways of manipulating this reproductive function, and despite the challenges of handling such subject matter on stage, playwrights have taken an avid interest in depicting contraception and abortion, even in the face of censorship laws or box office pressures.

Guillaume Apollinaire's play *Les Mamelles de Tiresias* (*The Breasts of Tiresias,* begun in 1903 and completed in 1917) satirically reflects the prominent place that reproductive issues were taking in the discourse on the declining birth rate that was affecting not only France but also many other European countries around the turn of the century. The play also forms an important precedent for Beckett's first play, *Eleuthéria*, as discussed in chapter 8. Apollinaire wrote it partly as a protest against realism, yet he chose as his theme a real social problem: "female emancipation, and its relation (which seems to have wholly charmed Apollinaire) to population decline."[118]

There are three striking things about the play's treatment of reproduction. First, he leaves the sexual act out of it, so that it is about "a man who makes children"; the emphasis is on manufacture, as if children are simply commodities. Second, there is a blurring of gender lines, a questioning of gender essentialism; Thérèse grows a beard and her breasts fly off her body (staged using balloons), and she refers to her husband as "less virile" than she is.[119] Third, despite the presence of exaggerated, grotesque breasts, they do not seem linked to anything either sexual or nurturing. In fact, it is the opposite: Apollinaire rather sneeringly notes in his preface that one critic "finds a ridiculous connection" between the rubber of the fake breasts and "certain articles recommended by neo-Malthusianism," and even provides a footnote to this comment on condoms "to clear myself of any reproach concerning the use of rubber breasts."[120]

Thérèse sings "let us get rid of our breasts," setting them on fire so that they explode, and her gender inversion becomes more pronounced as the play progresses (her husband does without testicles, she does without breasts; she trades clothes with him, puts on a mustache, and so on). In addition to destabilizing heterosexual norms and breaking down essentialist gender categories, the play satirizes feminism as incompatible with child rearing. Since the women of Zanzibar will not have children because they "want political rights," it is up to the men. As in Allan Neave's *Woman and Superwoman*, the play depicts women's desire for political rights as supplanting their natural reproductive role, rather than compatible with it.

Act 2 opens with the revelation that the Husband has just made 40,049 children in a single day, evidently by parthenogenesis. Along with lots of cradles, there is the continuous crying of babies on stage, in the wings, and in the auditorium throughout the scene. The stage directions indicate when and where the crying increases, not how this could possibly be staged. The Husband proclaims he has found "Domestic happiness / No woman on my hands" as "He lets the children fall."[121]

He tells a reporter (who is only a mouth, providing a notable link to Beckett's *Not I*, which similarly deals with reproductive issues) that he will bottle-feed the babies, something made possible and popular by the rapid advances in pediatrics and infant feeding at the time (pioneered in France). The incentive is wealth: the Husband says that having all these children will secure him immense riches. This is to counteract the pervasive fear that having children will make you poor.

Apollinaire belongs to the Continental avant-garde rather than the "new drama" that characterizes so many of the Anglophone plays under discussion here. This means that the play's modernist aesthetic draws more attention than the subject matter: his treatment of reproductive issues. By contrast, the American playwright Susan Glaspell has only recently begun to gain critical attention for the aesthetic innovation as well as the content of her plays. Like her compatriot Robins, Glaspell gives prominent place to vexed or failed parent–child relationships, especially to childlessness, and likewise has a more complex orientation toward feminism than has previously been recognized, now firmly endorsing female emancipation and free thinking about marriage, now retreating into reactionary anxiety about her childlessness and her need for conventional relationships. Glaspell, whose plays are deeply influenced by Ernst Haeckel, Jean-Baptiste Lamarck, and Herbert Spencer, shone a fierce light on parent–child relationships in the context of her broader evolutionary vision as exemplified by *Inheritors* and *The Verge*, which are explored in the following chapter. Here, it is *Chains of Dew* that has direct relevance, although already in *Bernice*, her breakthrough play, childlessness (and a possible miscarriage) defines the main character: "She would have made a wonderful mother, wouldn't she?"[122] There is also an implicit link between female activism and childlessness, just as in the Apollinaire and Neave plays.

Chains of Dew (1922) deals to some extent with the issue of birth control.[123] Though brisk and lighthearted in tone, it is a topical drama that reflects a contentious issue of the day: the campaign to legalize contraceptives and the dissemination of information about them, pioneered by Emma Goldman, Mary Ware Dennett, and Margaret Sanger.[124] This topic was not so unusual for a play; as we have seen with Brieux's and Apollinaire's plays, concerns over reproductive issues like these were common theatrical material, linked to and reflecting wider eugenic anxieties. What is different is Glaspell's attempt to strike a new tone, somewhere between the grave seriousness of Brieux and the surrealism of Apollinaire.

Chains of Dew opens in the New York city offices of an attractive birth control campaigner, Nora, who is having an affair with married poet Seymore Standish. There is little discussion of birth control itself, however; the play moves rapidly to the Midwest home of Seymore, where Nora suddenly arrives with a view to bringing her campaign to the wider population. The focus of the play shifts from Nora to Seymore's dutiful wife, Dotty, a self-effacing housewife reminiscent of both Vectia and Nora Helmer, who goes through a similarly life-changing awakening when Nora persuades her to campaign for birth control. Dotty pours her energies into this new cause. Her position is analogous to Vida's as a childless woman who devotes herself to women's emancipation.[125] Thus, *Chains of Dew* is part of a growing interest among playwrights in staging the tension between women's work and their reproductive lives—a new theatrical frontier, and an idea with controversial implications for human evolution if, as some feared, work supplanted childbearing.

Chains of Dew has a disappointingly nonfeminist ending, putting the doll back in the doll's house as Dotty sacrifices herself to what she perceives are her husband's greater needs. She returns to her normal life, with the one concession that Seymore promises to take her with him to New York occasionally; apart from this, the status quo is completely restored. It thus disappoints those expecting more of the radical feminist Glaspell of *Trifles* or *The Verge*. The play was a disaster on its premiere due to poor production values, poor acting, and bad timing, in that Glaspell and Cook were living in Greece and not able to exert any influence on the production in New York. Critics thought the characters were caricatures more than fleshed-out people and were disappointed in what seemed a thin and underwritten play after the powerful drama of *The Verge*. But, one critic, Maida Castellun, admired Glaspell's "ironic treatment of the theme and, especially, her extremely subtle satirical expose of the essentially conservative nature of men."[126]

One character critics particularly noted was Dotty's mother-in-law, who is central to the play both as its voice of reason and in symbolic ways through constant references to her avocation of doll making. The Seymore house is filled with her dolls, which pointedly look much more like people than playthings, and she constantly utters the pearls of wisdom expected of a wise older woman. Glaspell thus literalizes what was only a metaphor in Ibsen's *A Doll House*. There is a further parallel between this doll-making mother-in-law in *Chains of Dew*, whose dolls are obviously sublimations of her own repressed feelings (like the lame Jenny Wren in Charles Dickens's *Our Mutual Friend*) and Mrs. Solness's "nine lovely dolls" in Ibsen's *Master Builder*. What in Ibsen's play serves as a metaphor for Mrs. Solness's dead babies (and a constant reminder to Solness of his guilt for bringing this tragedy about) becomes in Glaspell's hands a powerful signifier of women's need to widen their sphere beyond motherhood. This connects with The Motherly One in Glaspell's short proto-Absurdist play *Woman's Honor*: "The prototype of the calm, wise, older woman who appears in many of Glaspell's later works and who is most aware of the workings of society; she voices Glaspell's contradictory feelings about individuality and evolution in traditional society."[127]

Glaspell's conclusions are as troubling and ambivalent as Robins's, with the birth control campaigner remaining single and childless while the mother who has taken up that cause with such zeal ends up abandoning it to preserve her marriage and family life. As with Robins, Glaspell's theatrical spectrum is vast and her political identity in constant renegotiation. This makes these writers harder to place than their male counterparts. Yet far from becoming dated, the play seems to work well in modern revival; the issues it addresses have by no means been resolved.[128]

Abortion

Another intervention in natural reproductive processes that gripped playwrights' imaginations in relation to evolutionary theory was abortion.[129] A surprising number of plays deal with this issue in these early years of the twentieth century, and they did not have an easy path to theatrical production, especially in England under the ongoing theatrical censorship. Granville-Barker's play *Waste* deals with a politician, Henry Trebell, whose lover, the married woman Mrs. O'Connell, has an illegal abortion and dies. The play was refused a license in 1907, the same year that Edward Garnett's *The Breaking Point,* a plea for more relaxed abortion laws, was banned. As a response to the decision on *Waste*, seventy-one playwrights sent a petition to the editor of the *Times* detailing why they objected to stage censorship: it was arbitrary, it undermined art and the craft of theatre, and it did not allow for any legal appeal process.[130]

But, for all its notoriety as an abortion play, Waste's main interest is in female sexual selection. The opening stage directions describe Mrs. O'Connell as "a charming woman, if by charming you understand a woman who converts every quality she possesses into a means of attraction, and has no use for any others." Amy O'Connell is driven by her life force (she is ovulating) to have sex with Trebell, who is helpless in its grip. There is a strong link here to Ann Whitefield, similarly portrayed as the predatory female turning all her charms on Tanner. The influence of Shaw's evolutionary theory (and perhaps also Henri Bergson's) on Barker was profound. The two men worked closely together on the groundbreaking 1904 to 1907 seasons at the Court Theatre (a partnership sartorially signaled in the tongue-in-cheek costuming of Barker to resemble Shaw when he played Tanner in *Man and Superman,* which he also directed, as shown in figure 1, chapter 5). The first version of *Waste* actually contains the term *life force*, and there are numerous other echoes of *Man and Superman*, though of a more somber nature, as *Waste* ends in suicide rather than the "bounding high spirits" of Shaw's play:[131] "In Barker's play the dominant image is that of barrenness (the word 'barren' itself occurs several times at crucial moments) and sexual barrenness and the wilful refusal of life is overtly and explicitly equated with spiritual barrenness."[132] I would argue that the word *Waste* recognizes not only the tragic waste of human lives (Amy's and the unborn baby's as well as Trebell's) but also the Malthusian excess of life that is required to ensure the continuation of the species—the downside being that some must "go to the wall" in the struggle for existence. The version in 1927 of the final speech by Trebell refers to nature as "spendthrift," then says: "Yet the God to whose creating we travail may be infinitely economical and waste, perhaps, less of the wealth of us when we're dead than *we* waste in the faithlessness and slavery of our lives."[133]

By contrast, Robins turns abortion-induced waste and barrenness to political efficacy. Her play *Votes for Women!* is in its treatment of abortion from the female perspective an exact inversion of *Waste*, though written in the same year and performed in the same context (the Court Theatre); indeed, the play was one of Barker's and the Court's greatest successes. In this play, Vida refers only obliquely to her abortion: "It was my helplessness turned the best thing life can bring into a curse for both of us [Vida and Geoffrey Stonor, her lover]."[134] As Eltis points out, this line comes after an exchange in which Vida has made it clear that her pregnancy came about through her own sexual desire and her recognition of a physical need as much as through her lover's advances; she was not just helplessly "succumbing to love," not taken advantage of as a stereotypically passive fallen woman. In this scene, "the tangential nature of the dialogue—necessitated by the constraints of theatrical censorship, if not by Robins's own artistic preferences—leaves it uncertain whether Vida's 'curse' refers to the aborting of the child or the reduction of her relationship with Stonor to a form of prostitution, or, perhaps, both."[135] This relates to evolution in the sense of directing it, being selective, making reproductive choices rather than letting nature take its course, as *Tares* put it. It also has to do with the impact on other lives, particularly the woman's, and the effect that in turn has on her life choices and her relationships.

By contrast, Eugene O'Neill's one-act play simply and shockingly entitled *Abortion* (1914) seems at first wholly uninterested in seeing the issue from the female perspective; the young girl who has had an abortion never appears, as she is already dead. Set on an unnamed university campus (probably Princeton), the play shows basketball star Jack Townsend being cheered by a crowd of classmates, his family, and his girlfriend. Moments later, he is visited by a scowling local youth suffering from tuberculosis who turns out to be the brother of a girl Jack got pregnant. He tells Jack that the girl has died as a result of a botched abortion that Jack and his father had financed. Jack must face the consequences of his affair with a girl from the other side of town, of a lower class. He realizes that not even money can clear his name, so he shoots himself, just as Trebell does in *Waste*.[136]

As with *Waste*, the sensational topic of abortion can too easily overshadow the real interest of the drama: its anatomizing of the sex instinct not only in terms of individual character but also in the broader context of evolution. *Abortion* likens the sexual urge to the vestiges of primeval mud on civilized man. When we give way to our sex drive, we act like savages. "We've retained a large portion of the original mud in our make-up," says Jack's father to him, "that's the only answer I can think of." Jack picks up on this in the next line, saying that he was not really himself when he slept with "this girl"; he was some form of early man, "the male beast who ran gibbering through the forest after its female thousands of years ago."[137] Jack's father calls this "pure evasion" and says we are responsible for both the Jekyll and the Hyde side of ourselves. Jack argues that in fact the sexual impulse he gave in to is more natural than "our ideals of conduct, of Right and Wrong, our ethics, which are unnatural and monstrously distorted."[138] This is the crux of the matter in evolutionary terms, raising again the issue of what is "natural" but framing it in a moral context (as J. M. Barrie does in *The Admirable Crichton* with a much lighter touch and within the context of the rigid British class system). O'Neill treats Jack sympathetically, as "a victim of both humans' natural, biological sex drive and the unnatural social result of puritanical morality."[139]

~

This chapter has shown three main things. One is that theatre's engagement with reproductive issues was deep, varied, and complex, and was often linked with deeply controversial areas such as eugenics and sexual selection. Two is the frequency with which the system of censorship and the need to please audiences prevented such plays being performed. Three is that the crux of the matter was, in many ways, the notion of what was "natural"—and this itself was a term of shifting semantic value at this time.

So far, much of the discussion in this book has centered on drama emanating from Britain, the home of Darwin, Huxley, Wallace, William Bateson, and so many other seminal evolutionists. We have seen some important early theatrical engagements with evolution by Americans Herne and Robins, but Herne is something of an isolated case, and some of Robins's work remains unpublished and unperformed, a treasure trove waiting to be fully mined. It is in plays by their successors Glaspell and Thornton Wilder that American drama fully confronts evolution in a distinctive style in the middle of the twentieth century.

7

Midcentury American Engagements with Evolution

We have almost become Darwinian in our playtaste.
Carol Bird, "Enter the Monkey Man"

The playwrights I consider in this chapter—especially Susan Glaspell and Thornton Wilder—take the theatrical engagement with evolution in radically innovative directions compared with their predecessors, and I explore what makes these innovations particularly American. Peter Middleton has noted a characteristically midcentury American attitude toward science, particularly physics.[1] The period (circa 1920–1955) also saw profound changes in evolutionary theory, from the gradual waning of the popularity of eugenics to the rejection of non-Darwinian alternatives to the consolidation of the genetics–natural selection camps into the Modern Synthesis. For almost this entire period, Glaspell experimented in a wide range of theatrical forms with both non-Darwinian and Darwinian thinking, and she, Wilder, and a range of other playwrights take the theatrical engagement with evolution in a variety of new and sometimes-startling directions and modes.

Although Paul Lifton argues that by the early 1940s "evolutionary theory was scarcely a subject of heated controversy . . . at least among aware, well-educated writers who were not religious fundamentalists or reactionaries,"[2] generally, Charles Darwin fared less well in America than in Britain in terms of broader cultural acceptance. There are strong resemblances, for example the enthusiasm for Herbert Spencer, the rise of eugenics, and the preference for "the Lamarckian view of human nature" around the turn of the twentieth century.[3] This is true also of developments in Continental Europe, where, particularly in Germany and France, genetics remained more open to Lamarckism and other non-Darwinian influences. The American resistance to Darwin was consistent with the optimism—albeit eugenic in its overtones—of the frontier hypothesis of Frederick J. Turner, with its notion that the American West was "a stimulating environment which worked directly on the constitution of immigrants to produce a superior form of humanity."[4] An important context here is the positive feeling about humanity's place in nature, so long as we can take care of it and be vigilant about limited resources—something that Glaspell touches on in several plays, long before other proto-environmentalist authors.[5] Plays like Hallie Flanagan Davis's $E = mc^2$ (1948) are typical of the time, balancing stark warnings about the danger of atomic energy with optimistic suggestions for the good that it can do, for example, for agriculture.[6] The theatre of midcentury America thus evinces a strong environmentalist streak, one that may be connected to the simultaneous romanticization of its wilderness (itself a problematic term since "wilderness" was often created by clearing land occupied by indigenous peoples) and steeped in the "vernacular modernism" of Georgia O'Keeffe, Wild West cinema, and the novels of John Steinbeck.[7]

Susan Glaspell's Theatrical Hybridity

Since Christopher W. E. Bigsby in 1984 described Glaspell as regrettably a mere footnote in the history of modern drama,[8] she has enjoyed a blossoming reputation: dozens of books and articles have been published on her work, her *Complete Plays* were published in 2010, a very active international Susan Glaspell Society now exists, and many of her works have enjoyed successful revivals, particularly at the Orange Tree Theatre in Richmond, England. Theatre critic Michael Billington has championed Glaspell as "American drama's best-kept secret," "an audacious pioneer whose voice cries out to be heard."[9] Tamsen Wolff writes that "the animation that important artists like Glaspell and [Eugene]

O'Neill brought to the vexed issues of biology and identity left a permanent stamp on American theatre."[10] Right to the end of her career, she was still showing this interest in biology and human destiny, in her collaboration with the Federal Theatre Project on the play *Spirochete* (tracing the evolution of syphilis) and in her last play, *Springs Eternal*, which continues her method of representing evolution through the depiction of multigenerational families.

Glaspell's first novel, *The Glory of the Conquered* (1909), reveals a wealth of intellectual influences, including George Bernard Shaw, Ernst Haeckel, and Darwin.[11] This first novel came out two years after Glaspell's partner, George Cram Cook, had published a paper on evolution first delivered in 1906 at the Davenport, Iowa, Contemporary Club (after *Scientific American* refused it).[12] Cook's paper referred to saltation as part of "the 'heroic history' of evolution, of one-celled plants and animals that had taken the leap and dared become more until some of them 'crossed the difficult gap from invertebrate to vertebrate life' and eventually developed a brain."[13] Perhaps this is the origin of Glaspell's use of this idea of the sudden leap in evolution in *The Verge* and *Inheritors*, both of which put forth saltation rather than natural selection as the preferred mechanism of evolutionary change. It offers a "heroic" vision of evolution that is fast, visible, and dramatic. Another distinctive quality in Glaspell's evolutionary vision is her focus on plants, drawing our attention back to botany, the foundation of evolutionary thought for Carl Linnaeus, Erasmus Darwin, Darwin himself, and so many other evolutionists before being overshadowed by the anthropomorphic appeal of mammals. Glaspell's *The Verge* and *Inheritors* (1921), as well as shorter plays like *The Outside*, foregrounds botany. In what follows, I trace her dramatic engagement with both of these concepts and show how innovatively she treats them in her work and discuss the other evolutionary element animating her plays, Haeckel's monism.

Bernice (1919) was her big breakthough, after which critics always reviewed her work, particularly in Britain.[14] The play contains a couple of direct references to Darwin, but they are not well integrated and therefore seem gratuitous; Darwin is just a label to slap on to a specific character, as with Robert Buchanan's references to evolution in *The Charlatan* or Henry Arthur Jones's use of Spencer in *The Dancing Girl*. Bernice's bereaved father is called "one of the wrecks of the Darwinian theory" by his son-in-law, Craig: "Spent himself fighting for it and—let it go at that."[15] The line is echoed a few moments later: "You said he was a wreck of the Darwinian theory. Then me—a wreck of free speech."[16] The line recurs when Margaret, "on the verge of being not herself" (a productive idea for Glaspell), calls Abbie "another wreck. It's your Darwinian theory. Your free speech."[17] This makes little sense given that Abbie has not been part of the previous exchanges regarding Darwinian theory and wrecks, and it suggests that Glaspell read Darwin and Spencer naïvely and did not seem to distinguish between them.[18] Certainly *Bernice* bears this out; Darwin is simply shorthand for free-thinking progressivism, just as he was in many of Glaspell's predecessor Herne's plays.

Inheritors still has this flavor but shows a much more thoroughgoing engagement with evolution, with its panoramic sweep encompassing the history of America, the pioneers settling and displacing Native Americans: the play stages the sheer force of history and offers as well a protest against World War I and its after-effects. The central figure is a jailed conscientious objector who never appears to the audience. As in *Bernice*, there are direct references to Darwin, but now better integrated and explained. "You haven't read Darwin, have you, Uncle Silas?" asks the student Felix, and he describes Darwin as "the great new man" with his theory of "the survival of the fittest," a term we know to come from Spencer rather than Darwin.[19] Both social Darwinism and Lamarckism crop up in this dialogue. Felix says that Darwin can make us "feel better about the Indians" because "in the struggle for existence, many must go down. The fittest survive. This—had to be." When Silas seems a bit shocked by such easy exculpation, Felix expounds: "[Darwin] calls it that. Best fitted to the place in which one finds one's self, having the qualities that can best cope with conditions—do things. From the beginning of life it's been like that. He shows the growth of life forms that were barely alive, the lowest animal forms—jellyfish—up to man." The exchange culminates in a lyrical explanation of evolution that is suffused with Lamarckism. Fejevary says "gently" that we should not be discouraged by the idea that we are descended from monkeys. Look at our hands, he says: "Why have we hands?" It is not because God gave them to us, but because "ages back—before life had taken form as man, there was an impulse to do what had never been done—when you think that we have hands today because from the first of life there have been adventurers—those of best brain and courage who wanted to be more than life had been, and that from aspiration has come doing, and doing has shaped the thing with which to do—it gives our hand a history which should make us want to use it well."[20]

We thus move from the ugly social Darwinism of the narrow-minded Felix to the Lamarckian vision of progress and human will equating to the pioneer spirit that Fejevary personifies. Silas is thrilled by this vision: "Think what it is you've said! If it's true that we made ourselves—made ourselves out of the wanting to be more—created ourselves you might say, by our own courage—our—what is it?—aspiration." Silas is sure that he has felt this thought emanating from nature itself: "The earth told me. The beasts told me." In fact, he says, even Fejevary's face has been revealing this thought: "In your face haven't I seen thinking make a finer face?"[21]

Geological time is acknowledged—human evolution has taken "many millions of years since earth first stirred"—but the role of the will is given pride of place in one of Glaspell's most syntactically stilted lines, characteristic of her approach to theatrical dialogue: "Then we are what we are because through all that time there've been them that wanted to be more than life had been."[22] Once he has realized this, Silas has a second revelation: that this progressive unfolding might easily be reversed, if the will has such a key part in it. In other words, we could regress. He determines to found a university as a means of keeping humanity on the steady evolutionary march forward. Silas clings to his idea that "we created ourselves out of the thoughts that came," rather than being created through natural processes. This is not far from Shaw's vision. Silas's university will be the culmination of "that thinking that breathes from the earth," those "dreams of a million years."[23]

J. Ellen Gainor likens *Inheritors* to *Angels in America* in its epic "scope, sweep, and political force."[24] I would suggest a kinship as well with Wilder's *The Skin of Our Teeth*, which is discussed further in this chapter. Wolff writes that *Inheritors* is about the American grand narrative of "self-production." The play is "an account of American history by way of an interpretation of Darwin." There is also sustained discussion of genetics in the play, connecting it to *The Verge*, written at about the same time, which likewise concentrates on genetic manipulation. But the difference between the two plays with regard to their engagement with evolutionary theory is vast. In *Inheritors*, Glaspell "allows Darwinian theory inaccurately to encompass a self-making (or neo-Lamarckian) myth: her version of evolutionary theory is one in which these individuals 'made ourselves—made ourselves out of the wanting to be more.'"[25] The hybrid corn that is being created in the subplot of the play by the farmer Ira Morton serves as the metaphor for this; it is now "best in the state. He's experimented with it—created a new kind. They've given it a name—Morton corn. It seems corn is rather fascinating to work with—very mutable stuff."[26] *The Verge* will take this concept further, staging it instead of describing it, and making the metaphor of mutation work on many levels.

Glaspell had spent some time around 1909 and 1910 working in Colorado for the United States Forest Service, "an experience that solidified her interest in ecology and sound environmental land use."[27] She uses trees, forests, and the unspoiled earth throughout her work as highly original and evocative metaphors. The setting of her one-act play *The Outside* (1917), for instance, is a barren and bleak Cape Cod coast where the sand meets the forest: "The rude things, vines, bushes, which form the outer uneven rim of the woods—the only things that grow in the sand. . . . The dunes are hills and strange forms of sand on which, in places, grows the stiff beach grass—struggle; dogged growing against odds."[28] The idea of these two environments meeting fascinates her, as it did François de Curel in the central speech about the forests and the seas in *Les Fossiles* that I discussed in the context of changing ideas about extinction. Glaspell uses the imagery of woods and sand in a more dramatic sense, visualizing the moment they run up against each other and the two distinct habitats collide. She stresses that they may be in constant competition but the sands and the forests coexist, in contrast to de Curel's sense of competition between these two environments, metaphors for incompatible kinds of humanity, unable to come together because they operate so differently and compete for supremacy. It is not inconceivable that Glaspell could have encountered de Curel's play when she lived in Paris in 1908; she saw all the new art that was emerging, as well as much European drama—Henrik Ibsen, August Strindberg, Gerhart Hauptmann, Maurice Maeterlinck. She went to the small theatres then leading the way in finding "new plays, new sight lines, cheap seats, ensemble acting."[29] Jacques Copeau and others were founding *Nouvelle Revue Française,* attacking popular theatre in ways Glaspell would also do in her attacks on Broadway.

Forests provide Glaspell with powerful metaphors relating to the human condition, and these metaphors run right through her work. In *Close the Book* (1917), she gives a pedigree-obsessed family the name of "Root" and makes them realize in the final moments of the play that their sacred family genealogy needs mixing up with new blood, new "air through our family trees," or else it will be "stifled."[30] This variation on the theme of the dendritic tree of life suggests that a healthy genealogy will not be pure and homogeneous but will follow irregular patterns. Glaspell's vision also encompasses a strong sense of agency in nature. In *Inheritors*, one character says, "Sometimes I feel that the land itself has got a mind," and that something that's "like thought" seems to be "coming up from" the earth.[31] In her last play, *Springs Eternal*, only recently published, she refers to the self-absorbed and blinkered family of the play as "a tree of many branches, and each twig attracts to itself people who will travel hundreds of miles—sleepless and hungry—that they may finally sit in this room. With other twigs."[32]

Glaspell transplants these ideas, and her European influences, into American contexts. She is prescient—like Ibsen's allusion to the filth-producing tanneries in *An Enemy of the People*—when in *Inheritors* she has a reference to industry encroaching on and polluting the land. The little town is growing. "It's grown so much this year, and in a way that means more growing—that big glucose plant going up down the river, the new lumber mill—all that means many more people," says Felix, and Fejevary adds that "they've even bought ground for a steel works." Yes, says Silas, "a city will rise from these cornfields—a big rich place—that's bound to be."[33] The land they are discussing was taken from the Native Americans, and that sense of wrongdoing haunts the play. No matter how hard the characters try, it cannot be expunged; even building Morton College, the Harvard of the Midwest, has not atoned for it because its original Arnoldian idealism of spreading the "sweetness and light" of culture has been undermined by political corruption and reactionary ideas on the part of the younger generation who were meant to be progressive pathbreakers.

Glaspell often reverses the expected pattern of increased enlightenment and progressivism, as in Horace, the racist son of the college's founder in *Inheritors*, and Elizabeth in *The Verge*, who is more conservative than her mother, Claire, who berates her for taking for granted all that women have fought for and for seeming to forget the struggle for the vote almost as soon as it was won. This maps directly onto the real-life backdrop for the play's first performance in Britain in 1925—a period of backlash and hostility to feminists and working women, as Susan Kingsley Kent points out.[34] *The Verge* is not only a key evolution-related drama but also an important precursor to the feminist approach to evolution in later plays like Bryony Lavery's *Origin of the Species*.[35] A number of closely related ideas about nature inform Glaspell's thinking in this play, among them saltation, Lamarckism, Haeckel's monism, the great chain of being, genetic hybridity. There is also Alfred North Whitehead, whose book *The Concept of Nature* was published in 1920 and to whom Glaspell refers in her notebook as having written of "the leap of the imagination reaching beyond what is then actual."[36] Claire is trying to make a great leap forward with her plant hybrid, breaking out of the old mold and creating a new form. This has always been seen as an analogy for what Glaspell is trying to do theatrically: breaking out of the realistic form and creating an expressionist hybrid whose dialogue is one of the first theatrical attempts at what we might identify now as "l'écriture feminine."

Glaspell's various evolutionary ideas were shot through with, and transformed by, a devotion to Friedrich Nietzsche, just as Shaw's were. Cook's greatest obsessions were the Greeks and Nietzsche. Margot Norris points out that Nietzsche used Darwin's ideas "as critical tools to interrogate the status of man as a *natural* being." But "Nietzsche misunderstands, rejects, and reappropriates an alienated version of Darwin's most radical thinking."[37] Discussing evolution, Nietzsche, and socialism, Glaspell and Cook and their circle seemed to conflate them into one monist, possibly eugenicist, vision of the eventual perfection of humankind akin to Shaw's Creative Evolution. Glaspell expressed in her writings her "need for oneness and the transcendence of humankind's limitations, which could only be achieved in a struggle that would not be 'woundless' but would bring those few individuals sufficiently courageous to risk pain and failure nearer to perfection"; this "Nietzschean idealism" was also present in *The Verge*, in which "a way to reach the perfection of humankind, to transcend man-made institutions, is signaled by woman."[38] Nietzsche's ideas of eternal recurrence and the will to overcome oneself are closely bound with the monism Glaspell and Cook embraced.

Glaspell's evolutionary mixture is thus a heady one, and performance is absolutely integral to it. Her texts contain numerous references to the evolutionary ideas described here, but it is the acting and staging that convey and clarify them effectively. In *The Verge*, the first thing the audience sees is a shaft of light coming from a trapdoor in the stage and striking "*the long leaves and the huge brilliant blossom of a strange plant whose twisted stem projects from right front. Nothing is seen except this plant and its shadow*."[39] Light then reveals that the whole room is full of strange plants. But this is no ordinary greenhouse: it is not plants but for experimenting with them. It is "a laboratory" belonging to botanist Claire Archer, whose work is analogous to Glaspell's use of theatre as a laboratory to experiment with new forms.

The oddest plant of all is the Edge Vine that Claire has been developing through experimentation with Hugo de Vries's theory of mutation. As Claire explains to her daughter, she is not attempting to create better plants, but different ones, new plants that have been "shocked out of what they were— into something they were not."[40] This is more than just an echo of the modernist demand, as expressed by Ezra Pound, to "make it new" or the feminist need to break out of patriarchal modes represented by the men who surround her, allegorically called Tom, Dick, and Harry. Claire explicitly links her project to evolution by couching it in terms that challenge assumptions about usefulness, the good of the species, and the upward drive toward perfection. The only character who comes close to understanding her project is Tom, her lover, whom she ends up strangling in a climactic frenzy.

The Verge has many themes, from the search for an authentically female language to a fierce questioning of biological determinism to the fragility of much-needed hybrid forms. The play has received thoroughgoing attention in terms of its science. In particular, Jörg Thomas Richter provides a detailed and insightful discussion of the play's use of mutation theory in relation to Glaspell's feminism.[41] Wolff also provides a searching analysis of Glaspell's use of genetics in the context of eugenics.[42] There are theatrical precedents as well as scientific influences; Glaspell's probing of plant life has an important precursor in the work of Strindberg, for instance, for whom, as we have seen, plants were "living beings with nerves, perhaps sense perceptions, and conceivably: consciousness."[43] Indeed, one recent revival of *The Verge* took the leap of having actors portray the Edge Vine and the other key plant in the play, Breath of Life. This gives added poignancy just for growing to Claire's interaction with her beloved creations, but it also makes her more of a Frankenstein who is simply playing with life.[44]

The play itself resists this kind of science-fiction reading, however, because the science being attempted is no fantasy but grounded in contemporary genetics and because the setting, for all its exotic feel, evokes a plausible botanical laboratory. Cook built an amazing greenhouse that must have influenced Glaspell in writing *The Verge*; she describes it as a "rampart against a thousand leagues of cold invading from the bitter polar night."[45] This contrast between extremes of heat and cold dominates the opening stage directions describing the greenhouse in *The Verge*, and it serves as further evidence of Glaspell's enduring interest in how incompatible environments coexist. The greenhouse has been likened to the remarkable loft space in Ibsen's *The Wild Duck*.[46]

At the beginning of the twentieth century, insects and plants were at the forefront of evolutionary science, as work with *Drosophila* and the foundation laid by Mendel's experiments with peas, respectively, made genetics the focus of new research on evolution. Jean-Baptiste Lamarck, Erasmus Darwin, and Linnaeus all began with plants as the foundation of their scientific investigations.[47] Indeed, the attention given humans and animals as the main focus of evolutionary science due to *The Descent of Man* and *Expression of the Emotions in Man and Animals* can obscure the fact that the bulk of Darwin's research was on plants, one of the main areas of interest and firsthand experimentation and observation (as opposed to the secondhand information he obtained about humans and animals). As Steve Jones has shown, England was Darwin's Galápagos and the garden at Down House his main center of experimental research.[48] He published numerous books and papers based on his botanical observations, and in his memoirs, he fondly recollects them: the *Fertilisation of Orchids* (1862), a paper on "the two forms or dimorphic condition of primula" (1862), a long paper on climbing plants (written 1864; expanded into book form and published 1875), a book on insectivorous plants (1875), the book *Effects of Cross and Self Fertilisation in the Vegetable Kingdom* (1876; expanded and revised edition 1977), the book *The Different Forms of Flowers* (1877; second edition 1880), the book *The Power of Movement in Plants* (1880), and "a little book on 'The Formation of Vegetable Mould, through the action of worms'" (1881).[49] Indeed, Darwin's enthusiasm for plant experimentation shines through his *Autobiography*; for instance, "No little discovery of mine ever gave me such pleasure," says the man who gave us natural selection and revolutionized our understanding of human origins, "as the making out the meaning of heterostyled flowers. The results of crossing such flowers in an illegitimate manner, I believe to be very important as bearing on the sterility of hybrids."[50] Claire's intense concentration on her plants evokes Darwin working away at his orchids or Mendel and his peas. *The Verge* also is timely for this reason, coming as Mendel's ideas were being compared but not yet synthesized with Darwin's.

This helps explain the presence of a theory that still held attraction although it was soon to be discarded by the New Synthesis. De Vries's mutation theory (published in 1901–1903; translated into English in 1910) became "the most popular theory of evolution in the early decades of the twentieth century."[51] Glaspell's most important full-length play directly reflects this influence. The idea of mutation appealed to biologists looking for an alternative to natural selection. It gained quite a following but was ultimately shown not to be able to cause the kind of long-term change or significant transformations in species that natural selection brought about. Darwin writes in *The Descent of Man* about the concept of "spontaneous variation," a term to denote that "large class of variations" that "appear to arise without any exciting cause." But he says that such variations, whether subtle and slight or "of strongly marked and abrupt deviations of structure, depend much more on the constitution of the organism than on the nature of the conditions to which is has been subjected."[52] He favored the principle of gradualism instead. Darwin's "deep faith that nothing in nature makes leaps" became absorbed as a political idea; "human society cannot make leaps either: evolution not revolution."[53]

The prominent geneticists William Bateson and Thomas Hunt Morgan embraced saltation as a more satisfactory explanation of how evolutionary novelties are produced than natural selection and Darwinian gradualism. Both were bitterly opposed to adaptation as the driving force of evolution. Like St. George J. Mivart before them, these men wanted "to believe that internal forces direct evolution along predetermined lines. Their saltation theory was an effort to suggest that evolution could become accessible to experimental study through direct observation of how new characters appear."[54]

Glaspell's *The Verge* is another example, then, of the *tværtimot* (contrarian) impulse among dramatists engaging with evolution. Saltation is inherently dramatic. The "sudden leap" was something that could be shown in the brief traffic of the stage and could signal evolution in a single bound, and hybridity was something visible and tangible, finding ready physical expression in an actor or through scenery. What is interesting is that the early discourse on mutation casts it generally in a negative light (it is "injurious," in Darwin's term, as quoted in my previous discussion of freakery). But by 1933, Richard Goldschmidt posited "hopeful monsters": those "rare but incredibly significant moments in the biological record when a species makes a random, radical change, a macromutation, taking it into a completely different set of survival circumstances."[55] This move toward a more positive kind of mutation is part of the cultural background informing *The Verge*. As Michael M. Chemers puts it: "Evolution is an unpredictable force. Every once in a while the great leap into the unknown actually works out for the better and, in some cases, generates a whole new line of species, a race of monsters that manages to survive while the rest of the species gets wiped out by predation, climate change, or getting hit by an asteroid. The monsters, then, become the new 'ordinary' species."[56] Claire's Edge Vine may well turn out to be such a monster; is this why she tears it from its trellis? Is she herself a monstrous freak who must be destroyed?

The limitations of language as a means of expressing the self are the key to these questions. But, whereas Claire's idiosyncratic and halting language is usually seen as an important early example of feminist writing, it also needs to be understood in evolutionary terms. Claire gropes toward self-expression through her many long speeches of disjointed, half-articulated thoughts that, paradoxically, are loaded with expression; we understand that they are the keys to her true self ("let her be herself," insists Tom). The paradox here is that while sophisticated spoken and written language is arguably what makes humans distinct as a species, Claire has reached a point at which she can no longer find a use for speech. Glaspell seems to be pointing out something along the lines of what Lavery would do in *The Origin of Species*: showing the bias of evolutionary discourse that takes the male as metonymic/

representative of the human race. Both seem to be saying that there not only is a single "human" evolution but also are distinctively male and female evolutions.

This state of speech abjection connects to Jean's elective muteness in *Alan's Wife*, as discussed in chapter 6, and it raises the question of why these female playwrights are showing female protest through evolutionary *regression* if language is the highest development of human evolution.[57] According to Mark Pagel, language is the key human innovation not only because it allows us to express ourselves most fully but also because it enables cooperation, which is essential to survival.[58] It allows us to operate as groups rather than simply as individuals, making the struggle for existence easier. But for Glaspell and Robins (as for later feminists like Hélène Cixous and Simone de Beauvoir, for that matter), the "group" is fundamentally patriarchal, and "cooperation" can mean collusion against women. So for them, the tactic is to refuse language. Is this regressive and counter-evolutionary, or is it simply an alternative strategy for survival? How are women to survive on their own terms if they conform to the evolutionary patterns and strategies of the group if these have been devised by men? This is the same kind of strategy as Nora leaving her family in *A Doll's House*. In all of these cases, playwrights are rejecting conventional assumptions about human evolution and seeking—if not actually providing—alternatives to it that are better for women.

In so many ways, *The Verge* radically departs from the kinds of plays we have seen in the Edwardian period focusing on gender issues relating to evolution, in particular motherhood. One of the great achievements of the play is that it explores femininity and creativity by replacing "the ubiquitous creative metaphor of motherhood" with scientific experiment.[59]

I want to conclude my discussion of *The Verge* with a brief look at its extraordinary fortunes in England, where Glaspell was held in higher esteem than O'Neill and where the play was embraced by a group of forward-looking female theatre practitioners who brought it to a highly appreciative audience. In particular, this is the story of one actress's instantaneous recognition of the play's evolutionary language and her ability to translate that into her performance.

Glaspell's fame had begun to spread to Britain after a collection of her plays was published in 1920, and she was enthusiastically embraced and "became the American playwright against whom other American writers were judged."[60] In 1924, *Bernice, Inheritors,* and *The Verge* were published in London, and in 1925, Emma Goldman provided an important link between the Provincetown Players and London's theatre world by giving lectures on Strindberg, German expressionism, O'Neill and his works, and Glaspell, in fact billing herself as "overseas representative of the Provincetown Players."[61]

Glaspell was immediately seen as the producer of hybrid forms. The *New Statesman* hailed *The Verge* as a mating of Pirandello and "the new German cinema," combining "Pirandello's pre-occupation with the possibility of the isolation of *thing*, the problem of reality" with the method of *Destiny* or *Dr. Caligari*, "the metamorphic method that is to say (as distinguished from the symbolist)."[62] The British also liked Glaspell's kinship with Shaw and Nietzsche; Claire was overwhelmingly received as an example of a female Nietzschean Superwoman.[63]

Even greater insight into Glaspell's appeal to British readers and audiences lies in the views expressed by her champions in the theatre. Sybil Thorndike saw the play instantly as a thrilling "mould-breaker" and related it to her reading of Pierre Teilhard de Chardin: "It was the first time I had come across the theory of the growth of the individual, the scientific concept which explains how certain unidentified creatures spring into another form of existence, and how it's always small, sensitive forms of life that make this leap, not heavily armoured things like dinosaurs which became extinct."[64] Teilhard's emphasis on the life force is closely linked to Cook's ideas and those of Gerald Heard, another influence on Thorndike later on, especially the notion of the Great Chain of Being, a concept that harks back to pre-Darwinian days yet held enormous popular appeal. James Harrison notes that Teilhard "at times reads like nothing so much as Pascal rewritten by Herbert Spencer and Bergson," giving full scope to the "covert teleology Bergson never acknowledges."[65]

Thorndike was fascinated by evolutionary theory and its possible links with acting. In a tract called "Religion and the Stage" (1928), she characterized theatre as a "microscope" as well as a vital evolutionary mechanism: "If, as we are taught, the history of the race is epitomized in the development of every child, then drama precedes music and painting, for nearly every child acts before it can sing or draw."[66] *The Verge* became a turning point in Thorndike's career: like

Glaspell, she began to use the theatre as "a laboratory for research and experiment in human nature and the possible breakthrough, over the horizon."[67] Through an exchange of increasingly excited letters, she implores Edith Craig to produce the play; she "adores" it and sees "wonderful awful things" in it.[68] She feels that Glaspell is voicing her own thoughts. Thorndike does not view Claire as insane but if anything a natural extension of the "Angel" St. Joan (a role Thorndike alternated with the role of Claire—playing Claire in the afternoons, Joan in the evenings), a saint with divine botanical visions.

This is remarkable, since Claire's murderous actions at the end of the play are almost invariably explained as manifestations of insanity, an instability of mind that has been there from the start and is shown in her crazy experiments. The trouble with the insanity plea as an explanation for Claire's behavior—just as with Jean's infanticide in *Alan's Wife*—is that it gives a rational cause for something that cannot be expressed or explained so easily and that is specifically related to being female. In fact, Thorndike's emphasis on Claire's otherness rather than insanity may well be connected to what she was doing prior to taking on the role. For two years (1920–1922), Thorndike had devoted herself to Grand Guignol, a form of performance that combined caricature, horror, melodrama, and the grotesque with ordinary domestic contexts. She loved it:

> With the intensest pleasure, Sybil Thorndike would endure mortal agonies on the stage of the Little Theatre, John Street, Adelphi . . . revelling in the Guignol's peculiar brand of horror. . . . Mad women in a lunatic asylum gouged out Sybil Thorndike's eyes with knitting-needles; she was a murdered cocotte stuffed into a trunk; as a war-widow she killed her scientist-brother, inventor of a bomb; she was neatly strangled; she did a little strangulation herself.[69]

Grand Guignol thrives on madness, but defamiliarizes it by taking it out of the realist acting context.[70] Wolff notes that Claire's strangling of Tom is "an unnatural act of strength, implausibly easily and rapidly [done], creating an almost stylized murder moment."[71] Such stylization is one of the defining features of Guignol theatre, which enjoyed a brief renaissance in the early 1920s (and earlier, in such plays as Alfred Jarry's *Ubu Roi*, 1896) as directors embraced its potential to disrupt realist

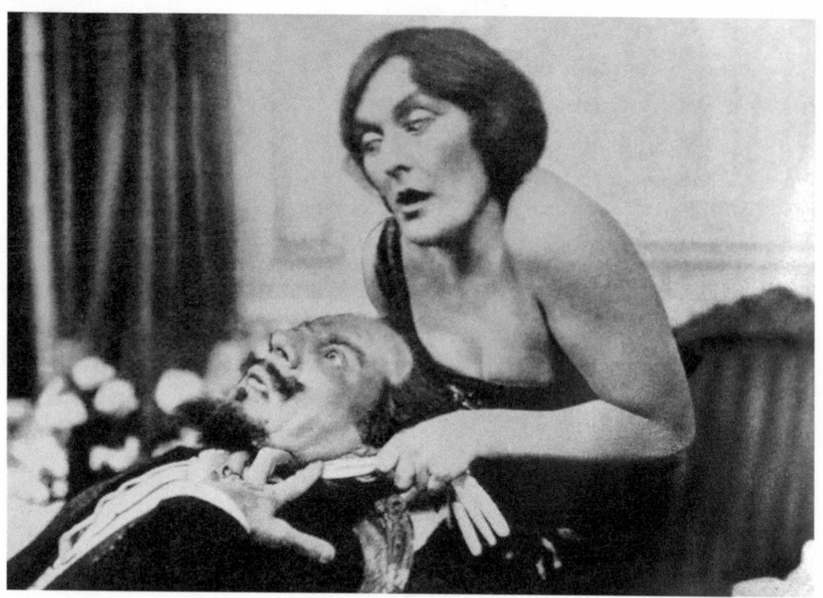

Figure 4 Lea (Sybil Thorndike) strangles Gregorff (George Bealby) in *Private Room No. 6* (1920), part of the Grand Guignol season at the Little Theatre, London. (© Thompson Theatre Collection/ArenaPAL)

expectations and conventions; Vsevolod Meyerhold, Aurélien Lugné-Poë, and Edith Craig's Pioneer Players are all working in this vein at this time.[72]

Thorndike, consciously or not, might have injected something of the Guignol spirit into her portrayal of Claire. A photograph of Thorndike in one of her Guignol roles shows her strangling a man (figure 4), an action she performed again just a few years later as Claire murdering Tom. Far from finding this kind of performance onerous, the Grand Guignol seasons were personally therapeutic for Thorndike: acting in these plays "did something for me, it was a kind of release," purging her fears of failure and her life-long nightmares.[73] I would suggest that Guignol gave Thorndike the theatrical idiom she needed for Claire's final actions. The Guignol connection is also persuasive given that critics still disagree on what form the play takes or what mode it is in—realistic, expressionist, or symbolist. The Grand Guignol, which has affinities with biomechanics in the stylized physicality and the grotesque features, might just be a plausible alternative.

This seems consistent with the critical response to the production and to Thorndike as Claire. *The Verge* was staged by the Pioneer Players (their last production) in March 1925 at the Regent Theatre for three matinee performances; it was seen as too "difficult" for any West End management.[74] Although reactions to the play were, in Julie Holledge's words, "extreme," it was a "success."[75] Lewis Casson, Thorndike's husband, played Tom—the man Claire murders. It was condemned by publications like *Vogue* and the *Daily Express*, but the more serious critics saw the play's pioneering quality: the *Weekly Westminster* calling it a brilliant psychological study, James Agate ranking it alongside Ibsen's plays, and the *Manchester Guardian* describing Glaspell as the greatest playwright writing in English since Shaw began.[76] Thorndike's acting drew ecstatic responses from audiences and critics alike, the *Lady* calling her "positively terrifying in her uncanny suggestion of evil and mental fury," while the *Era* wrote that "for the immense understanding and fine nervous force, Miss Thorndike's performance has rarely been equalled on the modern stage."[77] In the *Star*, A. E. Wilson thought it greater than her portrayal of St. Joan:

> The part of Claire might have been written for her, so closely did she enter into the mood and personality of this insane creature. Her own personality was absorbed in the part, and in her gesture, tone, facial expression, deportment, and everything that composes the art of acting she was precisely right. She delivered herself of the mad jargon with frenzied conviction. She permitted us an astonishing glimpse into a mind: no other actress could have created such an intensity of interest.[78]

This illuminating analysis suggests a split between Claire's "mad jargon" and her behavior. She is not simply dismissed as insane; instead, as Katharine Cockin notes, "Perhaps the most disturbing aspect of the play for many reviewers was the tremendous energy of Claire Archer."[79] We can never know exactly how Thorndike acted the role, but she clearly achieved a balance, a kind of theatrical hybrid echoing Glaspell's own experiment with form.

The bold evolutionary gestures of *The Verge* were unique. *Springs Eternal* (probably written in 1943) is about "the theme of evolution and achievement" in the face of an imperiled humankind.[80] Barbara Ozieblo characterizes the play's message as conservative in its rhetoric, arguing that women should remain at home, stay married, and look after the men. While the play may not be indicative of Glaspell's position on women, this conservative strain can be found in other works, like *Chains of Dew*, in which the potential of the wife to find fulfillment outside the home, pursuing a cause beyond raising children, is sacrificed to the perceived good of the husband. *Springs Eternal* focuses on a family line, using genealogy as the peg for larger evolutionary questions. "Are the Higginbothems unlike the rest of the human race?" asks Bill, the young doctor who is the moral center of the piece. This family is stuck in an evolutionary rut; they have not evolved.[81] Meanwhile, the war is going on far away. War as a catastrophic mutation or "sudden leap" in human evolution, analogous to a natural disaster, figures here as it does in Wilder's *The Skin of Our Teeth*. The returning soldiers will "be different," says Margaret, "so many will be different. That in itself will make a different world."[82] Later, Harry says "brightly" that "they say life is going to be entirely different after the war. In ten, twenty years you wouldn't know it for the same planet."[83] Glaspell's plays shift constantly between individual development and the evolution of species or the group as she seeks theatrically engaging and metonymic ways of conveying her ideas about human evolution through individual change; a single soldier returning changed by the war changes "the planet." Yet there is resistance to this kind of drastic change, even while it is deemed necessary, and we end up balancing the "Brave Old World" with the brave new one.[84]

Again, monism comes to the rescue, providing moral uplift and hope. Glaspell's monism is expressed in a long speech given by Bill when he explains that blood transfusions make some of the soldiers he treats "quite sentimental about the plasma. . . . They like to think of it as coming from a particular person . . . [and they] wonder whose it was—what they got, and what that person is doing now. How the rest of the blood is getting along. Maybe it's plowing—or it might be making love. Tending the baby—flipping the pancakes, any nice simple thing. It's quite a bond. I even think of it myself. Wouldn't it be funny, but perhaps we will become—one blood."[85] Owen, whose defeatist attitude has turned him into a recluse but who through the events of the play regains purpose and hope, echoes this "one blood" vision when he says: "This home of the human race is *one*. You've got to think of it as one—or be damned."[86] Thus, Glaspell's final play both revisits and expands on some of her evolutionary questions, and like Wilder's *The Skin of Our Teeth* (with which it is almost exactly contemporaneous), its stance is profoundly shaped by the war.

Glaspell's vision seems torn between two competing models, like the sand and forest of *The Outside* or saltation versus gradualism or progress versus degeneration. Like Robins, she confounds attempts to categorize and taxonomize her as writing a particular type of drama. Even her places within the historiographies of theatre and of modernism are ambiguous and seem to defy easy definition. Veronica Makowsky's essay on Glaspell and modernism, for instance, opens by acknowledging that modernism was for her, as for so many women writers, "at once a blessing and a curse. It stimulated her best work, and then scorned her aesthetic."[87] I would suggest that this ambiguity stems from Glaspell's sustained attempts to find an alternative to the male-dominated model of modernism that was emerging during her career, the model of "neo-catastrophism" rather than Darwinian gradualism, the "shock of the new" cliché of modernism that values sudden and drastic breaks with tradition over transitional work that gradually forges something new while retaining strong traces of the past. Modernist historiography values both the avant-garde "sudden leap" model of newness and neat categories of artistic definition, and neither Robins nor Glaspell fits easily within this framework.

Later in life, Glaspell came to consider motherhood as "the ultimate achievement for a woman."[88] Her plays oscillate between a radical feminist vision such as in *The Verge*, in which the scientist mother completely rejects the conventional, conservative daughter and kills her lover in her desperation to free herself from the bonds of traditional female life, and the vision of a traditional, self-sacrificing, motherly, nurturing woman such as the good wife Dotty in *Chains of Dew*. Glaspell was a woman of "vivid contrasts and sharp ambiguities" who defies easy pigeonholing into the "essentialized compartment" of feminism.[89] She was always torn: "Her firm belief in individual freedom of choice and freedom of speech was challenged by a belief just as firm in one's obligations and responsibilities to friends, family, and society, while her modernist desire to break out and seek new ways of life conflicted with her Victorian upbringing in a traditional, religious family."[90] Her solution to this conflicted state was to harness the science of genetics to the theatre to create new, hybrid forms and hope that they survived.

A Brechtian Bridge

Bertolt Brecht is a kind of counter-evolutionary force in the theatre, a lot like Shaw, to whom he paid tribute as a theatrical "terrorist" in his essay "Three Cheers for Shaw" (1926). He declared his theatre to be "theatre for a scientific age."[91] Not only his ideas but also his language are suffused with science; it is as if by emulating the methodology of science he lends his theatre its authority, continually referring to his "method," his "technique," the "model" he has set up, and his "experiments" in the theatre. In addition, his emphasis on imitation is similar to that of mimicry within Darwinian evolutionary theory; acting becomes for Brecht a process of imitation rather than embodiment, and such mimicry is shown to be a tool for survival, for enabling further processes to take place.[92] Although his initial love of science underwent drastic revision when he saw to what uses it was put in World War II, his work remains deeply connected to the science of his age. Most commentators have seen this in terms of his engagement with, and repulsion from, modern physics; but in fact, his work is no less concerned with evolution. Brecht's theatrical vision is founded on a particular conception of how human evolution works: he would "defy" naturalistic and Darwinian approaches to modernity.[93]

There is a profound and mutually dependent connection between Brecht's theatre and his views on evolution. His famous chart in "Notes on the Opera *Rise and Fall of the City of Mahagonny*" lists, under "Dramatic Form of Theatre," "evolutionary inevitability" as being replaced by "jumps" in the "Epic Form of Theatre." Although this last term is not defined further, *jumps* signals sudden and drastic changes rather than gradual transitions, a kind of punctuated equilibrium in evolutionary development. This chart contrasting the dramatic and the epic theatre is in fact mostly concerned with evolution: how far humans are alterable, how far life is teleological, "human nature as process" and whether there is growth through linear development or in "curves."[94] For Brecht, the emphasis is on "rupture" and how "historical trauma constitutes reality. . . . Experience is the awareness of violent rupture that destroys the consistency of the historical narrative."[95] Brecht had a "suspicion of the Enlightenment's philosophy of history, especially of its belief in progress."[96] Marxism would in any case have positioned him against natural selection and the way it came to stand, rightly or wrongly, for capitalism. He would be much more closely aligned with Lamarck or at least the way twentieth-century social reformers adapted Lamarckism. Brecht "treated nature as primarily a resource within an economy."[97] His plays reveal the insidious impact of survival of the fittest when turned to capitalist ends, most starkly illustrated in *Mother Courage and Her Children*.

Brecht's combination of saltation with epic form was highly suggestive to Wilder, who added a monistic element that suited his own theatrical and moral purposes.

Thornton Wilder's Epic Gradualism

One of Wilder's critical biographers writes that "all serious thought and art in the post-Darwinian age has been forced to take account, in one way or another, of the cosmological and religious implications inherent in *The Origin of Species*."[98] Wilder accepted Darwin's theories from an early age and expressed surprise that they were still questioned in China when he visited there in 1910, remarking (perhaps too optimistically) that Darwinian evolution was accepted without objection in the United States.[99] A well-established novelist as well as playwright, Wilder might have chosen to treat evolution through narrative means. But, as if answering Henry Arthur Jones's question about how to show in a brief evening's entertainment what in nature takes eons of time, he developed an extraordinarily ambitious project for putting evolution on stage. Like Strindberg's unrealized plan for an epic-scale depiction of evolution, Wilder's evolutionary vision is sprawling, encompassing many of Haeckel's ideas (recapitulation theory, monism) as well as a Spencerian and Lamarckian sense of progressive humanity that is constantly improving. In *The Skin of Our Teeth*, Wilder deftly telescopes vast stretches time into an evening's theatre-going through the device of anachronism, showing a modern American family living in New Jersey in the time of an encroaching ice age, with pet baby dinosaur and mammoth in their living room. The play goes on to highlight various milestones in human evolution, juxtaposing progressive events (the invention of the wheel, fire, the alphabet, the times tables) with destructive ones (the first murder, war, natural disaster) and using a biblical frame like that of *Back to Methuselah*. Act 2 ends with the flood, another catastrophic natural event like the ice age that engulfs act 1, and act 3 follows a seven-year war. Of all these destructive events, only the last is human-made (though in the Bible, humankind brings about the flood).

Wilder wrote in frustration in October 1940 that the play presented "problems so vast" because of the "gigantism" of what he was attempting: "to do a play in which the protagonist is twenty-thousand-year-old man and whose heroine is twenty-thousand-year-old woman and eight thousand years a wife."[100] He writes that "by shattering the ossified conventions of the well-made play the characters emerge *ipso facto* as generalized beings." He refers to his "myth-intention." He is trying to present Man and Woman; he is dealing in archetype. An early working title for this play was *The Ends of the Worlds*.[101] Wilder said it was "the most ambitious subject I have ever approached . . . the struggles of the race and its survival. . . . It is a subject as real as any other, as dramatizable as any other . . . [and] not so much a matter of emotion at all, as it is of seeing, knowing, and telling."[102]

Wilder's other way of showing evolution on stage was to juxtapose simultaneity with progression: as he writes in the preface to *Three Plays*, "The audience soon perceives that he is seeing 'two times at once.'"[103] Wilder sets everyday actions against "the vast dimensions of time and place." In an interview in 1957, Wilder noted how this was one of the prevailing ideas in all his art: "An unresting preoccupation with the surprise of the gulf between each tiny occasion of the daily life and the vast stretches of time and place in which every individual plays his role . . . [and] the absurdity of any single person's claim to the importance of his saying, 'I love!' 'I suffer!' when one thinks of the background of the billions who have lived and died, who are living and dying."[104] This is true of *Our Town* as well as of *The Skin of Our Teeth*. Ostensibly local and narrowly focused, *Our Town* actually offers two interconnected families in a small town in New Hampshire as metonymically spanning the same epic sweep as *Skin*. The audience is steered toward this view in many ways, always alert to the tension between the vast and the tiny.

The Skin of Our Teeth opens with Mr. Antrobus (the Greek word for "human"), having invented the wheel, now working on the alphabet and the multiplication tables. He is Adam ("He was once a gardener, but left that situation under circumstances that have been variously reported"), and Mrs. Antrobus is Eve. She asks the maid, Sabina, "Have you milked the mammoth?" Anachronism drives both the humor and the fundamentally serious side of this play and provides a shorthand way (theatrical equivalent to the time-lapse photography shown on nature programs) of indicating vast geological time as well as species transmutation. For example, Mrs. Antrobus tries to cheer her husband up as he despairs of the future, invoking harder times:

> *When the volcanoes came right up in the front yard. And the time the grasshoppers ate every single leaf and blade of grass, and all the grain and spinach you'd grown with your own hands. And the summer there were earthquakes every night.*[105]

At the end of act 1, Sabina appeals to the audience to help "save the human race" by passing their chairs up to be burned, to keep the fire from going out. Natural resources are limited; even the cuddly baby dinosaur and woolly mammoth are sent out into the cold to die. The comic and the tragic coexist. Wilder's depiction of these animals having to die (proleptically signaling their inevitable extinction) to make way for human survival echoes Lord (George Gordon) Byron's Cain when he laments the fate of animals forced into a fallen state by man's first disobedience, although they themselves were innocent. The connection, however suggestive, seems to be coincidental, as there is no evidence that Wilder knew of Byron's play.[106]

The play has no resolution. Early on, the exasperated Sabina states: "What the end of it [both the play and human evolution] will be is still very much an open question."[107] At the end, she breaks off in the midst of recapitulating the play's opening to tell the audience: "You go home. The end of this play isn't written yet."[108] Her reflections guide us constantly back to evolutionary concerns:

> *That's all we do—always beginning again! . . . How do we know that it'll be any better than before? Why do we go on pretending? Some day the whole earth's going to have to turn cold anyway, and until that time all these other things'll be happening again: it will be more wars and more walls of ice and floods and earthquakes.*[109]

Mrs. Antrobus will have none of this existentialist angst. She fiercely defends the purposeful quality of life, through work and through procreation. Adaptation is key, but Wilder does not call it that; instead, he goes in for more emotional and moral terms like perseverance and endurance. Of the human characters, Mrs. Antrobus is the most adaptive. "We can start moving. Or we can go on the animals' backs?"[110] Sabina just complains, or wants to give up and die; Mr. Antrobus thinks only of the present, glorying in his achievements but blind to impending disaster. Somehow, as the play's title suggests, we get by; perhaps the true evolutionary mechanism is sheer luck.

But Wilder's vision is anti-Darwinian with regard to human evolution in that, like Shaw, he refutes the blindness, randomness, and lack of purpose in the mechanism of natural selection. His faith in the human mind, enriched and sustained by knowledge and wisdom of the great thinkers he cites, supersedes religion, despite all the biblical references in the play. Wilder also finds much theatrical mileage in Haeckel's recapitulation theory. At the end of act 2, Mr. Antrobus talks of what Lifton characterizes as his own "prenatal history, during which he underwent all the traumas of evolution in his own person. . . . Not only is he Adam, Noah, and man in general, he also bears within him traces of earlier life forms." Wilder uses the biogenetic law to suggest at least a partial solution to the problem of individual significance "since it indicates that every human being not only is a distillation of influences from a myriad of historical and prehistorical human epochs but also contains physical or behavioral vestiges of even earlier, prehuman ages."[111] Lifton argues that Wilder also shows a general concern with harmony between humans and their environment, their "oneness with their surroundings"; Wilder is showing through this harmony a concern with "adaptation," which naturalistic works usually broach in terms of a *lack* of such harmony, a sense of conflict between individual and environment.[112]

The Skin of Our Teeth bears close resemblances to several works, from Shaw's *Back to Methuselah* to Brecht's theatre and the ambitious sprawl of James Joyce's *Finnegans Wake* (a debt that Wilder openly acknowledged). Wilder had met Shaw in 1928 but never commented much on his work, and he did not overtly draw a link between *Skin* and *Methuselah* despite their obvious similarities. Both depict biblical characters and events (Cain and Abel, Adam and Eve, Lilith, Noah's Ark), a vast sweep of time and place, the development of humankind telescoped into a few hours (or several days, in Shaw's case). Their Cains are virtually identical: both have turned their murderous bents to warfare and risen to the top of their ranks. But the two plays end radically differently: *Skin of Our Teeth* accepts our flaws and asserts human perseverance in the face of its own mistakes as well as natural disasters. It implies that there is something good in us that will keep surfacing and making us strive onward. *Back to Methuselah* turns its back on human imperfection, hoping that evolution will eventually result in an immaterial intellectual essence free from bodily limitations.[113]

Less well known is e. e. cummings's fragmentary *Anthropos or the Future of Art*, which contains a colorful reference to "prehensile precincts of predetermined prehistoric preternatural nothing" that is "fourteen million astral miles" away.[114] There is no evidence that Wilder knew of cummings's play, but the affinities are fascinating. Cummings relies on anachronism as a comic device, just as Wilder does, with cavemen speaking in modern advertising jargon and clichés, a contemporary artist sharing the stage with the "infrahumans," and a steam shovel acting as a woolly mammoth. (Byron also has a mammoth in his play *Cain*.) There is also the coincidence of cummings using Anthropos as his title character and Wilder using Mr. Antrobus.[115]

For all these possible or actual influences, Wilder claimed that *Hellzapoppin'* was the real inspiration for his play. This was a hugely successful musical revue in 1938 that ran for over three years and featured fast-paced comedy sketches constantly updated with new material to remain topical; circus-like in style, it used slapstick, mimicry, clowns, dwarves, and audience participation, with actors wandering through the audience. The zany influence of *Hellzapoppin'* would certainly explain why Wilder thought of *Skin of Our Teeth* as "a comic-strip play."[116] Its distinctly cartoonish quality nicely balances the stark subject matter. Wilder, in his journal, mentions *Candide* and Romain Rolland's *Liluli* as being of the "same category" as what he is attempting in *Skin*. His overall question was:

> What does one offer the audience as explanation of man's endurance, aim, and consolation? Hitherto, I had planned here to say that the existence of his children and the inventive activity of his mind keep urging him to continued and better-adjusted survival. In the Third Act I was planning to say that the ideas contained in the great books of his predecessors hang above him in mid-air furnishing him adequate direction and stimulation.

This, he notes, is in contrast to the "vast majority of writers hitherto" who would have turned to religion to explain man's endurance. He also notes that "the statement that the ideas and books of the masters are the motive forces for man's progress is a difficult one to represent theatrically. . . . The Hours-as-Philosophers runs the danger of being a cute fantasy and not a living striking metaphor."[117] Lifton characterizes Wilder's cosmology as "monistic" and says it is there in both *Our Town* and *The Skin of Our Teeth*.[118] The scene of the hours-as-philosophers includes Aristotle and Spinoza—key thinkers in developing preevolutionary explanations about the natural and spiritual worlds, Spinoza in particular founding the monist philosophy.

The play may have precedents in its epic structure, but its attempt to stage the natural order is unique. Act 2 imagines all the "orders" meeting at various conventions throughout the world, with the order of mammals' convention taking place in Atlantic City, New Jersey, the setting of this act. Wilder thus represents "the collected assemblies of the whole natural world" in a nonhierarchical Darwinian model. The order of Mammals, headed by their president, Mr. Antrobus, has "received delegations from the other rival Orders,—or shall we say: esteemed concurrent Orders: the WINGS, the FINS, the SHELLS, and so on. These Orders . . . have sent representatives to our own, two of a kind."[119] Antrobus addresses them, beginning by announcing that "the dinosaur is extinct," which garners applause, and "the ice has retreated." He then goes on to answer the accusation made during the last election that

> *at various points in my career I leaned toward joining some of the rival orders,—that's a lie. As I told reporters of the* Atlantic City Herald, *I do not deny that a few months before my birth I hesitated between . . . uh . . . between pinfeathers and gill-breathing,—and so did many of us here,—but for the last million years I have been viviparous, hairy and diaphragmatic.*[120]

The crowd cheers. Lifton notes that "the sense given by the entire act is of a deep bond of kinship uniting humans and animals."[121]

What is brilliant about this act is the staging, which situates the audience in the ocean; when Mr. Antrobus toward the end of the act addresses all the orders of animals, he is pointing to the audience and saying, "Look at the water. Look at them all. Those fishes jumping. . . . Here are your whales, Maggie!" In this zooistic vision, we *are* all the animals, and Mr. Antrobus begins his formal speech: "Friends. Cousins. Four score and ten billion years ago our forefather brought forth upon this planet the spark of life—." But thunder drowns him out, and moments later he is hustling all the animals two by two onto the ark, along with his family, to start a new world elsewhere.[122]

The play also focuses innovatively on women in an evolutionary context by exploring the troubling question of what has happened over the millennia to Mrs. Antrobus, whose power seems to have dwindled since act 1 rather than progressing: she's become harmless, merely interested in domestic matters, whereas before she was the anchor of the family who ensured its survival. She has become increasingly conservative; she invokes and lampoons the suffragettes' struggle for the vote when she recalls how she and her fellow woman fought for marriage ("at last we women got the ring") and enjoins her fellow mammals to "keep hold of that" and to "Save the Family."[123] Yet, toward the end of the act, as the deluge is about to break, she has a great speech championing women's unsung evolutionary roles. Taking a bottle out of her bag and flinging it into the ocean, she says that in it is a letter containing "all the things that a woman knows. It's never been told to any man and it's never been told to any woman, and if it finds its destination, a new time will come. . . . We're not what you're all told and what you think we are: We're ourselves. And if any man can find one of us he'll learn why the whole universe was set in motion."[124] This almost mystical invocation of women as the key to human evolution is akin to Ibsen's speech about looking to women to sort out humanity's problems. It is revealing to contrast this progressive male championing of women with the ambivalence of Robins and Glaspell, as well as many other women playwrights less sanguine about too easily finding such evolutionary hope in women. It would simply replicate the Victorian tendency to place women on a pedestal that the suffrage was supposed to help erase.

Glaspell and Wilder represent two of the most significant American interventions in the theatrical engagement with evolution in this period. They differ from the contemporaneous treatments of evolution in Britain, for instance in J. B. Priestley's *Summer Day's Dream* (1949), a virtually forgotten precedent for our current wave of climate change plays that amounts to an early, and prescient, call for environmental awareness and concern.[125] Priestley is usually noted for his "time plays" or for the expressionist interpretation brought to the revival of *An Inspector Calls* in the 1990s. He shares Wilder's idealism and theatrical techniques; a simple message like the social responsibility of each of us (*Inspector Calls*) is conveyed through didactic, semi-naturalistic theatre. It is illuminating to compare *Summer Day's Dream* with Glaspell's and Wilder's work. The futuristic play is set in 1975 in a kind of utopia after a Third War has wiped out civilization and the survivors have gone back to the soil and the old ways (using horses rather than tractors and living off the land in modest but complete self-sufficiency). The characters are "survivors from a wreck, from the war of split atoms and split minds—who crawled out of the darkness into the daylight."[126] Unlike most postapocalyptic plays (including Samuel Beckett's *Endgame*), this has all been to the good. The play shows "an England out of the rat race, reduced to third world status, but environmentally friendly" as it lives in an "ecological paradise."[127] Well before the current term *Anthropocène* came into vogue, this play suggests explicitly that the Industrial Revolution altered human evolution. "Hotter than it used to be, Fred," comments Stephen.[128] The staging visualizes this: the garden is "thick with climbing plants, lush flowers and foliage . . . all rich and grand in a natural fashion, and looking rather more sub-tropical than the usual English scene."[129] Far from being depleted, nature is simply changed; it is "slower" now, and in fact there are "more flowers, more birds, than ever before."[130] Chris tries to explain to the visitors the new and improved relationship to the land: "We're not just living off it. We're living with it. We love it—and we think it's beginning to love us."[131] This concern about how humans relate to the biosphere activates many Beckett's *Happy Days* and other works, as I will discuss in the next chapter.

Priestley's play also shows similarities with Glaspell and Wilder, challenging assumptions about what is "primitive" and what is civilized by inverting the two. In doing so, it also hints at ideas about altruism and cooperation, the group working together organically and only taking what it needs. As in many of his other plays, Priestley emphasizes "the group as much as the individuals."[132] And, like many of the generation of playwrights before him, Priestley seems to flirt with monism, especially through the rather mystical character of Margaret, a woman endowed with a kind of sixth sense. She says that "we men and women are part of a great procession of beings, many of them infinitely stronger and wiser and more beautiful than we are."[133] Priestley, Glaspell, and Wilder also share—with Brecht—an interest in showing how war can alter the course of human evolution.

Maggie B. Gale notes that "the play is replete with discussions about the merits of industry versus husbandry, of nature versus science, and the ending, whereby the interlopers, so moved by the rural utopia and so convinced by the dedication of the Dawlish family to this 'alternative' to their 'first world' existence, tell their respective superiors that the land will not provide the commodity they had originally thought it would."[134] But *Summer Day's Dream* is not just a "critique of capitalism."[135] It also muses seriously on humans' place in evolution and the relationship between humans and their habitats in terms of what we would now call sustainability.

Wrestling with the Impact of Darwin

In this final section, I want to turn to playwrights who explored the implications of Darwinism in very different ways from Glaspell and Wilder, beginning with Eugene O'Neill, Glaspell's close colleague and fellow playwright. Like her, O'Neill was a devotee of Nietzsche, a philosopher who rejected, even ridiculed, Darwinism; "[Nietzsche's] enthusiasm for strength of personality and its inspiration [came] from quite different sources."[136] O'Neill's earlier work contains evidence of Darwin's influence, or at least—as with his precursor, Herne—a watered-down Spencerian approach. As we have already seen, this was central to O'Neill's short play *Abortion.* In *The Great God Brown*, O'Neill refers to Darwin: "Watch the monkey in the moon! See him dance! His tail is a piece of string that was left when he broke loose from Jehovah and ran away to join Charley Darwin's circus!" *The First Man*, the play O'Neill wrote before *The Hairy Ape*, likewise carries evolutionary themes, with an idealistic explorer-anthropologist in search of human origins, partly as a way of running away from his domestic problems. His turn to anthropology—"the last romance"—has won him fame as "the most proficient of young skull-hunters," surely about to discover "the Missing Link," his friend says mockingly.[137]

Erika Rundle points out that O'Neill's "sea plays" are fundamentally concerned with "the discourse of species," and his "musings on adaptation and selection as mechanisms of survival (or extinction)" culminate in *The Hairy Ape* (1922; adapted from his short story of that title from 1917).[138] Already in his first sea play, *The Moon of the Caribbees*, O'Neill reorganizes "longstanding kinship models along the lines of species difference rather than those of race, ethnicity, or gender," as well as "invoking a vaudeville tradition involving a monkey who is forced to perform in order to survive." These elements become fully elaborated in *The Hairy Ape* a few years later.[139]

O'Neill was "concerned with the social ramifications of Darwinism," notes Rundle in her detailed analysis of the play's engagement with evolution.[140] *The Hairy Ape* follows the downward trajectory of Yank, who works as a coal stoker in the engine room of a ship. The stooping fire stokers with overdeveloped back and shoulder muscles resemble "those pictures in which the appearance of Neanderthal Man is guessed at"; one thinks instantly of Gabriel von Max's painting of *Pithecanthropus alalus* based on Haeckel's descriptions of prehistoric humans in *The Natural History of Creation,* a book still popular in O'Neill's time. According to O'Neill's stage directions, "all are hairy-chested, with long arms of tremendous power, and low, receding brows above their small, fierce, resentful eyes."[141] This grindingly hard labor has made Yank coarse of tongue and body, a huge, blackened figure of little obvious intelligence who is likened to an ape. The turning point of the play is his encounter with a spoiled millionairess who recoils in horror on seeing him. This sends him on a rampage through New York that culminates in a scene at the zoo, where he confronts—and is killed by—his close kin, the gorilla. Travis Bogard classifies this play as one of O'Neill's early "anthopological" dramas, "expressed in terms of a search for the origins of life and making reference to atavistic remnants of primitive man appearing in modern society."[142]

Rundle argues that *The Hairy Ape* helped to create a new genre, the "primate drama," which she calls "a twentieth-century American hybrid of classical and modernist structures that treats the subject of evolution, both explicitly and implicitly, through the disciplines of performance."[143] In addition to foregrounding and enacting ideas of regression, degeneration, heredity, and savagery, the play poses two challenging theatrical requirements. One is its invocation of a wide range of previous theatrical conventions and practices involving apes, such as the "missing link" show, the circus, the "menagerie," the zoo as spectacle, and the so-called ethnological exhibit. The other is the presence of a "real" ape on stage in the final scene of the play. There is much discussion of O'Neill's expressionism, of the play's obvious pull away from naturalism; the scene in which Yank attacks people on the wealthy streets of Manhattan yet they feel nothing, as if they are marionettes, is an example of this technique. O'Neill wrote after finishing the play that it did not fit "any of the current 'isms.' It seems to run the whole gamut from extreme naturalism to extreme expressionism—with more of the latter than the former."[144] Yet the final scene seems to demand realism in its representation of the real ape.

None of the discussions of the play in studies of O'Neill explains, in practical terms, *how* this was staged. Rundle does acknowledge that a real ape was not actually used, so it must have been a man in a monkey suit. There would have been strong evocations of the prehistory of such performances for a contemporary audience, not least the comical dimension as shown in farces like *The Missing Link*. So, for all the seriousness with which this scene is discussed, and all its tragic import, the reality is that it held potentially comic associations; it was daring, in other words, to disrupt this tradition and depart from the audience's expectations of men in monkey suits as farcical, associated with circus and vaudeville.

One way critics have dealt with this is to suggest that Yank's final visit to the zoo is merely in his mind—he is dreaming it.[145] Bogard sees the death in the gorilla cage as having "minimal plausibility. Although it realizes in a stage image the symbol of ape-in-cage which has been developed from the first scene, it does not entirely break from the context which the play has established as 'real.'"[146] Just as with Glaspell's plays, then, ambiguity of form—and the hybridity that results—generates sharp critical debate about what "kind" of theatre this is. Rundle points out that "O'Neill evokes the practices of nineteenth-century ethnographic display . . . [as Yank is] shackled to the violently constitutive performances" of so-called savages.[147] This is in stark contrast to the triumphant imperialism underpinning the popular Tarzan stories of the time, showing the savage becoming civilized; here, the savage regresses even further, dying in the gorilla cage at the hands of his seeming kin and equal, who then tosses him aside after killing him. Una Chaudhuri usefully contextualizes the play in terms of the rise of the modern zoo but likewise refrains from providing a sense of how this final scene was actually staged.

Bogard finds the ending "ambiguous" and unclear:[148] "Yank is destroyed . . . by a figure out of his own 'racial' past, by a gorilla in the Zoo."[149] Just when the audience thinks that Yank is successfully ascending the evolutionary ladder, he slides back down it. O'Neill discards that upward and predictable trajectory, instead dropping Yank "back into darkness by suggesting that he can only belong to a force of simian brutality."[150] Rodin's *The Thinker* is repeatedly invoked as the paradoxical model for how Yank (and eventually the gorilla) should sit and attempt to think.[151]

Kenneth Macgowan was a decisive influence on O'Neill, and his book *The Theatre of Tomorrow*, which O'Neill read in 1921, makes a clear link between theatre and evolution: "The grandeur of the play of the future must lie not in a superhuman figure, but in the vast and eternal forces of life which we are made to recognize as they play upon him." The drama of the future will, he predicts, show us "the things that, since Greek days, we have forgotten—the eternal identity of you and me with the vast and unmanageable forces which have played through every atom of life since the beginning." We must "recover the sense of our unity with the dumb, mysterious processes of nature," and we must apply our scientific knowledge to drama.[152]

A play that attempts to do this explicitly is *Inherit the Wind* (1955), which uses the clash between creationism and evolution as a means to criticize McCarthyism.[153] This pseudodocumentary play by Jerome Lawrence and Robert E. Lee dramatizes the notorious Scopes trial of 1925, in which a schoolteacher in Tennessee was prosecuted for teaching evolution rather than creationism. Just as Arthur Miller did with his adaptation of Ibsen's *Enemy of the People*, Lawrence and Lee used the harassment of the scientist-figure to make a strong statement about free speech during the peak of McCarthyism (the blurb on the paperback edition of the book reads: "The spectators sat uneasily in the sweltering heat with murder in their hearts, barely able to restrain themselves. At stake was the freedom of every American"). It is pure courtroom drama depicting the trial in which two of the country's greatest lawyers, William Jennings Bryan (prosecution) and Clarence Darrow (defense) went head to head over evolutionary theory.

Inherit the Wind has renewed political relevance now with the emergence of intelligent design as a pseudoscientific version of creationism. Like Miller, Lawrence and Lee were harnessing the persecution of progressive thinking to political purpose. They wrote in their original prefatory note that the play is timeless: "The stage directions set the time as 'not too long ago.' It might have been yesterday. It could be tomorrow." But the play is not particularly interested in science trumping religion; it ends with the Darrow figure (pro-Darwin) alone on stage after the trial has ended, and as he packs up his briefcase, he takes the Bible in one hand and *On the Origin of Species* in the other and, "balancing them thoughtfully, as if his hands were scales," he then "slaps the two books together and jams them in his brief case, side by side."[154] The world can accommodate the views contained in *both* these books, the playwrights seem to be saying, but it cannot tolerate suppression of ideas. Clearly, what is at stake in this play is freedom of speech and resistance to oppression, exactly the same subject matter as in Brecht's *Life of Galileo* a few years earlier. (When Brecht was questioned by the House Un-American Activities Committee, one observer remarked it was a case of "the biologist being examined by the apes.")

Edward J. Larson, in his study of the Scopes trial, reveals the extent to which Lawrence and Lee distorted the facts of the case in their supposedly documentary approach. He shows in detail how the two playwrights oversimplified and made reductive some of the complexities of the trial and also made up things, like the hostile attitude of the crowd (actually, the atmosphere in the town was more like a circus or festival than a witch hunt with an angry mob) and the placement of Cates/Scopes in jail when in reality he was not jailed and it was highly unlikely that he ever would be.[155] Larson refers to Lawrence and Lee's influence (and the highly successful film version that was later made of the play) as contributing to an ongoing myth about this "trial of the century": the myth of a resounding and definitive defeat of religious fundamentalism in America. Running for three years on Broadway, and going on tour across the United States, the play had a remarkable impact as a piece of theatre, regardless of the film version that was later made. But most reviewers at the time scorned both the play and the film; the *New Yorker* wrote that "history has not been increased but almost fatally diminished" by it, and *Time* characterized the film as "wildly and unjustly" lampooning the fundamentalists as "vicious and narrow-minded hypocrites . . . [and] just as wildly and unjustly idealizes their opponents, as personified by Darrow."[156] Stephen Jay Gould emphasized the falsehoods perpetuated by both play and film: "John Scopes was persecuted, Darrow rose to Scopes's defense and smite [*sic*] the antediluvian Bryan, and the antievolution movement then dwindled or ground to at least a temporary halt. All three parts of this story are false."[157]

It is striking that, in 1955, the most prominent American theatrical engagement with evolution was a historical one, set decades earlier, though encapsulating a debate that would dominate the popular reception of evolution in America for decades to come. *Inherit the Wind* looked backward, while in the same year in Britain, Beckett's *Waiting for Godot* signaled a very different kind of encounter between evolution and the stage, opening up new possibilities and reflecting an acceptance of Darwinism on a scale not seen in the theatre since Ibsen. The broader cultural and scientific landscape had changed utterly as the New Synthesis took hold, conveying

> certain moral and political lessons. . . . Man, too, was integrated in the synthesis—as a rational, responsible being, the pinnacle of evolution. Evolution was connected to progress, and knowledge of evolution would be crucial for the future of man. [Ernst] Mayr presented the new "population thinking" as bringing new respect for the uniqueness of the individual, as opposed to earlier essentialist, typological thinking. In the post-World War II climate, the modern synthesis represented optimism and hope.[158]

Progress, optimism, hope; this is hardly a Beckettian landscape and it warrants further discussion.

8

Beckett's "Old Muckball"

Samuel Beckett's *Waiting for Godot* and John Osborne's *Look Back in Anger* are typically cast in theatre histories as signifying, virtually simultaneously, two new and clashing paradigms in drama.[1] But the contrast between the seemingly apolitical, meaningless world of Beckett and the ostensible social protest mode of Osborne has distorted our understanding of Beckett's work, relegating the dramatic milieu of his plays to some kind of alien, airless, and moribund world divorced from our own. Beckett scholars have offered many ideas about what kind of "world" Beckett depicts and indeed creates in works like *Waiting for Godot* with its lunar landscape or *Endgame* with its "corpsed" setting, but a common assumption is that it is somehow different from our own. This reaches back to the very first scholarship on Beckett, as Linda Ben-Zvi in her introduction to *Beckett at 100* points out, quoting Ruby Cohn in 1962 characterizing Beckett's world as a different "planet" where

> matter is minimal, physiography and physiology barely support life. The air is exceedingly thin, and the light exceedingly dim. But all the cluttered complexity of our own planet is required to educate the taste that can savor

> the unique comic flavor of Beckett's creation. Our world, "so various, so beautiful, so new," so stingily admitted to Beckett's work, is nevertheless the essential background for appreciation of that work.[2]

Cohn clearly marks Beckett's and our own world as *materially* distinct entities, with our own world merely a "background" to his, the stage scenery for the main drama.

Cohn's was one of the earliest scholarly assessments of Beckett's work, and it arguably set the tone for the subsequent reception of his works as "other worldly" (yet without the tinge of supernaturalism that term often implies). This approach persists. Downing Cless refers to "the wastelands and voids of Beckett's settings."[3] Reviewing JoAnne Akalaitis's production of Beckett shorts in 2007/2008 starring Mikhail Baryshnikov on a sand-strewn stage, Ben Brantley noted how the sand slowed him down: "This grounding of a winged dancer poignantly captures the harsh laws of Beckett's universe, where Mother Earth never stops pulling people toward the grave."[4] Brantley's phrasing is confusing; how can Beckett's universe be so different from our own as to have its own "harsh laws" yet at the same time encompass "Mother Earth" and her relentless birth-to-grave march?

Beckett's dramatic world has been understood in a vast array of contexts spanning existentialism, Friedrich Nietzsche, the cataclysmic impact of two world wars, ancient philosophy, antitheatricality, Irishness, feminism, and so on. Although his interest in science more generally has been acknowledged, the nature and extent of his engagement with evolutionary theory has not been fully explored, particularly in his plays. Evolution has profound relevance to Beckett's drama and raises significant questions: How might his repeated motif of giving birth astride a grave, or his sense that Godot randomly may or may not come, be understood in the context of a Darwinian world? How do the common critical descriptors of Beckett's works as about death, endlessness, and meaninglessness relate to characterizations of his vision as "unsentimental" and "harsh," terms often used to describe the nature of Darwinian evolution by natural selection? What is his theatre saying about the interplay between the organism and its environment?

Waiting for Godot introduced audiences to a playwright whose dramaturgy would be underpinned by the fundamental ideas of evolution, paradoxically foregrounding nature through its seeming absence; a lone tree in a stark and empty landscape can make a greater environmental statement than a rich natural setting. *Happy Days* is built around the dualism of earth and air.[5] Willie crawls around on all fours; likewise, in *Waiting for Godot*, Estragon echoes the primordial: "All my lousy life I've crawled about in the mud! . . . Look at this muckheap! I've never stirred from it!"[6] Beckett wrote to Thomas MacGreevy of John Keats's "thick soft damp green richness," presumably finding Keats's language itself an embodiment of the moss he squats on.[7] The imagery is palpable, tactile, sensory; it evokes the physical world directly. Indeed, Beckett thought elementally. He told the cast of his production of *Waiting for Godot* in 1974: "Estragon is on the ground; he belongs to the stone. Vladimir is light; he is oriented toward the sky. He belongs to the tree."[8] Beckett thus conceives of this play—and indeed many others—in terms of the irreducible, fundamental qualities of nature, rather along the lines of Empedocles's four elements. Contemplating their situation, Estragon decides, "We should turn resolutely towards Nature."[9]

A brief overview of the critical interpretations of Beckett's depiction of the natural world shows how recent scholarship is beginning to open this line of thought. Chris J. Ackerley finds that although geology is fundamental to Beckett's work, and though traces of Karl Popper and Thomas Kuhn on scientific method can be found, an attempt "to saddle Beckett with a scientific temperament, let alone a scientific methodology, runs into an impasse generated by Beckett's deep distrust of the rational process. His rejection of rationalism entails a rejection of scientific methodology, less the process of uncertainty and fallibility . . . than the capacity of reason and its handmaiden hypothesis to shape a sufficient understanding of the natural world."[10] Ackerley concludes with a piercing insight about the relationship between Beckett's characters and their environments: "It is not the universe that is the absurdist structure . . . but rather the agency of the human mind that must accommodate itself to the only possible, the demented particulars of quotidian reality, in rare moments only being able to intuit the deeper laws that frame them."[11]

An ecocritical approach to Beckett offers some key insights along similar lines. Paul Davies characterizes Beckett's preoccupation with climate and weather as generating "a disparity of climates [that] sets the planetary biosphere directly at odds with the environment of the artificial interior," and he argues that Beckett's "deeply symbolic penchant" was "to articulate the hell that is caused by abstraction from the biosphere, from the living environment."[12] This hell manifests itself in its harmful, painful effects on the body, from dried-out eyeballs and skin to the feeling of "something dripping in my head" (this obsession with the fontanelles not having fused properly is a recurring theme). Davies also suggests that Beckett deliberately disrupts circadian rhythms as he plays with the natural cycles of light and dark, sleep and wakefulness, dawn and dusk. Why does he, as Davies puts it, continually "tell the condition of human alienation from the biosphere"? It is because "*clearing* equates to *space made*."[13] Destruction allows rebuilding, reconceptualizing. Along with Ackerley's conception of something amiss between humans and their frame, we might take Davies's argument a step further and suggest that perhaps what Beckett's work is really suggesting is that in fact it is the biosphere that is being alienated from the human. His plays again and again suggest this turn, perhaps most strongly in *Happy Days*. Worrying over the loss of God seems like misplaced energy when we are actually losing our whole planet. As Cyril Darlington writes in the conclusion to his epic study of the evolution of humankind and society published in 1969, "Every new source from which man has increased his power on the earth has been used to diminish the prospects of his successors. All his progress has been made at the expense of damage to his environment which he cannot repair and could not foresee. Surely this is the most practical of all the lessons of history."[14] This tale of destruction and environmental alienation, told only a few years earlier by Thornton Wilder in *The Skin of Our Teeth* and J. B. Priestley in *Summer Day's Dream*, is retold by Beckett in shorter, sharper form in *Happy Days* in particular, as I will show. In short, Beckett's theatre dramatizes the process of ecocide; there is "no way out of a denuded nature—just an endlessly denatured void, but with a few tiny signs of new or continued growth."[15]

Like Davies, Joseph Roach refers to "topographical alienation" between Beckett's characters and their environments.[16] Roach calls the "natural-historical landscape" of *Waiting for Godot* "desolate but not empty" since it has a tree with five leaves, a population of five people, and a lot of disembodied voices—"in other words, it is haunted."[17] What Roach then movingly goes on to show is that Beckett is summoning a specifically Irish context here; it is a landscape haunted by the famine that emptied Ireland of so many of its people through starvation and emigration, rendering whole villages extinct.[18] Roach sees *Godot*'s country road with a tree at evening as emblematic of that empty landscape, the "profundity and duration of that silence" brought about by the famine: "the sparsely peopled countryside, the wind in the ruins, the rocks scattered like bones under an indifferent sky."[19]

Following Roach's grounding of Beckett's work in a specific landscape, I would like to pursue this line of inquiry and extend it to consider the ways in which Beckett's theatrical work tells an evolutionary story. The earth and its processes are the mainstay of his theatre: Winnie's "earth ball," and "earth, you old extinguisher," Krapp's "old muckball."[20] And Time is "that old fornicator," according to Murphy, a word on which Winnie and Willie relish punning.[21] In these cases, the phrasing is affectionate, not alienated; "old" indicates something familiar, predictable, and oddly intimate even though it is monumental and macrocosmic. This suggests a deep *connectedness* to the familiar, everyday world we live in, to the random physical processes of the Darwinian universe: birth, death, natural selection, adaptation, deterioration and decrepitude, extinction, mutation, and so on. The world Beckett continually depicts in his plays is our own. He is obsessed with natural phenomena and how people interact with their natural surroundings when both are reduced to their bare minimum.

Cohn notes that with *Happy Days*, Beckett abandons thick, dark mud for "desiccated earth."[22] But Steven Connor counters the emphasis usually put on earth, "muck," and mud in Beckett's work. Although it may be "less well-ventilated than that of almost any other writer," air and breath are still everywhere in Beckett's work, as they are essential to sustain life.[23] Connor charts a development from the "searing mistral" of the earlier works, like *Dream*, to "the almost windless calm of the later works."[24] Beckett took notes on the chemical components and geographical features of the planets, including earth, in his "Whoroscope" notebook, for instance observing that Venus's makeup is similar to that of the start of life on earth and calling the atmosphere of Mars "tenuous."[25]

The "Whoroscope" notebook is just one of many unpublished archival sources (most of them at the University of Reading) that, along with the several volumes of his letters now in print, provide fascinating insights for anyone wanting to trace the sources of Beckett's thought.

They help to ground his works in all kinds of contexts, from the political to the scientific to the philosophical, enabling new interpretations and deeper understandings of Beckett's works. Yet the wealth of new studies building on such material is sometimes accompanied by attempts to police the boundaries of Beckett studies, to adjudicate which readings are valid and which are not, or as Matthew Feldman puts it, "to quietly negate overarching readings of Beckett that attempt to say what he (or 'it') actually means."[26] There is a repeated emphasis on the idea of scholarly responsibility, as if studies of Beckett have crossed some imaginary line; Shane Weller warns that "if one wishes to contribute to the understanding of Beckett's œuvre, then it is not enough to detect more or less striking resemblances between his works and those of a range of philosophers, theologians and literary figures picked to suit a particular commentator's intuitions or predilections."[27] Paraphrasing Beckett, Feldman and Weller warn against a "neatness of identifications."

Feldman himself, however, notes that Beckett avoided "the systematic harnessing of knowledge."[28] As Rita Felski points out in her exuberantly polemical piece "Context Stinks!" finding intertextual (and interartistic) connections is hardly spurious since works of art are autonomous and speak to one another and make connections across periods well beyond the strict boundaries of historical context; this is precisely where the most exciting interpretative possibilities lie.[29] The "labour of empirical groundwork"[30] is admirable and necessary, but it should not prevent the more speculative work of linking Beckett, Henrik Ibsen, Susan Glaspell or indeed any other author to his or her wider spheres even if those links might not have been intended by the authors or were even dreamed of in their philosophies. "A commitment to the empirical and to responsible literary scholarship does not preclude an adventurousness that is itself perhaps the most effective provocation to other scholars to enter into the task of trying to 'make sense' of Beckett."[31] So we are supposed to be "adventurous" while also being "responsible," to sense the difference "between ungrounded speculation and well-evidenced argumentation."[32] While this is indeed a desirable balance, there is an implicit assumption that suggesting influence "where none intended" is out of bounds. Perhaps the term *interlocutor* best captures the essence of the kind of interaction with Charles Darwin and other evolutionary thinkers that I am exploring here.[33] Rather than a question of direct influence, they act as a springboard for Beckett's own ideas about geology, evolution, and the natural world generally.

Although there is undoubtedly much to explore in Beckett's fiction as well as his drama, I focus on the latter because of the link between evolution and performance that has been so compellingly explored by Jane Goodall.[34] The actor's body is central to Beckett's evolutionary vision and is the main means of exploring it. Much is inscribed into performance that does not register fully enough in written accounts of Beckett that are still focused on how little is said and how much silence there is—not taking into account the rich and complex nonverbal expression that goes on throughout those silences and often destabilizes the apparent meanings of the words.

Ben-Zvi argues that after 1946 Beckett changes his focus to the body, to impotence, decrepitude, and general physical malfunction.[35] There is a concomitant celebration in Beckett of the materiality and corporeality of the theatre; the physical, the everyday, the material, and the "gross" are constantly present for the audience through the actor's body. Beckett is immersed in biological nature, fascinated by physiological processes. This is consistent with the celebration of the natural in all its aspects that one finds in Darwin.

Beckett and Darwin

Allusions to Darwin, usually indirect and oblique, can be found throughout Beckett's writing (perhaps most amusingly in Estragon's comment in *Waiting for Godot* that "people are such bloody ignorant apes").[36] One of the first to trace the Darwinian connections was Frederik N. Smith, who found in the "Whoroscope" notebook several quotations from *On the Origin of Species* and at least one reference to *The Voyage of the Beagle*.[37] Intriguingly, the notebook also contains a possible allusion to *The Descent of Man* for, as Smith notes, "among the many polysyllabic rarities listed in the notebook, one finds 'steatopygous'; it is interesting to note that the *Oxford English Dictionary* gives only a single example of this word, which is derived from the Greek roots meaning 'fat buttocks,' and that is from Darwin's *Descent of Man*. Could Beckett have discovered this word in that book?"[38]

Steatopygia was what attracted so many spectators to the Hottentot Venus in the early nineteenth century, as discussed in chapter 1. Why would Beckett jot down this word? The speculation that he might have read, at least in part, *The Descent of Man* is highly suggestive, for this is the text in which Darwin confronts the issue of human evolution most fully, while also drawing on unusual animals behavior to address topics such as sexual selection. Beckett's notation of the term also attests at this early stage to his interest in bodily dysfunction, the extremes of the human condition, the full spectrum of human physicality.

Beckett seemed utterly dismissive of Darwin. Of *On the Origin of Species*, he wrote to his friend MacGreevy that he had "never read such badly written cat lap."[39] This single line has been taken by scholars to mean that Beckett did not think much of Darwin, although he may well be criticizing the prose style rather than the ideas. In fact, the famous "cat lap" dismissal was not all he said about Darwin; further in the same letter, he volunteers: "I only remember one thing: <u>blue-eyed cats are always deaf</u> (correlation of variations)."[40] This is from an early section of *Origin* dealing with the "laws regulating variation," which, as Ernst Mayr points out, Darwin little understood, as he was unaware of Mendel's discoveries, even though he was fascinated by "the mysterious laws of the correlation of growth." What Darwin wrote was that "some instances of correlation are quite whimsical: thus cats with blue eyes are invariably deaf; colour and constitutional peculiarities go together."[41] Further in the *Origin*, he again refers to the "relation between blue eyes and deafness in cats," asking, "What can be more singular" than this correlation?[42]

Two things stand out about this part of Beckett's only direct pronouncement on evolution. One is that what he remembers comes from the beginning of *Origin*, suggesting that he perhaps did not read the entire work; Michael Beausang speculates that he read up to chapter 8.[43] Another is that he remembers something odd about the natural world that Darwin himself found "whimsical"; a quality of studied whimsicality is certainly a defining feature of Beckett's theatrical vision.[44] "Whimsicality" goes directly against teleology, purpose, order, predictability, and progression; in fact, it would jeopardize all of these. Blue eyes recur in Beckett's works, as in *Embers* ("the old blue eye" repeated several times). He was not the first writer to be drawn to this detail in

Origin; as Valerie Purton notes, there is "one small but telling allusion in both [Tennyson and Darwin to] . . . a certain blue-eyed cat."[45] It is perhaps when Darwin is at his most whimsical that Beckett connects with him; it is tempting to think that they share a sense of wonder and enchantment at the natural world more than the harsh, random, and blind force of nature usually associated with both writers (if we speak of nature at all in Beckett's plays, which are often erroneously read as denatured). But the wonder never becomes sentimental; plays like *Act Without Words I* show that force as not just whimsical but dark and malignant.

There is actually another direct reference to Darwin in Beckett's writing, and that is in his interwar notebooks. It is a rich, but perplexing, reference. He writes that in the "ceaseless transformation of all things nothing individual persists, but only the order, in which the exchange between the contrary movements is effected—the law of change, which constitutes the meaning and worth of the whole."[46] Life is a process of endless transformation subject to a "law of change," and the greater "order" supersedes the individual. Beckett's notes indicate a strong sense of an overarching evolutionary vision that puts the individual human being in his or her place:

> Living [is] an expiation of the arrogant desire for individual existence. From plant through to animal to man, who is finally worthy to return to primal unity. Propagation is an evil, because it retards reorganisation of primitive unity.
>
> Precursor of Darwin and Schopenhauer.[47]

Is the ultimate end of evolution paradoxically a "return to primal unity"? This would be a twist on the progressive view of Herbert Spencer, Jean-Baptiste Lamarck, and others. And, where is the will in this process—is it simply the "arrogant desire" for individual life? Though Beckett's lifelong interest in Arthur Schopenhauer is well documented,[48] I would suggest that Beckett's idea of the will in nature leans more toward Lamarckism precisely because it is continually invoked and framed within a specifically natural context concerned with individuals in relation to an environment that is blind to their needs.

An important indirect route to Darwin was through Fritz Mauthner, on whose work Beckett took extensive notes. Mauthner's admiration for Darwin "shows even in his writing style," notes Dirk Van Hulle.[49] He was taken with Darwin's idea of contingency, arguing that "the evolution of our senses is just the work of chance, which implies that the 'laws' of nature have the status of a law only because human beings' senses have developed in more or less the same way." By the same token, history is equally contingent and nonteleological.[50] Most significantly, Mauthner condemns the social Darwinists such as Spencer and (in his view) Ernst Haeckel, whose distortion of the term *evolution* has reinstated exactly the teleological element that Darwin rejects, and insists that evolution can be rescued only if it is liberated from the two main ideas of these so-called Darwinians: purpose and progress.[51] Beckett's interest in Mauthner thus further aligns him with Darwinian evolution.

Another possible reference to evolution, and another, if tenuous, link with Darwin, can be found in Beckett's poem "Serena I," in the line "Our world dead fish adrift" (rendered in Beckett's notes as "Earth afloat [dead fish]"). Peter Fifield interprets this image not as a fish floating on top of water but as an early geological state (glacier) that left seashells embedded in mountain rocks. Beckett drew on the discussion of the Greek philosopher Thales in the first volume of Friedrich Ueberweg's *History of Philosophy*, in which Ueberweg observed, "Aristotle reports that Thales represented the earth as floating on the water. It is possible that geognostic observations (as of sea-shells in mountains) also lay at the bottom of Thales' doctrine." This is Beckett's first direct incorporation of his philosophy notes into his work.[52] Fifield reads Beckett's dead fish as actually a seashell, "unmistakably the remains of marine crustaceans found in the mountains due to tectonic movements."[53]

In 1838, Darwin made one of the great mistakes of his career when he tried to solve the puzzle of Glen Roy, which had mystified so many geologists at the time. The glen was miles long and had three parallel paths running along its sides, which were really 60-foot-wide slanted shelves tilted at twenty degrees. Standing in the middle of one of these shelves, Darwin saw that it was "the same for the other terraces, 200 feet below and 100 feet above him. These 'parallel roads' went right around Glen Roy as far as the eye could see, while behind him, twelve miles away, Ben Nevis, Britain's tallest peak, completed the panorama." Had there been, as many posited, an ancient lake there that had dropped three times, "each time leaving a shore-line cut into the mountain side?" If so, perhaps the valley "had once been dammed to hold the lake water." But Darwin had a theory about the earth's crust as "oscillating," based in part on what he had observed on his voyage on the *Beagle*. "He had seen terracing in Chile; there the 'roads' were littered in shells—they were obviously old beaches," so he assumed this was the case with the roads in Glen Roy, even though they lacked shells or barnacles. "Proving the parallel roads to be sea margins would in turn confirm his global geological theory" based on floating land masses.[54]

One of Darwin's early ambitions had been to "create a 'simple' geology based on the Earth's crust."[55] This desire crowded out the lingering doubts and questions; Darwin published his theory about Glen Roy, and it increasingly came under fire from colleagues like Louis Agassiz and David Milne, who asserted that it was an ancient lake formed by glacial meltwater, and not the sea, that had caused those roads. "But Darwin clung to his seashores, holding tight because of his *idée fixe* with bobbing landmasses."[56] I am not suggesting that Beckett knew of Darwin's Glen Roy mistake and deliberately alluded to it. The point is that both in their very different domains seem taken with the concept of "earth afloat" in a specific and simple geological vision, a "global geological theory." Beckett took detailed notes on geology. He made a chart of the key eras from the Precambrian to the present day, and he jotted the emphatic phrase "the *geology* of conscience—Cambrian experience, Cainozoic judgments" in the "Whoroscope" notebook.[57] He included many references to geology in his plays. Yet in our haste to read this as "a metaphor for the prehistorical landscape of consciousness . . . suggesting the strata of guilt and repression," we rush past the fact of the interest in geology in the first place, for its own sake.[58]

In addition to a shared inclination to conceptualizing geological and temporal vastness,[59] two further affinities between Beckett and Darwin are particularly worth noting. One is their love of popular forms of performance: Beckett famously draws on Charlie Chaplin's and Buster Keaton's silent film work, Darwin preferred entertaining theatrical fare such as the equestrian melodrama he took his young cousins to see. Along with her observation about the link between evolution and acting, Goodall notes that "the experimentalism of minstrel entertainers, actors, acrobats and dancers often mirrors that of natural scientists in its exploration of the limits and modalities of the human body, but it tends to be a burlesque mirroring."[60] These shared lowbrow tastes thus relate to important epistemological processes, experiment, and burlesque, each enabling distinct kinds of knowledge. The esoteric and intellectual Beckett, the arcane, inaccessible, and deeply learned side, can overshadow his immersion in the popular culture of his time—vaudeville, cinema, music hall—and the pleasure he took in the mundane and quotidian aspects of life, especially its bodily processes. Even his use of the figure of the tramp connects him to evolutionary theory by the unexpected route of eugenics, a topic I return to in considering *Eleuthéria*, Beckett's first full-length theatrical work. Tramps, as William Greenslade notes, potently signified humanity's refuse, the undesirables whom eugenicists wanted expunged in the early decades of the twentieth century.[61]

Finally, on a biographical note, both Beckett and Darwin suffered from chronic physical conditions, the causes of which have never been concretely established. This link is so striking that Goodall even suggests a Beckettian reading of Darwin, "a doubt-ridden and generally troubled occupant of his own mental world" who genuinely doubted the foundations of his own argument and suffered physically from it in ways that parallel Beckett's experience.[62] Darwin's chronic nausea, retching, and vomiting drove him to try water treatments and all kinds of other remedies, without much success, and kept him a recluse. People have posited a range of causes, from a mysterious, tropical bug picked up on his voyage on the *Beagle* to psychosomatic illness caused by the social and theological implications of his work. Beckett had a succession of similarly crippling gastroenterological problems, from vomiting, diarrhea, and excessive wind to boils under his testicles and in his anus, all described in meticulous detail in his journals.[63] These problems arose during his trip to Germany in 1936 as he witnessed firsthand the effect of the Nazis on Jews, artists, and students. "Beckett's body starts to mimic the disorder he perceives around him," writes Andrew Gibson, but it is more than a psychosomatic condition; "his symptoms seem more like manifestations of an almost terrifying susceptibility to the world around him."[64] This visceral response to the world around him indicates a heightened awareness of the physical processes of nature and their impact on humans that both Beckett and Darwin so deeply felt.

Non-Darwinians

A key point about Darwinian evolution is that the basic condition of life is variation, not stability. It was revolutionary to suggest that the natural state, which seems to be defined by its equilibrium and balance, is actually in constant flux and that species always vary rather than remain fixed, or they would not survive. We can see this in Beckett, who often depicts the same circumstances or environment with slight variations and looks at how characters respond or adapt. In fact, what we often see in Beckett's plays is a single moment in the "endless repetition of the same, simple physicochemical laws" that govern life, rather than its "cumulative effect."[65]

The discourse on Beckett's plays has arguably made too much of their "cyclical" patterns, assuming that *Waiting for Godot* and *Endgame*, for instance, simply recapitulate the main action. This cliché has stuck stubbornly to Beckett studies and prevents us from considering alternatives. There is certainly truth in it: not only the full-length plays but also the shorts, like *Play* and *Come and Go*, do emphasize the cyclical format, with its repetition-with-variation structure. The plays are not strictly repetitive or cyclical, though, because they do not come back to the exact place where they began. They progress, usually showing small changes leading to fundamental differences rather than simply repeating the exact same words and actions. In this way, the plays enact the minute changes that constitute biological variation: the tree in *Godot* changes (in one act there are leaves; in another act there are none); Krapp's recordings are iterations of his evolving self; Winnie gradually becomes subsumed by the rising earth. This is all part of both the action and the language of evolution, which emphasizes the cycle of life and death, transformation and transmutation, and infinitesimal changes leading to bigger shifts—and, ultimately, extinction for some.

The now-famous pronouncement that in *Godot* "nothing happens—twice" thus fails to account for the variation within such repetition. A more apt analogy lies in Haeckel's recapitulation theory, a motif that intrigued Beckett, who was well aware of Haeckel; he mentions him in *How It Is* and makes the recapitulation motif central to *Watt*.[66] Indeed, the image of Darwin's caterpillar is often taken as a description of the structure of the novel.[67] In *On the Origin of Species*, Darwin observes how remarkable it is that people and animals are such creatures of habit that when interrupted in a song, for instance, or in repeating anything from rote memory, a person

> is generally forced to go back to recover the habitual train of thought: so P. Huber found it was with a caterpillar, which makes a very complicated hammock; for if he took a caterpillar which had completed its hammock up to, say, the sixth stage of construction, and put it into a hammock completed up only to the third stage, the caterpillar simply re-performed the fourth, fifth, and sixth stages of construction. If, however, a caterpillar were taken out of a hammock made up, for instance, to the third stage, and were put into one finished up to the sixth stage, so that much of its work was already done for it, far from feeling the benefit of this, *it was much embarrassed*, and, in order to complete its hammock, seemed forced to start from the third stage, where it had left off, and thus tried to complete the already finished work.[68]

Note this anthropomorphized sense of the caterpillar experiencing embarrassment; this is typical of Darwin's writing, particularly in *The Descent of Man*, yet something that Beckett avoided. Where they do connect is in the mutual interest in the force of habit as a natural phenomenon, strongly embedded in our evolution. Habit is embodied recapitulation.

Beausang argues that the caterpillar's circular behavior is a metaphor for the way in which Western thinkers become prisoners of their own logical constructions. He suggests that Beckett saw in Darwin's caterpillar a kind of narcissism, the possibility of "metamorphosis against nature."[69] The motif also signals an equality between before and after: origins and terminations cannot be identified because they are a continuum. This principle of recurrence then shapes Beckett's interest in the conflation of birth and death.[70]

With their intense focus on the end of life, it is sometimes easy to miss the emphasis in so many of Beckett's plays on begetting, origins, and transmission from one generation to the next, which counterbalances the motifs of decay, death, and annihilation. "Progenitor" is a frequent term in Darwin's works, and it recurs in Beckett. "Accursed progenitor" in *Endgame* echoes the sense of a single originator of life in chapter 9 of *Origin*: a "group of forms" all descended from "some one progenitor." Embryology is key in Darwin's thinking. Beckett also invokes the fetus—whether born, stillborn, aborted, miscarried—and the womb, often in strange formulations.[71] What does Winnie mean when she refers to "the womb, where life used to begin"? Does it not begin there anymore? If not, where does it begin? This is almost as odd as Mrs. Rooney's comment in *All That Fall*: "The trouble with her was she had never really been born!"[72] Mouth in *Not I*'s refusal to acknowledge her abortion is similarly rendered in a linguistic construction, in a simple shift of pronoun. Already in 1930 in his first published work, *Whoroscope*, Beckett alluded to Descartes's ideas about the formation of the fetus as relayed by John P. Mahaffy's book on the philosopher. Like other animals, humans originate in "the fermentation produced in generation, which causes heat and expansion, so forming the heart, and next producing a motion of the subtler matter there found towards the locus which becomes the base of the brain, with a consequent return of the grosser matter into the places thus vacated."[73] Beckett had a conviction that he remembered being in the womb—a "uterine reminiscence"—that completely goes against scientific evidence.[74] He was also fascinated by the idea that consciousness begins with conception.

All That Fall centers on reproduction and childlessness: procreation, conception, and the prevention of children through various means, from hysterectomy (Mr. Tyler's daughter) and menopause (Mrs. Rooney imagines that her lost daughter, Minnie, would be in her forties or perhaps fifty, "getting ready for the change") to abortion, miscarriage, and death either accidental or intentional. Mr. Rooney asks (presciently), "Did you ever wish to kill a child? Nip some young doom in the bud"—and this links to the many references to miscarriage and abortion, to conception gone wrong, in Beckett's other plays, such as *Not I* and *Rough for Theatre II* ("five or six miscarriages").[75] The dialogue of the radio play *All that Fall* is set against a soundscape of, on the one hand, animal noises evoking nature and all its irrepressible urges and, on the other, the sounds of vehicles representing, in Katharine Worth's view, "a whole history of human transport," from the cart full of manure, to the bicycle Mr. Tyler rides, to the car with its new tires driven by Mr. Slocum, and of course to the train that Mrs. Rooney's husband is on.[76] This representation of technological development parallels human evolution, which in Mrs. Rooney's view seems to be teleological: as she says, we are all going in the same direction.[77]

As noted throughout this book, one of the challenges for playwrights engaging with evolution has always been how to represent vast eons of time and imperceptible change in the brief two-hour traffic of an evening's theatrical entertainment. Beckett confronts this head on in several ways. One is his sense of time's elasticity, signified by the tree's overnight sprouting of four or five leaves in *Waiting for Godot* (at which the men marvel) and in the play's metatheatrical acceleration of the diurnal cycle ("In a moment it is night. The moon rises at back, mounts in the sky, stands still," and then, "The sun sets, the moon rises. As in Act One").[78] Another is Beckett's compression of birth and death, the sense that "the gravedigger puts on the forceps."[79] In his study *Proust* (1930), Beckett notably does *not* conflate them: "By no expedient of macabre transubstantiation can the grave-sheets serve as swaddling clothes." But by the time of *Waiting for Godot*, humankind is giving birth astride a grave, and this motif permeates his subsequent drama. *Krapp's Last Tape* refers to a nurse pushing a black pram, a "most funereal thing."[80] And what is life anyway? What does it all amount to? "The same old moans and groans from the cradle to the grave," says A in *Rough for Theatre I*.[81] This image is made literal in *Breath*, which represents a whole life span in seconds: vagic cry, inhale, exhale, silence. In the midst of this arc, the "miscellaneous rubbish" strewn on the stage is significant. It may well be taken as a pessimistic dismissal of what lies between birth and death (life is just so much rubbish). But it may also symbolize not detritus but accumulated experience; not what human life produces, but what the self encounters and accrues in its engagement and struggle with its environment. There is also the sense of waste implied in Thomas Malthus, whose argument that the reproductive excess of nature is necessary for the survival of species deeply influenced Darwin's thinking, offsetting death, famine, starvation, and other depleting mechanisms. Yet between these extremes an equilibrium does prevail, and perhaps this also is true of Beckett's theatrical "world" where there is no such waste, in the Malthusian sense. His achievement, especially in the short plays, revolves around a paring down of the inessential: even "story itself is often occluded or subverted and so often secondary."[82]

Entropy and Adaptation

Evolution takes place on a vast scale of geological "deep" time and is subject to the laws of physics—entropy and the second law of thermodynamics—which ultimately project the death of the sun. One can see this paradox in Beckett's vision: We are evolving, but for what? Only to die out in millions of years? James Harrison, for example, writes that "the overall slide toward stasis and silence" in Beckett's plays "is almost a staged enactment of entropy."[83] There are indeed moments, as in *Rough for Theatre I*, that invoke stasis, entropy, and the cooling of the earth: "It seems to me sometimes the earth must have got stuck, one sunless day, in the heart of winter, in the grey of evening."[84] John Calder sees both *Godot* and *Endgame* as projecting a sense of "the great cooling and the end of all life."[85] *Happy Days* can be seen as entropic. Lucky's speech in *Godot* was originally written in a single block, as Cohn notes, "without the three-part division to which the author later called attention—indifferent deity, dwindling humanity, and stone-cold universe."[86] From this perspective, Lucky's speech gives Beckett's Darwinian view in a nutshell: there is no God; humanity is moving toward extinction; and the universe is succumbing to entropy.

Goodall offers a sharp and subtle reading of Lucky's speech in terms of entropy. She also briefly considers reading the speech in Darwinian terms, the idea that "he speaks here to bear desperate witness to something that has somehow been registered, 'beyond all doubt': that the shrinking and the dwindling happens. It happens before your eyes, in spite of the tennis and all those other activities any good Darwinian would confidently presume to be working toward the selection of improved human specimens."[87] Beckett's plays chart this entropic movement toward shrinking and dwindling, and the theme of extinction haunts them as it does all of humanity.

Extinction hovers over *Endgame, Krapp's Last Tape*, and *Happy Days*. It is also mentioned in the Nobel Prize speech for Beckett: "In the realms of annihilation, the writing of Samuel Beckett rises like a miserere from all mankind, its muffled minor key sounding liberation to the oppressed and comfort to those in need."[88] Such terms as "complete annihilation" and "a universal winter" abound in the writings of Darwin and his contemporaries, presaging a Beckettian view of the world as obliterated, entropied, with humankind at its last moments before extinction.[89] Total extinction is, in Roach's view, the most powerful intellectual contribution of Darwinian modernism to postmodernism.[90] This helps to explain why Beckett so often deals with extinction and why he gives it a humorous edge, so that his characters see themselves going extinct yet quip and dispense black humor. In fact, it may explain Beckettian humor and the "tragicomedy" with which he is associated, why he would call *Breath* "a farce in five acts."

All we have to cling to in the face of this inevitable final demise of life is the body and the natural environment that sustains it. Although the idea still reigns that Beckett is a dramatist of the end of nature, of our great alienation from our natural surroundings, recent work on Beckett and the body has gone some way toward complicating this approach. Herbert Blau says that Beckett depicts the "brutal material world," while Katharine Worth alludes to Beckett's "cosmic scenery."[91] Kathryn White focuses on the central notion of decay, whereby the exterior deterioration mirrors internal decay.[92] Ben-Zvi has explored how deeply interested Beckett is in the physical and how he grapples with the material world through the body's experience of it. An obsession with the body and an exploration of stylized alternatives to the natural body is surely one of the hallmarks of theatrical modernism, furthered by Edward Gordon Craig and the metatheatrical bents of Luigi Pirandello and Bertolt Brecht, among many others. But with Beckett, whom Joseph Roach calls "this most physical of playwrights," something new happens.[93]

After decades of plays that explored and debated single ideas or aspects of evolution, such as maternal instinct, natural selection, adaptation, mutation, and heredity, Beckett returns the theatre to thinking about humans as a species, our place within a natural order, mechanisms of species evolution and whether it occurs on the group or individual level, and the interconnections between us, the animal kingdom, and the environment. *Waiting for Godot* invokes "the natural order."[94] The nonappearance of the presumed "savior" Godot would seem to indicate that such a being lies outside the natural order, that waiting for some kind of external salvation is a waste of time and energy. "Species" occurs frequently in Beckett's plays, as in *Rough for Theatre I* (B wants to "die reconciled, with my species") and *Rough for Theatre II* (the subject wished for "the extermination of the species").[95] In *Waiting for Godot*, when Pozzo first sees Didi and Gogo, he pronounces them "of the same species as myself. [*He bursts into an enormous laugh.*]

Of the same species as Pozzo!"[96] A little further, this seems questionable as Didi and Gogo are horrified at Pozzo's inhuman treatment of Lucky: "To treat a man . . . like that . . . I think that . . . no . . . a human being . . . no . . . it's a scandal!" Pozzo admits a few lines later: "I am perhaps not particularly human," which Estragon's later comment, "He's all humanity," flatly contradicts.[97] It is significant that only when Pozzo has become abject, blind, and dependent on others can he become "all humanity."

One of the central speeches of *Waiting for Godot* passionately foregrounds this question of what defines the human "species," with Shakespearean echoes:

> [VLADIMIR:] *Let us not waste time in idle discourse!* [Pause. Vehemently.] *Let us do something, while we have the chance! It is not every day that we are needed. Not indeed that we personally are needed. Others would meet the case equally well, if not better. To all mankind they were addressed, those cries for help still ringing in our ears! But at this place, at this moment of time, all mankind is us, whether we like it or not. Let us make the most of it, before it is too late! Let us represent worthily for once the foul brood to which a cruel fate consigned us! . . . It is true that when with folded arms we weigh the pros and cons we are no less a credit to our species. . . . But that is not the question. What are we doing here,* that *is the question.*[98]

Fundamental to our species is cooperation: not only the capacity to help one another but also the compulsion to do this. This has been the subject of intense debate and experiment across many scientific fields (in addition to its much longer presence within philosophical discourse). Beckett's writing for the theatre, beginning already with *Eleutheria*, coincides with this developing discourse on altruism and cooperation as evolutionary mechanisms. "Christ what a planet!" Mrs. Rooney "explodes" violently in *All That Fall*, triggered by Miss Fitt ("misfit," echoing the eugenic concerns about the unfit), who has not been quick enough to proffer "your arm! Any arm! A helping hand! For five seconds!"[99] It is possible to read *Waiting for Godot* as a sustained meditation on this theme, intensifying in Didi and Gogo's second encounter with Pozzo and Lucky, which prompts a series of reflections about the extent to which they are "friends" through "helping" one another, an idea on which the central speech quoted previously so eloquently elaborates.[100]

The perpetuation of "the species the human" relies on procreation, which centers on the female body. Beckett marks a distinct shift from previous drama not only in his unflinching focus on ailing and decrepit bodies but also more specifically in his sustained exploration *without hostility* of the female body and its reproductive and sexual functions, whether in the compassionate auditor of *Not I*; the "vagic cry" of *Breath*; the continuous talk of birth and "forceps," childlessness, and barrenness; the powerful presence of women and their physicality. This is all done in a context of a fully Darwinian world that is being embraced and explored in Beckett's works. The catch is that sex is responsible for life as well as for the inevitable suffering this entails, which leads Beckett to juxtapose, in plays like *Happy Days*, references to sex and sexuality with sterility or "discreation."[101] For example, Willie is, of course, a child's word for "penis"; thus, procreation is underscored even while it is shown to be pointless, as Winnie (who is of menopausal age anyway) can hardly have sex while encased in earth from the neck down. Although she is delighted to see new life as symbolized by an ant carrying its egg, Winnie says it is a "blessing" that "nothing grows. . . . Imagine if all this stuff were to start growing." Likewise, in *Endgame* the idea that the human species might be regenerated by the flea in Clov's pants causes panic ("Catch him, for the love of God!"); it does indeed seem to be a "blessing" that things are winding down. This moment in *Endgame* is also significant because, as Ulrika Maude points out, Beckett "collapses the neat categorical distinction between the lowliest insect imaginable, a parasitic flea, and the highest mammal," thus questioning "the fault line between the animal and the human."[102]

Beckett conveys the blindness, randomness, and chance of evolution; the lack of teleology and will: and the fact of death and extinction as just that—facts, not tragedies. Change, chance, and lack of human agency are the drivers of Beckett's evolutionary vision: "Just chance, I take it, happy chance," in *Happy Days*.[103] Both chance and change are central to the seemingly static situation Winnie finds herself in: "For however unchanging and apparently endless Winnie's existence might appear to be, change *is* present in the shape of decline, degeneration and deceleration. For Winnie and Willie are not so much the last survivors of a giant holocaust, such as a nuclear explosion, or of a sudden and dreadful natural disaster, as the victims of a slow process of running down: dentifrice, universal 'pick-me-up,' lipstick and Willie's vaseline are all depleted and in danger of running out."[104] This entropic process has even threatened natural laws. "Is gravity what it was, Willie, I fancy not," says Winnie, wondering if one day the earth will simply crack open and she will float upward.[105] "Ah well," she says, "natural laws, natural laws, I suppose it's like everything else, it all depends upon the creature you happen to be."[106] *What* depends? Since when are natural "laws" relative? How could they be "laws" if they are not absolute?

Happy Days contains one of Beckett's most direct evolutionary statements: "That is what I find so wonderful. [*pause*] The way man adapts himself. [*pause*] To changing conditions."[107] Yet, like so many of Beckett's characters, Winnie simultaneously recognizes the marvel and necessity of adaptation as the key to survival and its tragic elusiveness in the face of a world gone to environmental extremes: the scorching heat that causes her parasol to explode and is literally baking her, the sense that very few humans are left on earth (references to a couple having passed this way, possibly the "last human kind"), her frequent references to past weather conditions, and her alertness to the state of nature (unusually high temperature, level of earth) all give the play a new relevance to climate change. She realizes what will no doubt happen to her: "Shall I myself not melt perhaps in the end, or burn, oh I do not mean necessarily burst into flames, no, just little by little be charred to a black cinder, all this . . . visible flesh."[108] The question is rhetorical; whether she burns or melts or is swallowed alive by the rising mound of earth, she will die soon. Or, she imagines, she might die of extreme cold; but, whichever it is, it will be some extreme of climate, and it will be horrible. The play is a meditation on our relationship to nature generally and to our specific environment, and while humans' ability to change and adapt to circumstances is lauded, it is also futile.

For Winnie, the earth/her surroundings are both familiar and "strange." She used to perspire a lot; now, even in this extreme heat, she hardly sweats at all. How can this be? Another natural law is being broken. The contrast between this lively, persevering character and her probably hopeless condition is heartbreaking. Yet her exclamation "Ah earth you old extinguisher" is closer to affection than bitterness.[109] By the second act, she is asking, "Do you think the earth has lost its atmosphere?"[110] And the aforementioned reference to "the womb, where life used to begin" likewise seems to indicate some fundamental shift in the natural order.

Monkey Men

Happy Days manages to encompass some of the major tendencies of thought within the field of evolutionary theory itself, from the Victorians to Beckett's own times. The play both visually and textually telescopes human evolution by juxtaposing references to our evolutionary past directly with allusions to our highly evolved, refined current state and by juxtaposing the "hairy" and silent (except for occasional grunts and monosyllables) male Willie with the primping and chatty female Winnie. Willie may represent our primitive past, the ape-like attributes made all the more pronounced when he appears incongruously wearing tails and crawling on the ground, evoking chimps put on show in circuses: "He is on all fours, dressed to kill—top hat, morning coat, striped trousers, etc., white gloves in hand. Very long bushy white Battle of Britain moustache. He halts, gazes front, smooths [*sic*] moustache. . . . He advances on all fours towards centre"[111] and proceeds to mime in a way suggestive of a circus chimp.

It is worth noting here a likely connection with Franz Kafka, another biocentric writer fascinated by the fluid line between animal and human and an author in whom Beckett had a lifelong interest. Willie's ape-like guise evokes Kafka's story "A Report to an Academy" (1917), which features a chimp named Red Peter giving a talk to an academic audience about how he achieved his accelerated evolutionary ascent.[112] In a kind of dramatic monologue, Red Peter explains how he trained himself to become a successful vaudeville performer; he knows how to win an audience, and he foregrounds the performative nature of identity. He is, as Margot Norris puts it, acting as "the successfully adaptive Darwinian organism."[113] Kafka's monkey, like Beckett's characters, sees music hall and vaudeville as his survival strategy; he must learn to perform to survive. We are back to that fundamental link between evolution and "low" forms of theatrical performance.[114]

Red Peter and Willie stand at opposite ends of the evolutionary spectrum: the chimp made human and the human made chimp, each realizations of "metamorphosis against nature." Willie represents human regression, or evolution reversed; he hardly speaks, he is living in a hole, he crawls on all fours, and he even, at one point, crawls backward. Winnie tries to guide him back into his hole, warning him, "Keep your tail down, can't you!"[115] Toward the end of the play, he

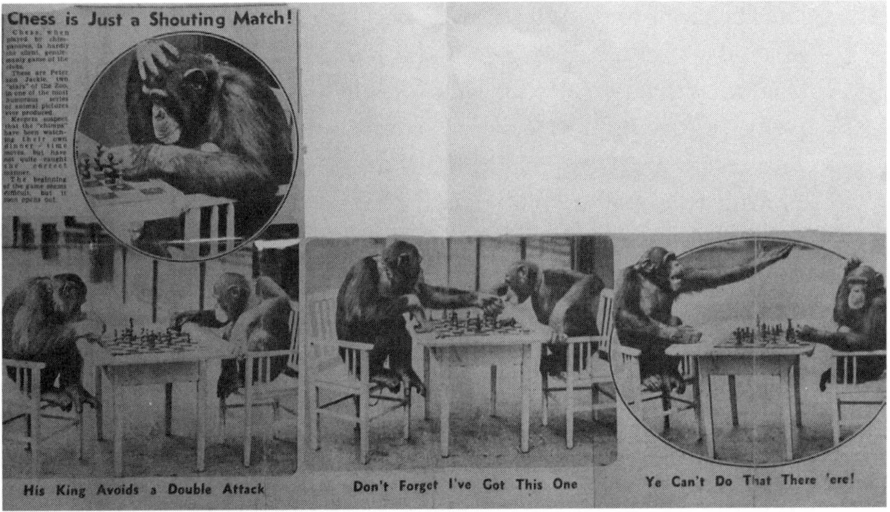

Figure 5 Chimps playing chess, from an unidentified newspaper clipping in Samuel Beckett's "Whoroscope" notebook. (Courtesy of Beckett Collection, University of Reading)

slithers down the mound. Regression, or reversion, was of course an early fear prompted by evolution—the idea that man might degenerate to an earlier state and become an ape again.

Apes clearly fascinated Beckett. In the "Whoroscope" notebook, there is a cutting from a newspaper in 1936 showing four photographs of two chimps playing chess (figure 5). One is a close-up of one of the chimps scratching his head and looking like he is really thinking. The three others are in a sequence showing a chess game in progress. The first shows them each contemplating the board, the next shows them each making a move, and the last shows the chimp on the left comically extending his arm in outrage while the one on the right scratches his head and looks a bit like Laurel when Hardy gets angry at him. This is the only newspaper clipping in the notebook, and it is further evidence not only of Beckett's fascination with ape–human proximity but also of his particular interest in the comical side of that relationship.

Apes and regression also crop up in *Act Without Words I* (1956). Interested in gestalt psychology, Beckett in his "interwar notes" had referred to Wolfgang Köhler's study of apes on Tenerife (published 1917) and their behavior as evidence that, as Beckett concluded, "the brain works in large patterns, closing gaps." This has had a range of analyses. Feldman notes that twenty years later, "Köhler's study acts as the setting and events in Act Without Words I," contrasting Köhler's apes as "can-ers" with regard to "conceptual learning, and Beckett's mime as a 'non-can-er,' with relief ever beyond reach."[116] Ulrika Maude provides a compelling case for seeing *Act Without Words I* as a critique of gestalt theory. "Beckett subjects the human animal to a set of experiments," echoing Köhler's with the apes on Tenerife but challenging his conclusion that "animals will learn if they are given the full picture." Neither Köhler's apes in their laboratory environment nor Beckett's characters and audience (including readers) are "in a position in which a full Gestalt—a full picture could emerge." Beckett takes this idea further in his later writing, which is "notoriously and deliberately devoid of contextual definition."[117] Seeing Beckett's work in the light of such evolutionary discourse is more suggestive and illuminating than the traditional recourse to seeing his "universe" as other worldly, meaningless, and static.

The allusion to apes in *Happy Days* also stimulates reflection on what it means to be human: "'Tis only human. [*pause*] Human nature. . . . Human weakness. . . . Natural weakness."[118] What is "only human," "natural weakness"—is it to do with her relationship with Willie? Why, for instance, does he not dig her out of her mound? And why is there a gun lying nearby? While she delights in the unexpected appearance of the ant carrying its egg, signifying "life of some kind," she herself is powerless, however much she wants to survive. There is no trace of Lamarckian will directing the course of evolution. If anything, it is the environment that seems to exert a will of its own, to possess some agency. Whether the mound is rising up to swallow her gradually or she is sinking more deeply into it, the outcome seems hopeless as no rescuer appears and there seems to be no way out. The bleakness of this vision of life on earth is enhanced by Winnie's memories, "giving oblique expression to the fear, violence, and suffering that, in Beckett's view, seems to be an unavoidable accompaniment to procreation, birth, and being."[119]

Beckett's world is Darwinian and not Spencerian or Lamarckian—a significant departure from the drama and performance I have been charting throughout the foregoing chapters. Beckett is the post–New Synthesis playwright par excellence. Gone is Lamarck: Beckett's theatrical universe has no room for the inheritance of acquired characters, the notion of progress, the idea that the will could ever enter into species transmutation. *Act Without Words I*, for example, is a microscopic exploration of how impossible it is for the will to have anything to do with human evolution, for the man is entirely at the mercy of his environment—and the tables have been turned, for it is now the *environment* that seems to have a will, not the person. Seeing the play in this evolutionary light contextualizes it and seems more productive than the self-limiting reading of it as signifying the "meaningless" human condition.[120] So Darwin is inverting the Lamarckian idea, showing how futile the will is and that humans must constantly struggle to adapt to the changing environment or perish.

This affects possible interpretations of the ending of *Happy Days*. Winnie does all she can to adapt to alarming, and rapidly worsening, environmental change. Depicting as it does the relationship between humans and their dynamic environment, *Happy Days* takes on new meaning in the context of contemporary concerns about global warming. The earth is suddenly and catastrophically hot, and we do not know what has happened to the rest of its inhabitants. The world around her is still recognizably the "earthball," though it is entropic: "It has almost dried up in the fierce heat of the sun and, by the second act, might conceivably have lost its atmosphere."[121] The worsening situation in which she finds herself seems to point toward death, though nothing is certain—there is no third act. Perhaps the mound of earth that gradually swallows Winnie is neither a malignant force (like that of *Act Without Words I*) nor a benign one, but simply a fact of life. It could thus be something to be reckoned with and altered, rather than inevitably killing her. This makes Winnie's cheerful insistence on everyday rituals, and her reliance on the objects of normal life despite her increasingly desperate situation, not only endearing and admirable but also necessary. Her resilient refusal to give in to despair resonates with environmentalist debates over the earth's resilience in the face of.

Paley 's Watch and Darwin's Finches

In *Happy Days*, perhaps more strongly than in his other plays, Beckett resists the idea of a creator. No one is watching or indeed watching over Winnie—except the audience. For Beckett, questions of origin, creation, and teleology are inherent to the theatre: the very act of creating and staging a play mandates not only an invisible designer but also live witnesses to that act, all pretending that there is no creator. Although his plays sometimes contain metatheatrical moments, it is striking that he remains largely uninterested in Brechtian techniques that shatter the fourth-wall illusion, such as on-stage narrators, placards to announce scenes, and an episodic structure. Such techniques would give the feeling of an offstage creator who knows more than the audience, akin to an omniscient god, whereas Beckett's plays (with striking exceptions like *Act Without Words I*) posit no such being; as he insisted in the introduction to *Waiting for Godot*, "All I knew I showed." Indeed, *Act Without Words I* suggests how sinister such a force can quickly become.

Likewise, the fundamental differences between prose fiction and theatre may explain why Beckett's explorations of evolution are so different in both forms: saltationist in his fiction but hewing to a more Darwinian line in his theatre.[122] Beckett's novels feature sudden physical changes in the characters, whereas the plays show gradualism at work everywhere: we watch as Winnie gradually is buried; we wait with Didi and Gogo in surroundings that change minutely. The nature of how time works in the theatre and the coexistence of "real" time with "stage" time allows Beckett to experiment with a different vision of evolution from that of his novels. What is most striking about this is that it directly contradicts what one might expect: that the stage would lend itself more readily to saltationism than to gradualism, as Susan Glaspell's plays demonstrate. Beckett did a television play, *Quadrat*, which he then wanted to do again in a way that seems to indicate this gradualist evolutionary perspective: "'I think I have an idea—*Quadrat II*—the same action 1000 years later, monotone.'"[123]

A teleological approach to evolution posits a designer at work organizing chaotic nature into an ordered and progressive system. William Paley, the most famous exponent of the teleological argument, maintained that nature, like an intricate watch, could never have come into being without a creator (the watchmaker) behind it. Paley's watchmaker analogy is one of the emblematic tropes of the nineteenth-century debates about the transmutation of species.[124] There is a moment in *Rough for Theatre II* (written in French in the late 1950s, around the much-celebrated centenary of the publication of

Origin of Species) when Paley's watch is invoked obliquely to rebut the idea of design. "B" reads in the file that the subject once won in the lottery a "high class watch . . . solid gold, hallmark nineteen carats, marvel of accuracy, showing year, month, date, day, hour, minute and second, super chic, unbreakable hair spring, chrono escapement nineteen rubies, anti-shock, anti-magnetic, airtight, waterproof, stainless, self-winding, centre seconds hand, Swiss parts, de luxe lizard band," a description that highlights the craftsmanship, sophistication, and complexity of this marvelous instrument. Beckett's excessively detailed description of the watch (almost a parody of Joyce's parodic chapter of lists in *Ulysses*) seems pointless and extraneous—and that is precisely the point. The verbosity itself shows how useless this is. Even as it engages it, the play thus simultaneously invokes and destabilizes the watchmaker analogy.

But Beckett does not leave it at that. Not only is the question of a designer irrelevant, but also the watch was won by lottery—a chance win. "A" notes that the subject was "chancing his luck. I knew he had a spark left in him."[125] It then turns out that the chance was all the greater because the lottery ticket that won him the watch was a gift—a random act of kindness, an impulse of sympathy by someone who saw him sitting on a bench, "down and out." This combines chance with altruism; and it shows Beckett again foregrounding cooperation as an evolutionary mechanism. The idea is also signaled in *Rough for Theatre II* in its touching allusion to a pair of birds that seem mutually dependent, in stark contrast to the pair of human characters, who show utter disregard for human life as they assess the worth of a third character, C, and come to the conclusion that he should commit suicide. A and B peer through the cage at the two "lovebirds," and A identifies them as finches:

> *Look at that lovely little green rump! And the blue cap! And the white bars! And the gold breast!* [Didactic.] *Note moreover the characteristic warble, there can be no mistaking it.* [Pause.] *Oh you pretty little pet. . . . And to think all that is organic waste! All that splendour!*"[126]

This brief passage contains several interesting elements. The fact that they are finches[127] associates the birds with Darwin, who in his travels on the *Beagle* noticed that finches from different areas and islands had significant physiological divergences (i.e., in their beak structure and plumage) that could not be explained by existing theories. Darwin's finches became synonymous with his argument for natural selection as they demonstrated adaptation, or descent through modification. There is also the self-conscious assumption of a "didactic" voice, as if A is a scientist instructing onlookers about the ethology of finches, contrasted with the warm, affectionate voice of the animal lover ("Oh you pretty little pet"). Which voice is the more natural, which one conveys the true nature of human–animal connection? The final lament for the dead bird as "organic waste" may be read as an acknowledgment of the Malthusian excess of nature as well as the inevitability of death, for the female bird has already died; the male, barely alive, is warbling—uselessly—beside her.

Symbiosis and Antagonism

The two finches perhaps signify symbiosis, another key motif in his work and his evolutionary vision. Beckett once told an actor that *Godot* was all about symbiosis, and his plays often feature complementary or interdependent pairs of characters: Didi and Gogo and Lucky and Pozzo in *Godot*, *Endgame*'s Hamm and Clov and Nell and Nagg, Winnie and Willie in *Happy Days*, Krapp's current and older selves, Mouth and Auditor in *Not I*, *Rough for Theatre I*'s B and A, who (though they have only just met) must "join together, and live together, till death ensue" as "we were made for each other."[128] There is a clear debt here to the conventions of vaudeville and music hall and to silent film's comic teams—Laurel and Hardy, Abbot and Costello, the Marx Brothers. Given the innate hostility and resentment underpinning so many of Beckett's pairings, they also seem to suggest the inevitability of the master–slave paradigm in all human relationships (one of the pair must dominate). If his pairs represent both physical and psychological symbiosis, this would provide an interesting twist on Beckett's Darwinism by endorsing cooperation, not competition, for resources.

Beckett's pairs have an almost-biological need for each other, yet usually these pairs are same sex, therefore biologically sterile, and no help to species perpetuation. His plays show this biological underpinning and inspiration—not only this idea of symbiosis in his frequent use of interdependent pairs but also such aspects as his emphasis on anatomy and physiology, animal husbandry/domestication, breeding, and larger environmental processes; his preoccupation with the process of birth; and the idea of play/playful aggression and feigned hostility as an evolutionary "rehearsal" and therefore a strategy for survival in many species. For example, the symbiotic pairing in *Endgame* is echoed in *Théâtre I,* first published in French in *Minuit 8* (1974) but originally written in English in December 1956—in between the French and English versions of *Endgame*, which explains their close similarity. Character A is blind and seated on a folding stool playing a violin. Character B is crippled and in a wheelchair, getting about by means of a pole that he pushes. He seems to signify "a point prior to a state of interdependence such as that in which Hamm and Clov are found. In this case, the mutual needs rapidly become clear and the advantages of a symbiotic relationship seem obvious. . . . But when put into practice, however tentatively, the theory conspicuously fails to lead to any satisfactory state of companionship," as the characters alternate between irritation and altruism, almost randomly—and these seem to cancel each other out.[129]

Extinction

Beckett's plays seem repeatedly concerned with death not only of the individual but also of groups ("loss of species"). I would like to take a closer look at this in the context of the role of extinction in evolutionary theory that I discuss in other chapters in this book as a major preoccupation of dramatists and indicative of their engagement with evolution. Beckett's interest in early Greek philosophy provides valuable insight into his ideas about extinction. In the notes he took as he was reading Ueberweg's translation of the Greek philosophers, Beckett wrote: "definite individual existence constitutes an injustice & must be atoned for by extinction," suggesting a view that extinction "is scripted into existence as a just facet of individual being."[130]

Happy Days strongly suggests that we are witnessing—as Winnie says several times—"the last human kind." As I discussed in chapter 2, extinction is an interesting aspect of evolution in terms of how our view of it has changed over time. As Gillian Beer has pointed out, what we regard with alarm and horror Darwin viewed as a natural, everyday occurrence.[131] He scoffs at people's melodramatic reactions to the idea of extinction, but "because his thinking is extended backward over vast aeons, he is able to invoke the long processes of extinction without qualm. Time dilutes terror."[132] The idea of extinction has particularly interested dramatists, from Ibsen, Henry James (*Guy Domville*), and François de Curel (*Les Fossiles*) to Wilder (*The Skin of Our Teeth*) and Beckett right up to contemporary playwrights like Terry Johnson (*Cries from the Mammal House*) and Steve Waters (*The Contingency Plan*). They show extinction as a current event, happening in real theatrical time, putting back all the shock that Darwin took out. They can thus focus on the drama (and trauma) of "sudden" extinction, and in fact, they create this counter-evolutionary idea for the stage; in reality, dinosaurs apart, most extinctions happen over time, a slow death.

Repeatedly, Beckett's plays posit this idea of the "last human kind." Krapp comments that "the earth might be uninhabited," so still is the night around him. In *Endgame*, the world outside is "corpsed." *Endgame* is arguably more final than cyclical; even the title suggests there is no repetition here, this is the last set of moves in an extended game.[133] Three generations come to an end in the course of one evening's drama: the binned parents die or disappear; Hamm, their son, is blind and confined to a wheelchair and will die if left by his servant and possible son, Clov; Clov freezes in the midst of deciding whether to go or to stay, but he is dressed in his overcoat and on the way out. James Knowlson and John Pilling point out that in *Happy Days*, Winnie's world (here again the idea that her world is somehow different from ours) is full of light, but "the light is just as hellish as Krapp's darkness had come to be and evokes extinction just as unequivocally."[134] In fact, the dualism of light and dark in *Krapp's Last Tape* is paralleled in the "elemental contrast of earth and air" in *Happy Days*.[135] "Ah earth you old extinguisher," says Winnie. "The eternal dark . . . night without end."

In *Endgame*, Clov announces periodically the death of various things that Hamm mentions: there are no more bicycle wheels, pap, nature, sugarplums, tide, navigators, rugs, painkiller, coffins.[136] Soon, if Clov carries through his threat, there will be no more Clov either, and by implication no more humans at all. The characters seem not only to accept but also to be complicit in this sense of dwindling. "In an ending world," writes Cohn, "generation is an ever-present danger; the human race may evolve from the flea in Clov's crotch; a rat or a boy may germinate life. Plant life seems to be more acceptable, since Hamm and Clov hope forlornly for seeds to sprout. In sum, we hear about a moribund world outside the refuge."[137] Cohn refers to the play's "anti-creation theme."[138] I would suggest that this play in particular stages Beckett's sense that human beings are at odds with the earth. "You're on earth, there's no cure for that!" shouts Hamm. There is also no alternative planet. But for Beckett, the most important line in the play was "Nothing is funnier than unhappiness."[139]

There is an often comical mismatch between population and available sustenance that pushes the characters to the edge of existence, flirting with extinction. In *Godot*, the tree and its barren setting cannot give nourishment to the obviously hungry men; and a carrot, a turnip, and a radish will not last long. The carrot in all its sparseness appears also in *Act Without Words II*, a metonym for dwindling resources and likely starvation, just as the spindly tree is a metonym for denatured nature. Roach points out the pathetically small size of the carrot in *Godot*: "Beckett's personal selection of the properties for his own stagings of *Godot* is very suggestive in this context [of hunger and famine]: 'the carrot was usually pitifully small, another example of diminishing resources.'"[140]

These "diminishing resources" can be found throughout the play and are set against Pozzo's abundant chicken and wine picnic, suggesting a political comment on the unequal distribution of material goods. Vladimir's little song, "A dog came into the kitchen," "heightens the perception that deprivation, violence, and punishment are the normal expectation for those whose physical needs transgress against the prevailing maldistribution." It is as if "fighting over the last scrap were a daily ritual" just as the waiting for Godot.[141] Pozzo blames the blindness of nature, cruelly indifferent: "That's how it is on this bitch of an earth."

Death is a great leveler. But, far from taking meaning away from life or making living pointless, it reinforces the importance of quite the opposite idea—reveling in meaning and doing so through words. The paradox for Beckett is that language is at a crisis of meaning, yet it is the main thing that separates us from the animals; so, throughout his plays, people talk and talk in the face of possible extinction, as if afraid that if they give up on language they give up on life. There is a possible link here between evolutionary psychology and Beckett's suggestion of storytelling as serving some kind of biological function. We seem to be hardwired to make ourselves (in Beckett's words) "mean something," even in a seemingly meaningless void, and one way we do that is by telling stories about ourselves.[142] On the face of it, Beckett's characters confirm what the literary Darwinists argue: that stories serve an evolutionary role in helping us to survive. Hamm and Winnie are the supreme examples of this; Winnie "survives by talking her way through each day," and Hamm "hams" it up as self-conscious actor and metatheatrical commentator of events on stage (and the extended joke told by Nagg further exemplifies the need for such storytelling).[143] But at no point does Beckett evince the hierarchical view implicit in literary Darwinism. For all his characters' storytelling compulsion, they may well simply die out, subject to environmental pressures beyond the control of language. Animals, earth, the landscape, the elements, indeed time itself all threaten to overtake human existence, which is shown to be expiring.

Change

But while they do live, they can change, and change is the key to ongoing, sustainable, and even meaningful existence. Hamm celebrates the fact that "we breathe! we change! we lose our hair, our teeth, our ideals!" Such loss affirms life. He comments on the odd fact of "something dripping in my head"—a foreboding of death, perhaps, but equally of the earliest days of infancy, when the fontanelles are not yet closed. That fusing of the skull's plates is an important anatomical milestone signaling a transition from the most vulnerable state of the newborn to its next phase of development. Growth and change are synonymous, and they are "paramount" in Darwin's "portrait of nature" as well.[144] The key to an organism's survival is its ability to change.

Yet early in his playwriting career, Beckett shows an interest in human beings "in stasis," as he puts it in the preliminary note to *Eleuthéria*, his first play.[145] Here, nature is very different from what it becomes in *Godot* and *Happy Days*: it is something to be controlled, to manipulate. It is a reminder that just as Ibsen's engagement with evolution changed over time, there is a significant shift in Beckett's view of human evolution from his early to his late theatrical work. In this early work, Beckett engages with one form of evolutionary change that was not natural: eugenics. *Eleuthéria* revolves around procreation and breeding, and its language reflects "the contemporary discourse of race and racial purity" but falls short of embracing it.[146]

A key idea in *Eleuthéria* is to tamper with evolution by medical intervention, as the evil Dr. Piouk ("puke") explains:

> *I would ban reproduction. I would perfect the condom and other devices and bring them into general use. I would establish teams of abortionists, controlled by the State. I would apply the death penalty to every woman guilty of giving birth. I would drown all newborn babies. I would militate in favour of homosexuality and would myself set the example. And to speed things up, I would encourage recourse to euthanasia by all possible means, although I would not make it obligatory. Those are the broad outlines.*[147]

This eugenic program can be achieved by taking a pill. And it is all calculated to go against the natural processes of procreation and death by which nature replenishes itself and humankind goes on; Henri Krap at one point recalls "with some tenderness" his wife's attempted abortion of their son.[148] So Beckett in his earliest play (written 1947) is thinking about disrupting nature and evolution, in contrast to his later theatrical work which accepts and indeed celebrates the laws of nature as they are, although the central issues of *Eleuthéria*—the value of life, the morality of euthanasia—also find expression in *Embers*.

There is of course a strong satirical bent to Piouk's speech—it echoes Jonathan Swift's *Modest Proposal* and is hardly prescriptive.[149] A possible theatrical precedent, and one that is much closer both temporally and geographically, is Apollinaire's *Les Mamelles de Tiresias*, discussed in chapter 6. Both plays deal with birth control, eugenics, and reproduction in grotesque and cartoonish ways, yet few critics have seen a connection here.[150] As mentioned earlier, Mouth in *Not I* may also echo Apollinaire's play, which has a reporter whose "face is blank; he has only a mouth."[151]

Eleuthéria is the Greek word for "freedom," and the paradox—the reason why Beckett's characters again and again reject suicide—is that one can enjoy freedom only by being alive (i.e., having consciousness of the state of freedom)—consciousness being, many argue, not only the key difference between humans and animals but also the elusive "holy grail" of science. As tempting as it is to see *Godot* as the product of a more mature theatrical mind, chronologically at least this is simply not the case. Seeing *Eleutheéria* as Beckett's theatrical intervention in eugenic discourse gives it a new meaning and relevance rather than dismissing it as inferior to Beckett's other theatrical work. Despite his decades-long suppression of *Eleuthéria*, the play shows Beckett stretching the way the stage could explore sexual and reproductive issues—much as his American contemporary Edward Albee would do in dramas that include "false" pregnancy, childlessness, abortion, miscarriage, and even cross-species sex.

~

Beckett's later work (after the Nobel Prize) has often been seen in the light of compassion and empathy with human suffering. His plays do indeed show not only the need for these qualities but also the extraordinary capacity of human beings, against all the odds, to extend them toward one another. But this emphasis on humanity in Beckett can also help to explain why his evolutionary vision has been relatively little considered, his seeming dismissal of Darwin taken at face value. Beckett's empathy is "studiedly diverse" and extends well beyond the human, indicating an "abject fascination" with those lowliest of beasts, the flies, hens, and toads of the animal kingdom.[152] He depicts not so much the "entangled bank" but a tiny fragment of it, gesturing toward the interconnected life that is teeming somewhere off stage, as when the apparent isolation and desolation of Didi and Gogo are shattered by the sudden appearance of Pozzo and Lucky, laden with good food, signaling abundance elsewhere. Although Beckett embraces the inherent bleakness of Darwin's vision with its underlying competition, entanglement, ruthlessness, and randomness (rather than determinism), he also recognizes that such bleakness is not all there is to it. Winnie's attitude of resolute optimism in her dire circumstances chimes with one of the perplexing things about human beings that Darwin tried to figure out—what is the evolutionary purpose of some of our most seemingly useless yet utterly defining characteristics as a species, such as altruism, cheerfulness, humor? These also have a place in Beckett's drama and in his evolutionary vision.

Beckett's engagement with evolution is unrivaled in modern playwriting for its depth and range. He finds new dramatic subject matter in the age-old cycle of birth and death, extinction, adaptation, natural selection, and change without turning it into a social Darwinist "survival-of-the-fittest" paradigm. He attacks anthropomorphism, eschews "facile speculations about a creator whose signature might be inscribed" on beast and man alike, and fiercely questions the biblical teaching that "grants the human species dominion over the beasts of the field and the fowl of the air."[153] Rather than simply reflect evolutionary ideas, he transforms them into something uniquely his own and uniquely theatrical, as Ibsen had done decades earlier in his own encounter with evolution. They are separated, of course, by the Modern Synthesis.

Although it is wise to be wary of too great a fondness for paradigm shifts, epochal changes, and watersheds, both Darwin and Beckett in their respective fields ushered in new ways of thinking and, most important, doing. Theatre has never been the same since Beckett, biology likewise since Darwin, Wallace, and their colleagues. The epilogue to this book looks briefly at what happens to theatre and evolution after Beckett and what these developments suggest might happen next.

Epilogue

Staging the Anthropocene

Theatre continues to interact with evolutionary ideas, and it remains a valuable site in general for artistically and intellectually wrestling with scientific ideas through its intimacy, immediacy, and communality. It can engage the mind and senses in a way that is unique and that is becoming, if anything, more vital in a fragmented digital culture. In fact, recent plays that engage evolution are pretty bleak about where we have arrived as a species and where we are headed. "Why is it that everything that humans touch turns to shit?" asks a character in Richard Bean's climate-change play *The Heretic.*[1] Plays are collectively asking not whether evolution is valid but what all these eons of struggle have amounted to, certainly not Bernard Shaw's vision of metaphysical perfection at the end of *Back to Methuselah* or the sense of communal uplift and striving that triumphs over war-wrought devastation in Thornton Wilder's *Skin of Our Teeth*. Playwrights are depicting human evolution in varying degrees of crisis, regression, and stasis, with the added element of threat posed by climate change. It makes the earlier plays looked at in this book seem positively cheerful.

But is this really so new? Theatre has always unflinchingly shown the worst sides of humanity as well as the best: from *Oedipus Rex* through the medieval morality plays, from Jacobean tragedy through nineteenth-century naturalism, from the baby killing of *Saved* to the baby eating of *Blasted*. As Michael Billington notes in *State of the Nation*, theatre bears witness; it also articulates a consciousness. The consciousness theatre is articulating now about evolutionary ideas seems tacitly to be in agreement with the scientists who are proposing a new epoch: the Anthropocene, dating from the onset of the Industrial Revolution and marking a period when humans are having a direct, and largely adverse, impact on the lithosphere and atmosphere of the earth. This helps to explain that nostalgia for nineteenth-century meanings of evolution that I mentioned at the beginning of this book—and a concomitant retro-Victorianism in theatre's engagement with evolutionary thought.

Retro-Victorianism

An entire subgenre of plays re-creates the Victorian freak show, much as Arthur Wing Pinero did in *The Freaks* in 1918 (and e. e. cummings in an obscure play *Him*, 1946).[2] But this is no sudden revival of interest. The nineteenth-century fascination with freakery in relation to human evolution found renewed force in the mid-twentieth century, partly through the phenomenon of Ripley's Believe It or Not! franchise of cartoons, products, and live shows. Robert Ripley trafficked in "staged eccentricity, freak tourism, and eye-popping spectacle" that reached millions in America from 1930 onward. His Odditorium, on Broadway in New York, boasted "curriodities from 200 countries," such as a man with no stomach, "a man who drove spikes into his head, and a woman who swallowed a neon tube bolted to a rifle, which she then fired." A master of ceremonies would announce these incredible displays, and young women dressed as nurses (with "Ripley" embroidered onto their hats and jackets) would circulate, helping audience members who fainted.[3] Recent plays show a continued need to revisit this troubled legacy of human display and exploitation and its specific links to the stage; among them are Bernard Pomerance's *Elephant Man*, Suzan-Lori Parks's *Venus*, Mary Vingoe's *Living Curiosities; or, What You Will*, Shaun Prendergast's *True History of the Tragic Life and*

Triumphant Death of Julia Pastrana, the Ugliest Woman in the World, and Robin French's *Gilbert Is Dead.* In these new shows (as in Pinero's play), the freak is the object of empathy, and Otherness provides the moral center. Tony Kushner notes how daring it was for Parks to make the Hottentot Venus the heroine of a serious tragedy:

> *Venus* expresses both a global empathy, a mourning for all of suffering humanity, and at the same time an anger at oppression and oppressors. . . . The play places human paradoxes of love and loathing, attraction and revulsion, pleasure and denial in a historical context of racism, sexism, exploitation, voyeurism and colonialism. By contextualizing these paradoxes the play places the historical in dialogue with the eternal (if anything is eternal) . . . acknowledging the tragic, the immutable, while not extinguishing the possibility of mutation, of change.[4]

Similarly, Mary Vingoe makes a heroine out of a "freak" when she recreates Phineas T. Barnum's American Museum circus performances of 1863 in her play *Living Curiosities* (2010). Using the device of metatheatre (the "freaks" rehearse a Shakespeare play), Vingoe shows the redemptive power of theatre to bring dignity and hope to society's castoffs.

The retro-Victorian fascination with the circus and its suggestive stagings of evolutionary themes can also be seen in Cirque du Soleil's *Totem,* which toured internationally (2011/2012) and played at huge venues like the Royal Albert Hall. *Totem* invokes distinctively Victorian associations of the kind I discussed in the first two chapters: a love of spectacle and extravaganza, a seemingly insatiable curiosity about the natural world and displays of animals and nature scenes, an interest in origins and descent, and a preoccupation with the extremes of human capability, especially physical strength, stamina, and precision. Written and directed by Robert Lepage (channeling Barnum), this high-tech modern circus visually celebrates the rich diversity of nature and purports to be about the evolution of human life from the water to the air. *Totem* boasts a team of distinguished designers and theatre artists—everything except, apparently, a scientific adviser, despite a two-year development period. For a show that stakes so much on a single scientific concept, this is a strange omission, especially at a time when it is standard practice for theatre practitioners to bring scientists into the process not only to ensure accuracy but also to develop ideas.

Given its immense popular success, it is worth analyzing *Totem*'s use of evolution rather than simply dismissing it as erroneous because of what it reveals about the assumptions that still dominate our understanding of evolution in the modern world—a catastrophic world of global warming, climate change, and accelerated species extinction in which, one might argue, a deep understanding of evolution is essential for Earth's survival. The program explains that *Totem* is about "the odyssey of the human species," tracing "humankind's incredible journey" from amphibian to bird. It explores "our dreams and infinite potential, and the ties that bind us both to our collective animal origins and to the species that share the planet with us." But this circus foregrounds exceptional, not "normal," human beings; the artists we see represent the pinnacle of human achievement in terms of physique, balance, coordination, flexibility, and sheer hard work and endless practice. This is drummed home to the audience with every new feat it witnesses: a team of unicylists juggling golden bowls from their feet to their heads; two sisters whirling pieces of sequined fabric on their feet and hands while balancing precariously on chairs, with heads down and feet up; a roller-skating couple spinning around on an impossibly small, leather-covered, drum-like platform. Most acts revolve around this crucial combination of balance and teamwork. Interdependence, a very Darwinian concept, bumps up against a non-Darwinian reliance on absolute perfection and repetition that does not allow for variation. Each routine must follow the exact choreography laid out for it. The acrobats and other performers start with an already-difficult task and then repeat it with increasingly complicated variations so that by the end of their act the audience is astonished at a display of human ability mastered by only a handful of people on the planet. We see human beings at their most spectacular, performing acts mere mortals can only dream of, and suggesting even greater possibilities for the future (e.g., human flight) if we continue to evolve along these lines. We are right back in Shaw territory, evocative of the final lines of *Back to Methuselah*. If the performers deviate from the rehearsed choreography, however, with even a slight shift in weight or a missed cue, someone could be killed. If anything, what we take away from this show is humankind at its most robotic; these people resemble machines or cyborgs more than humans.

The show moves quickly and randomly from one act to the next, with no apparent logic regarding their sequence; most of the amazing acts are "complicatedly futile," and the entire cast is "crazily miscellaneous."[5] But the title of the show points to its hierarchical vision of evolution: life is one long column and we are at the top of it. The production claims to show "scenes from the story of evolution randomly linked together in a chain." *Totem*, the program goes on to state, "depicts a world of archetypal characters who, in their own way, witness and act out the perennial, existential questions of life. Alternating between primitive and modern myths, and peppered with aboriginal stories of creation, *Totem* echoes and explores the evolutionary process of species, our ongoing search for balance, and the curiosity that propels us ever further, faster, and higher." This implies agency and will in the process of evolution, as if we can direct its course, despite the assertion that the "story" of evolution is shown here "randomly."

The program notes situate the show "somewhere between science and legend." Much like its Victorian predecessors, this circus invokes the concept of evolution without showing any awareness of its scientific basis and implications, its historical context, or its complexity; over a century's worth of further knowledge about evolutionary mechanisms seems erased. The show simply tacks a vague idea of "evolution" onto its series of amazing acrobatic acts to provide a unifying theme and coherence. It purports to show how far we have come since we came out of the primordial slime: we now "break free from gravity." Intellectually, the entire thing is shaky and confused.

Why should this matter? Why should this show be chastised for its distortion of evolution when so many others in this book have not, even though they likewise confuse issues, are too reductive, or are misleading? The answer is that we have moved on so far, and have so much more understanding of how evolution works, that it is strange to (inadvertently) reflect such dated ideas and package them as new. In addition, given the rise of creationism and intelligent design over the past century, *Totem* is a missed opportunity to face down some of the ignorance about evolution. As Michael Billington points out in his review, the concept "that we are watching the evolution of species . . . doesn't make much sense. . . . The evolutionary theme is largely window dressing. . . . It is visually impressive without making logical sense."[6] Lyn Gardner simply dismisses it as a "mind-bogglingly daft show on the theme of evolution."[7]

Although Charles Darwin makes an appearance—an amiable white-bearded man shuffling around the stage performing experiments and occasionally walking hand in hand with his chimpanzee assistants—the real, yet unacknowledged, stars of evolution here are Jean-Baptiste Lamarck and Ernst Haeckel. The inheritance of acquired characters, the role of the will, progressivism, and recapitulation theory are emphasized not only visually but also in Lepage's program notes:

> The different stages in the development of a human being in many ways encapsulate the evolution of species. As fetuses, we float weightless like fish. As infants, our first attempts at self-propelled motion resemble an amphibian creeping along the ground. As we grow, we crawl like four-legged mammals, then climb like monkeys and finally walk assuredly on two feet in preparation to join the adult world.

Haeckel's biogenetic law was already questioned in the Victorian period, but it makes a handy visual metaphor for a theatrical entrepreneur needing a shorthand and instantaneous way of signifying the whole of evolution in a single moment or gesture.

The radical changes in the biological sciences wrought by modern genetics and molecular biochemistry and the rapid advances in the technological capabilities of science have, if anything, sharpened the epistemological nostalgia for such nineteenth-century meanings of "evolution." This is partly shown by the rising interest in the scientists who shaped it, a strong biographical inclination in large part generated by the steadily growing interest in Darwin from the mid-twentieth century onward, culminating in the Darwin celebrations of 2009. A growing number of plays depicting Darwin includes *Trumpery* (Peter Parnell), *Re: Design* (Craig Baxter), *Darwin in Malibu* (Crispin Whittell), *Darwin's Flood* (Snoo Wilson), and *After Darwin* (Timberlake Wertenbaker). But, for all this interest in Darwin, the most compelling contemporary theatrical engagements with evolutionary theory move away from biography and instead probe the tensions inherent in the evolutionary worldview that have never been resolved. Catherine Trieschmann's *How the World Began* shows the painful consequences of the ongoing clash between pro- and antievolutionists in America. Caryl Churchill (*A Number*) and Bryony Lavery (*Origin of the Species*) continue the exploration of gender questions by challenging the concepts that seemed central to evolution yet at the same time most problematic for women, such as maternal instinct, nurture, and biological determinism.

Genetics and Epigenetics

A key development is the interest in questions raised by genetics. Churchill probes the nature-versus-nurture debate in *A Number*, coming down firmly in the nurture camp, while *An Experiment with an Air-Pump* by Shelagh Stephenson dramatizes the ethical problems posed by genetic manipulation. More recently, *Ex Vivo/In Vitro* by Jean-François Peyret and Alain Prochiantz, the director–scientist team behind many other devised works relating to scientific ideas, shows how increasingly knotty the kinds of reproductive issues discussed previously in this book have become in light of reproductive technology. This piece also brings epigenetics explicitly into the theatre, another link with the Victorian scientific past as it gives renewed interest and credibility to Lamarckism.

August Weismann had rendered the inheritance of acquired characters untenable: as Francis Galton put it,

> As a general rule, with scarcely any exception that cannot be ascribed to other influences, such as bad nutrition or transmitted microbes, the injuries or habits of the parents are found to have no effect on the natural form or faculties of the child. Whether very small hereditary influences of the supposed kind, accumulating in the same direction for many generations, may not ultimately affect the qualities of the species, seems to be the only point now seriously in question.[8]

These words were prescient, as the field of epigenetics is suggesting that this is indeed highly likely to occur. Epigenetics suggests that it is not only through DNA that characteristics are transmitted, but also that genetic makeup can be affected by external factors. A well-known example is the finding about pregnant Dutch women during the famine winter late in World War II, which showed that those who were undernourished only in the first trimester went on to have infants who were normal in weight, whereas those women who were undernourished only in the last trimester had small babies.[9] There is ongoing debate about the significance of such findings. "Epigenetically acquired characteristics generally do not get inherited, and therefore do not have much significance for evolution," writes Jonathan Hodgkin, although there are occasional, striking exceptions. In fact, "much of epigenetics is standard fare in molecular biology, and scarcely revolutionary."[10] What does seem certain is that the inheritance of acquired characters can be seen as complementary to natural selection and it is enjoying renewed interest.[11] But why would epigenetics hold particular appeal for theatre makers? One of the vectors in theatre's engagement with evolution that is explicitly linked to the act of performance is the need to render observable what is either microscopic or too gradual to be seen. Perhaps epigenetics has caught the fancy of dramatists as well as audiences not because we are all closet Lamarckians but because it is so well suited to the power of theatre to render the microscopic and the gradual.

Relating modern advances in treatment by in vitro fertilization (IVF) to fundamental questions about identity, *Ex Vivo/In Vitro* updates the nature-versus-nurture debate (as we have seen, a theatrical mainstay ever since naturalism) and, without lecturing the audience, discusses pluripotent and totipotent stem cells and powerfully evokes Conrad H. Waddington's epigenetic landscape in references to "redescending new valleys" and "epigenetic reprogramming" (figure 6).[12] It also puts a modern twist on the theme of maternal instinct by musing on the relationship between carrier mother and the baby she is carrying that is not her own but that she is still biologically tied to in subtle ways.

The piece muses on what a person really is if he or she is conceived artificially; indeed, it goes to the crux of what conception really is if not done naturally. *Ex Vivo/In Vitro* explores the paradox that the most personal of all experiences, conceiving a child, can be so public. The opening line of the play asks: Is in vitro fertilization the triumph of biology over love as well as dissociating sex from procreation? This was, as we have seen, positively portrayed in Shaw's *Back to Methuselah* (women no longer have babies; instead, children hatch full grown from giant eggs), negatively in Neave's dystopian *Woman and Superwoman*. Maybe Shaw was not so far-fetched after all. Nor was Apollinaire: cloning is very much present in the intellectual terrain of *Ex Vivo/In Vitro*, and there are moments that strongly recall *Les Mamelles de Tireseas*, a male character, for example, talking about having dozens, indeed millions

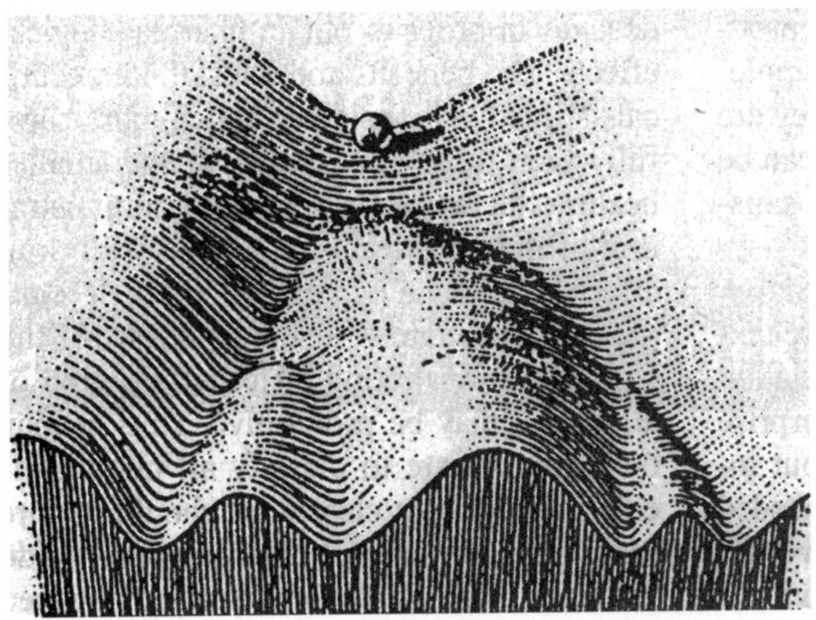

Figure 6 Conrad H. Waddington's epigenetic landscape, visualizing the totipotent stem cell's journey toward specialization and differentiation as it matures. (From Conrad H. Waddington, *The Strategy of the Genes: A Discussion of Some Aspects of Theoretical Biology* [London: Allen & Unwin, 1957])

and millions, of babies. This is not just textual but visual. *Ex Vivo/In Vitro* uses over a thousand "cordes de chanvre" (thick hemp ropes) hanging from the ceiling like a forest of umbilical cords through which the actors walk and intertwine themselves. These ropes also signify the increasingly complex "filiations" among us, our tangled bank of relationships never imagined possible before intracytoplasmic sperm injection (ICSI) and not yet fully understood.

Ex Vivo/In Vitro is only the latest of several theatrical engagements with artificial fertilization, following on from Carl Djerassi's *Immaculate Misconception* and Anna Furse's *Yerma's Eggs* (2001–2003), works representing vastly divergent approaches and conclusions. One of *Ex Vivo/In Vitro*'s questions is whether reproductive technology affects the human species collectively. ICSI is held up as a triumph over natural selection because it involves "a SINGLE spermatozoa instead of the usual 150 million! Technology takes the place of nature, of natural selection. Can you hear me Darwin? We did it!"[13] Eugenics, though not named as such, has renewed relevance now in terms of genetic counseling and medical genetics. It is not so much a question of individual identity (who are you?) as of the group (what are you?). Alarmist concerns over the results of artificial reproduction are bringing us full circle back to the days of the freak show and the "missing link." Philip Ball, in his book *Unnatural*, explores this subject and exposes the damage done by the tabloid media in misrepresenting scientific advances in reproductive technology as leading to horrific human anomalies, or "Frankensteins."[14] This is still an endlessly riveting subject for theatre, and it also relates to a concomitant fascination with uselessness. In *Origin*, Darwin defines a monstrosity as "some considerable deviation in structure, generally injurious, or not useful to the species." Many recent theatrical encounters with evolution seem to reject the valorization of or bias toward all things "useful to the species"—the implicit, unspoken pressure put on all of us to live well, multiply, prosper, and be useful not only to society but also to our species. They seem to sense the irony that, far from sweeping away religion with its moral teachings, the biological sciences through the rise of evolution were simply replacing them with equally stern commandments about how to live.

Zooësis, Mimicry, and Interspecies Performance

It is therefore no surprise to see an avid theatrical interest in animality, the nonhuman, and interspecies performance.[15] The frontier between human and nonhuman has always held a deep appeal to playwrights, and this is only increasing, even as ethology and sociobiology, among many other fields, are constantly enhancing our knowledge of animal life. What is new is the seeming desire to stage the *becoming* of the animal Other, giving renewed meaning to the concept of mimicry in nature. In *Totem*, several acts simulate animal behavior while maintaining a human guise, invoking courtship rituals in birds and primates through astonishing acrobatic feats. A male–female trapeze duo rehearses a mating ritual that strongly evokes bird behavior and departs from the stereotypical coy female/aggressive male dichotomy that often prevails in representations of animal courtship. As the couple hangs in midair, they use the only props they have at their disposal—the trapeze bar and the ropes that hold it—as well as their own bodies in extraordinarily complex and inventive ways. The key thing about the scene is that it is not clear who is leading and who is following; gender roles are challenged, and the markers of gender are physical, not behavioral. There are some really surprising, almost disturbing gestures, as the male and female shift seamlessly from tenderness to violence, from gentle caresses to fierce pecking, gripping, throwing, and clamping onto each other, echoing the many instances of similar behavior in animals engaged in mating rituals as the males try to win the females to select them (as Darwin documents in *The Descent of Man* in examples referred to throughout this book). This happens as well in Terry Johnson's *Cries from the Mammal House,* David Greig's *Outlying Islands,* and Peyret and Prochiantz's *Les Variations Darwin. Outlying Islands* (2002), for instance, features humans enacting bird behavior, blurring the human/nonhuman boundary, and weaving many other evolutionary motifs into the drama, such as adaptation, survival in hostile habitats, Malthusian culling (natural death making way for the fitter specimens to survive), mimicry, sexual selection, and maternal instinct. The overriding question guiding the play is what is "natural" in the Anthropocene age. This and many other plays adopt what Cary Wolfe has described as "zoontologies," recognizing the important role played by the animal in decentering the figure of the human, and Una Chaudhuri's "zooësis," the discourse of animality in human life.[16]

An important forerunner in theatricalizing zooësis is Edward Albee, who was once called a "playwright of evolution" because of the overt concern with evolutionary ideas in his play *Seascape*.[17] Albee's zooëtic concerns begin with his early play *Zoo Story* in Jerry's histrionically narrated encounter with the dog, and quickly broaden to embrace other evolution-related issues raised by Jerry's suicide (altruistic; yet where does altruism fit within the natural world?) and the genetic dead-endedness of all the characters, both those on stage and those Jerry recollects in the tenement house as isolated and miserable. Albee continues probing evolutionary ideas in *Who's Afraid of Virginia Woolf?* in the focus on reproductive issues: the obsession with procreation, especially stillborn procreation; the hysterical false pregnancy of the younger woman coupled with the illusory child who then has to be "exorcised" (so twice dead). But, with *The Goat; or, Who Is Sylvia*, his evolutionary vision crystallizes in his treatment of the taboo topic of interspecies sex. One of the key things about this play is that, much like the elusive dodo in *Cries from the Mammal House,* we do not see the animal that is central to the play until the final moments. We are entirely reliant on Martin's description of the goat Sylvia and must imagine her. This makes it all the more shocking when his wife Stevie carries the bloody carcass of the goat on stage at the end, dumps it there in full view of the audience, and exits.

Another interspecies performance mode that deserves mention in this context is a modern version of the device of the "man in the monkey suit" that has been a favorite from the Victorians through Eugene O'Neill and beyond, given a new twist in Kathryn Hunter's haunting portrayal of Red Peter, the chimp addressing the academicians in a stage adaptation of Franz Kafka's short story "A Report to an Academy" as I discussed in the previous chapter. (This story also provided source material for *Des Chimères en automne*, by Peyret and Prochiantz, performed in 2003.) Colin Teevan's adaptation garnered wide acclaim, largely due to the feat of Hunter's performance, "perhaps the most physically remarkable I've ever seen on a stage," writes Charles Isherwood in the *New York Times*. There is no monkey suit, and it is a woman enacting chimphood, not a man, "scampering up to a woman in the front row and picking through her hair for tasty morsels of lice, which he then offers to share with others, as a delicacy," or clambering up a ladder to hang by one leg "as he casually continues his narration. But this is much more than a feat of actorly athleticism. Ms. Hunter imbues Red Peter with a wry wisdom, a touch of cheeky humor and, above all, a sense of dignity just slightly tinted with melancholy."[18] The performance indicts the audience: "all humanity" has done this to the chimp. Productions like this increasingly point the finger at the audience, increasingly imply, either directly or indirectly, some shared blame for things as they are in our current state of evolution, which some are calling "posthuman."

Climate Change Drama

The most obvious way in which theatre represents the Anthropocene epoch is through its increasing engagement with climate change, giving new meaning to the vexed issue of how individuals relate to their environments. Some see theatre as playing an almost-salvationist role, getting us "back to nature in an authentic way," because as a live experience,

> theatre of any kind in any space is ecological in a social way, even if removed from nature. It is always a niche in an ecological sense, what [Baz] Kershaw calls an ecotone. In addition to that, plays can tell powerful tales of human relationship with the earth, its creatures, and its endangerment. More than literature or art, theatre deeply connects nature back to humanity, because it can combine all of Felix Guattari's "three ecologies"—mental, social, and natural.[19]

Climate change dramas like Richard Bean's *Heretic* and Steve Waters's *Contingency Plan* are part of what Julie Hudson calls "a rapid escalation of interest in climate change on the stage" since 2008.[20] Such plays echo themes dealt with in earlier drama that I have discussed, such as extinction and catastrophism, giving them a sharper urgency in light of current concerns about global warming. They recall Victorian stage spectacles of nature and geology, but cast in a different light, as global, man-made disasters.

Extinction also features in plays less interested in climate change than in how species simply die out and in fantasizing about what might happen if they could be recovered. *Cries from the Mammal House* imagines a scientist finding examples of living dodos sustained and protected by enlightened "third world" humans. Robin French's *Gilbert Is Dead* (another retro-Victorian play) explores the connection between the rise of the zoo and the nineteenth-century craze for taxidermy. The main theme of *Gilbert Is Dead* is the idea that evolution by natural selection can be entirely discredited by locating a single organism, a rare loris that survives despite having all kinds of physical, nonadaptive disadvantages. A common thread in these plays is the fantasy of undoing or reversing evolutionary processes such as natural selection and extinction. The very presentation of such ideas as dramatic subject matter indicates their thorough absorption into the public consciousness.

Supercooperation

If this truly is the Anthropocene age, what will the next geological epoch be called? Will there even *be* a next epoch? Perhaps the answer is affirmative only if we cease to think that competition is at the heart of survival and instead embrace our cooperative side. Some scientists are skeptical of the selfish gene theory, of scientists who see all human behavior in terms of "reproductive ambition and the aggression this instils," and culture as just "a set of strategies to promote reproductive viability." Instead of Spencer's "dog-eat-dog survivalism," there is now a greater belief that "reproductive viability requires teamwork and cooperation."[21] This emphasis on cooperation puts Darwinism in a far more positive light: evolution as a collaborative effort rather than a competition, with humans working together rather than against one another, something for which theatre has renewed relevance through its central quality of empathy. Just as humans living in small groups depended for survival on cooperation and this strategy along with the quality of empathy gave an evolutionary advantage, theatre, argues Bruce McConachie, "trades on these proclivities [for cooperation and empathy], through the co-operation of actors, and of the audience, in the production of the theatrical event, and we spectators take pleasure in this experience of co-operative flow" as well as in empathizing with the characters we see on stage.[22] Critics may object that the idealization of "theatre as a badge of the human achievement of co-operative culture" ignores the even more important aspect, which is that theatre not only depicts but also enacts conflict.[23] We are back to the basics of theatre as bearing witness and articulating a consciousness. The discovery of mirror neurons suggests a way of probing the exact nature of the empathy that is so crucial to encounters between performers and audiences. Just as it did 150 years ago, theatre is responding to these new ideas in innovative and provocative ways, but it is also continuing to play a constitutive role in our ongoing search for answers to the fundamental questions posed by evolution about our origins, our development as a species, and ultimately, our place in the universe.

~

While this epilogue has explored some recent developments and new directions in theatre's engagement with evolution, the book as a whole suggests one abiding, eternal theme: the family. From mate selection and the relation between the sexes, to breeding and reproduction, to the raising of young, theatre has always foregrounded the family constellation, right from its origins to the present. Evolutionary ideas collectively proffered powerful new theory that changed the way we think about that core unit, and theatre is still probing the aftermath and repercussions as they continue to play out.

This book has explored the two-way conversation between evolution and theatre, and the story that has emerged points to nothing less than the birth of modern drama from the spirit of evolution. All along, the underlying mandate has been the question of what theatre takes from evolution, which aspects of this vast and complex set of scientific ideas the theatre—as opposed to other art forms or modes of public discourse or expression—seizes on. I hope the book has gone some way toward answering this and opening up new areas of investigation as we continue to try to address it fully.

Notes

Preface

1. Peter Allan Dale, *In Pursuit of a Scientific Culture: Science, Art, and Society in the Victorian Age* (Madison: University of Wisconsin Press, 1989), 32.

2. Neil Vickers, "Literature and Medicine: A Snapshot" (keynote address, meeting of the British Society for Literature and Science [BSLS], April 2012); Stuart Firestein, "What Science Wants to Know," *Scientific American*, April 2012, 10. See also Stuart Firestein, *Ignorance: How It Drives Science* (Oxford: Oxford University Press, 2012), and "The Pursuit of Ignorance" (TED talk, presented February 26, 2013, at TED2013, Long Beach, Calif., February 25–March 1, http://www.ted.com/talks/stuart_firestein_the_pursuit_of_ignorance [accessed June 5, 2014]).

Introduction

1. "Darwinism in Literature," *Galaxy* 15 (1873): 695, quoted in Cynthia Eagle Russett, *Darwin in America: The Intellectual Response, 1865–1912* (San Francisco: Freeman, 1976), 11.

2. Michael M. Chemers, *Staging Stigma: A Critical Examination of the American Freak Show* (London: Palgrave, 2008), x.

3. Gillian Beer, "Darwin and the Uses of Extinction," *Victorian Studies* 51, no. 2 (2009): 323.

4. According to Jean-François Peyret, "Amusons-nous pendant que le pape et notre législateur ont le dos tourné et promenons-nous dans les lois pendant que le comité d'éthique n'y est pas" ("En avoir ou pas," program notes for *Ex Vivo/In Vitro* [Théâtre de Colline, Paris, 2011], 7).

5. Jane R. Goodall, *Performance and Evolution in the Age of Darwin: Out of the Natural Order* (London: Routledge, 2002), 5.

6. Henry Arthur Jones, *The Relations of the Drama to Real Life* [lecture presented at Toynbee Hall, London, November 13, 1897] (London: Chiswick Press, 1897), 8.

7. See, for example, Bernard Lightman and Bennett Zon, eds., *Evolution and Victorian Culture* (Cambridge: Cambridge University Press, 2014); Gillian Beer, *Darwin's Plots: Evolutionary Narrative in Darwin, George Eliot and Nineteenth-Century Fiction* (Cambridge: Cambridge University Press, 1983), and *Open Fields: Science in Cultural Encounter* (Oxford: Oxford University Press, 1996); George Levine, *Darwin and the Novelists: Patterns of Science in Victorian Fiction* (Chicago: University of Chicago Press, 1991), *Realism, Ethics and Secularism: Essays on Victorian Literature and Science* (Cambridge: Cambridge University Press, 2008), and *Darwin Loves You: Natural Selection and the Re-enchantment of the World* (Princeton, N.J.: Princeton University Press, 2008); Gowan Dawson, *Darwin, Literature, and Victorian Respectability* (Cambridge: Cambridge University Press, 2007); and John Holmes, *Darwin's Bards: British and American Poetry in the Age of Evolution* (Edinburgh: Edinburgh University Press, 2009). On the widespread engagement with evolutionary theory in nonfiction writing, see Louise Henson, Geoffrey Cantor, Gowan Dawson, Richard Noakes, Sally Shuttleworth, and Jonathan R. Topham, eds., *Culture and Science in the Nineteenth-Century Media* (Farnham, Eng.: Ashgate, 2004); Geoffrey Cantor, Gowan Dawson, Graeme Gooday, Richard Noakes, Sally Shuttleworth, and Jonathan R. Topham, eds., *Science in the Nineteenth-Century Periodical* (Cambridge: Cambridge University Press, 2004); and Geoffrey Cantor and Sally Shuttleworth, eds., *Science Serialised: Representations of the Sciences in Nineteenth-Century Periodicals* (Cambridge, Mass.: MIT Press, 2004).

8. Gillian Beer, "Science and Literature," in *Companion to the History of Modern Science*, ed. Robert C. Olby, Geoffrey N. Cantor, John R. R. Christie, and M. J. S. Hodge (London: Routledge, 1990), 787.

9. Michael H. Whitworth, "The Physical Sciences," in *A Companion to Modernist Literature and Culture*, ed. David Bradshaw and Kevin J. H. Dettmar (Malden, Mass.: Blackwell, 2008), 46.

10. Goodall, *Performance and Evolution in the Age of Darwin*, 6.

11. Ibid., 7.

12. Diana Donald and Jane Munro, *Endless Forms* (pamphlet for the exhibition presented at the Yale Center for British Art, New Haven, Connecticut, February–May 2009, and the Fitzwilliam Museum, Cambridge, June–October 2009). See also Diana Donald and Jane Munro, *Endless Forms: Charles Darwin, Natural Science, and the Visual Arts* (New Haven, Conn.: Yale University Press, 2009).

13. Quoted in Patrick McGuinness, *Maurice Maeterlinck and the Making of Modern Theatre* (Oxford: Oxford University Press, 2000), 7.

14. Simon Stephens, comments at Panel on Theatre and Politics, with Vanessa Redgrave, Ralph Fiennes, Michael Billington, and Simon Stephens, Humanitas series, University of Oxford, February 10, 2012.

15. James Harrison, "Destiny or Descent? Responses to Darwin," *Mosaic* 14, no. 1 (1981): 112.

16. Gillian Beer, "The Challenges of Interdisciplinarity" (speech presented at the Institute of Advanced Study, Durham University, April 27, 2006), https://www.dur.ac.uk/ias/news/annual_research_dinner/ (accessed June 5, 2014).

17. Ibid.

18. Luca Ronconi, "Movement as a Metaphor of Time" [interview with Maria Grazia Gregori], trans. Bruno Tortorella. The manuscript was kindly given to me by Pino Donghi.

19. Geoffrey Cantor, Gowan Dawson, Richard Noakes, Sally Shuttleworth, and Jonathan R. Topham, introduction to *Culture and Science in the Nineteenth-Century Media*, ed. Henson et al., xx.

20. Whitworth, "Physical Sciences," 47.

21. Beer, *Open Fields*, 171.

22. Tiffany Watt-Smith, "Darwin's Flinch: Sensation Theatre and Scientific Looking in 1872," *Journal of Victorian Culture* 15, no. 1 (2010): 101–17. See also Iwan Rhys Morus, "Worlds of Wonder: Sensation and the Victorian Scientific Performance," *ISIS* 101, no. 4 (2010): 806–16.

23. Nessa Carey, *The Epigenetics Revolution: How Modern Biology Is Rewriting Our Understanding of Genetics, Disease, and Inheritance* (London: Icon Books, 2012), 2; Robert P. Crease, *The Play of Nature: Experimentation as Performance* (Bloomington: Indiana University Press, 1993), 4, 6.

24. Joseph R. Roach, "Darwin's Passion: The Language of Expression on Nature's Stage," *Discourse* 13, no. 1 (1990–1991): 41.

25. Rita Felski, "'Context Stinks!'" *New Literary History* 42, no. 4 (2011): 573–91.

26. Eve-Marie Engels and Thomas F. Glick, eds., *The Reception of Charles Darwin in Europe*, 2 vols. (London: Bloomsbury, 2008); Thomas F. Glick and Elinor Shaffer, eds., *The Literary and Cultural Reception of Charles Darwin in Europe*, 3 vols. (London: Bloomsbury, 2014).

27. Elin Diamond, "Beckett and Caryl Churchill Along the Möbius Strip," in *Beckett at 100: Revolving It All*, ed. Linda Ben-Zvi and Angela Moorjani (Oxford: Oxford University Press, 2008), 286.

28. Paul White, quoted in Gowan Dawson, "'Like a Megatherium Smoking a Cigar': Darwin's *Beagle* Fossils in Nineteenth-Century Popular Culture," in *Darwin, Tennyson and Their Readers: Explorations in Victorian Literature and Science*, ed. Valerie Purton (London: Anthem, 2013), 83.

29. See, especially, Peter J. Bowler, *Evolution: The History of an Idea*, rev. ed. (Berkeley: University of California Press, 2009), and *The Non-Darwinian Revolution: Reinterpreting a Historical Myth* (Baltimore: Johns Hopkins University Press, 1988); and Rebecca Stott, *Darwin's Ghosts: In Search of the First Evolutionists* (London: Bloomsbury, 2012). Bowler argues that the emphasis on Darwin as the "pivot around which everything else moved" has distorted his impact on both science and intellectual history(*Non-Darwinian Revolution*, 93, 4).

30. Goodall, *Performance and Evolution in the Age of Darwin*, 177. See also Chris Fleming and Jane Goodall, "Dangerous Darwinism," *Public Understanding of Science* 11, no. 3 (2002): 261.

31. Bowler, *Non-Darwinian Revolution*, 5.

32. Ibid.

33. Ibid.

34. Richard D. Altick, *The Shows of London* (Cambridge, Mass.: Belknap Press of Harvard University Press, 1978), 287, 484. On the problems with this "triumphalist" narrative of Darwin's reception, see also Joe Cain, introduction to Charles Darwin, *The Expression of the Emotions in Man and Animals* (London: Penguin, 2009), xxxiv.

35. Fleming and Goodall, "Dangerous Darwinism," 269.

36. Daniel Pick, *Faces of Degeneration: A European Disorder, c. 1848–c. 1918* (Cambridge: Cambridge University Press, 1989); William Greenslade, *Degeneration, Culture and the Novel* (Cambridge: Cambridge University Press, 1994).

37. Charles Darwin to George Henry Lewes, August 7, 1868, Darwin Correspondence Project, http://www.darwinproject.ac.uk (accessed June 10, 2014).

38. Charles Darwin, "Autobiography, May 31, 1876," in *Charles Darwin and T. H. Huxley: Autobiographies*, ed. Gavin de Beer (Oxford: Oxford University Press, 1983), 54.

39. Ibid., 52.

40. Bowler, *Evolution*, 358.

41. Thornton Wilder, *The Journals of Thornton Wilder, 1939–61*, ed. Donald Gallup (New Haven, Conn.: Yale University Press, 1985), 24.

42. Londa Schiebinger, *Nature's Body: Sexual Politics and the Making of Modern Science* (London: Pandora, 1993). See also Cynthia Eagle Russett, *Sexual Science: The Victorian Construction of Womanhood* (Cambridge, Mass.: Harvard University

Press, 1989); and Mike Hawkins, *Social Darwinism in European and American Thought, 1860–1945* (Cambridge: Cambridge University Press, 1997).

43. Darwin, "Autobiography," 15.

44. Angelique Richardson, ed., "Essentialism in Science and Culture," special issue, *Critical Quarterly* 53, no. 4 (2011); Elizabeth Grosz, *Time Travels: Feminism, Nature, Power* (Durham, N.C.: Duke University Press, 2005), 1–42; Levine, *Darwin Loves You.*

45. See, for example, Elaine Aston and Ian Clarke, "The Dangerous Woman of Melvillean Melodrama," *New Theatre Quarterly* 12, no. 45 (1996): 31–32; and Sos Eltis, *Acts of Desire: Women and Sex on Stage 1800–1930* (Oxford: Oxford University Press, 2013).

46. For a penetrating analysis of this movement, see Jonathan Kramnick, "Against Literary Darwinism," *Critical Inquiry* 37, no. 2 (2011): 315–47.

47. Seamus Perry, review of *Why Lyrics Last: Evolution, Cognition, and Shakespeare's Sonnets*, by Brian Boyd, *Times Literary Supplement*, May 25, 2012, 8.

48. William Flesch, "Acting Together," review of *Theatre and Mind,* by Bruce McConachie, *Times Literary Supplement*, September 20, 2013, 29.

49. The anthropological roots of theatre as ritual were long ago established by Victor Turner, Richard Schechner, Eugenio Barba, and many others, but that is a separate issue.

50. Marco Iacobini, *Mirroring People: The New Science of How We Connect with Others* (New York: Farrar, Straus and Giroux, 2008); Giacomo Rizzolatti, Corrado Sinigaglia, and Frances Anderson, *Mirrors in the Brain: How Our Minds Share Action, Emotion, and Experience* (Oxford: Oxford University Press, 2008). See also Bruce McConachie, *Engaging Audiences: A Cognitive Approach to Spectating in the Theatre* (New York: Palgrave Macmillan, 2008); Bruce McConachie and F. Elizabeth Hart, eds., *Performance and Cognition: Theatre Studies and the Cognitive Turn* (London: Routledge, 2006); and David Z. Saltz, ed., "Performance and Cognition," special issue, *Theatre Journal* 59, no. 4 (2007).

51. James M. Harding, ed., *Contours of the Theatrical Avant-Garde: Performance and Textuality* (Ann Arbor: University of Michigan Press, 2000).

52. Tamsen Wolff, *Mendel's Theatre: Heredity, Eugenics, and Early Twentieth-Century American Drama* (New York: Palgrave Macmillan, 2009), 7–8.

53. James Secord, *Victorian Sensation: The Extraordinary Publication, Reception, and Secret Authorship of "Vestiges of the Natural History of Creation"* (Chicago: University of Chicago Press, 2003).

54. Fleming and Goodall, "Dangerous Darwinism," 265.

55. Ibid.

56. James Moore and Adrian Desmond, introduction to Charles Darwin, *The Descent of Man, and Selection in Relation to Sex*, ed. James Moore and Adrian Desmond (London: Penguin, 2004).

57. Bowler, *Non-Darwinian Revolution*, 5.

58. Ibid., 24. But Darwin did not derive his idea of natural selection from Thomas Malthus: "Far from it. Darwin already knew, months before he read Malthus, that selection was the key to man's success in breeding and improving cultivated plants and domestic animals" (Gavin de Beer, introduction to *Charles Darwin and T. H. Huxley*, ed. de Beer, x–xi). Ironically, it is Darwin's own *Autobiography* that overstates his debt to Malthus.

59. Bowler, *Non-Darwinian Revolution*, 25.

60. M. J. S. Hodge, "England," in *The Comparative Reception of Darwinism*, ed. Thomas F. Glick (Chicago: University of Chicago Press, 1988), 15.

61. Bowler, *Non-Darwinian Revolution*, 5.

62. Harrison, "Destiny or Descent?" 113.

63. Russett, *Darwin in America*, 16–17.

64. Hodge, "England," 15 (emphasis added).

65. Rosaura Ruiz, "Lamarck," in *Encyclopedia of Evolution*, ed. Mark D. Pagel, 2 vols. (Oxford: Oxford University Press, 2002), 1:601. See also Pietro Corsi, *The Age of Lamarck: Evolutionary Theories in France, 1790–1830*, rev. ed. (Berkeley: University of California Press, 1988), and *Evolution Before Darwin* (Oxford: Oxford University Press, 2015).

66. Wolff, *Mendel's Theatre*, 40.

67. Eva Jablonka and Marion J. Lamb, "Lamarckism," in *Encyclopedia of Evolution*, ed. Pagel, 1:602.

68. Russett, *Darwin in America*, 10–11.

69. Ibid., 10.

70. Harrison, "Destiny or Descent?" 113.

71. The story of how Joseph Hooker and Thomas Henry Huxley found a compromise so that Wallace's work received some recognition while still foregrounding Darwin's is well known and has formed the basis for several books, films, and at least one play: Peter Parnell's *Trumpery* (2007). See Kirsten Shepherd-Barr, "Darwin on Stage: Evolutionary Theory in the Theatre," *Interdisciplinary Science Reviews* 33, no. 2 (2008): 107–15.

72. Alfred Russel Wallace website, http://wallacefund.info/faqs-myths-misconceptions (accessed June 5, 2014).

73. Robert J. Richards, *The Tragic Sense of Life: Ernst Haeckel and the Struggle over Evolutionary Thought* (Chicago: University of Chicago Press, 2008), 2.

74. Bowler, *Non-Darwinian Revolution,* 162.

75. Robert J. Richards, "Ernst Haeckel's Alleged Anti-Semitism and Contributions to Nazi Biology," *Biological Theory* 2, no. 1 (2007): 97–103, and *Tragic Sense of Life*, 506–12.

76. Nigel Rothfels, "Aztecs, Aborigines, and Ape-People: Science and Freaks in Germany, 1850–1900," in *Freakery: Cultural Spectacles of the Extraordinary Body*, ed. Rosemarie Garland Thomson (New York: NYU Press, 1996), 165.

77. Ibid., 166.

78. Keith R. Benson, "Recapitulation," in *Encyclopedia of Evolution*, ed. Pagel, 2:985.

79. Bowler, *Non-Darwinian Revolution*, 88.

80. Ernst Mayr, introduction to Charles Darwin, *On the Origin of Species by Means of Natural Selection, or the Preservation of Favoured Races in the Struggle for Life*, facsimile of the first edition, ed. Ernst Mayr (Cambridge, Mass.: Harvard University Press, 1964), xvi–xvii.

81. Darwin, *On the Origin of Species*, 74.

82. Charles Darwin, *On the Origin of Species*, 4th ed. (London: Murray, 1866), 504–5 (emphasis added). This passage does not appear in the first edition of *Origin*.

1. "I'm Evolving!"

1. Robert Altick, *The Shows of London* (Cambridge, Mass.: Belknap Press of Harvard University Press, 1978), 2.

2. According to George Rowell, "The late Victorian theatre remained fundamentally a popular theatre" (*Late Victorian Plays, 1890–1914* [Oxford: Oxford University Press, 1968], vi). Jane Goodall discusses Barnum's definitive importance for the relationship between science and performance, his shows turning "the whole notion of species into a vast programme of entertainment," in *Performance and Evolution in the Age of Darwin: Out of the Natural Order* (London: Routledge, 2002), 21–45. On popular theatre, see especially Jacky Bratton, *The Making of the West End Stage* (Cambridge: Cambridge University Press, 2011), and *New Readings in Theatre History* (Cambridge: Cambridge University Press, 2003).

3. Iwan Rhys Morus, "Worlds of Wonder: Sensation and the Victorian Scientific Performance," *ISIS* 101, no. 4 (2010): 814.

4. Quoted in Bernard Lightman, *Victorian Popularizers of Science: Designing Nature for New Audiences* (Chicago: University of Chicago Press, 2007), 457. Lightman comments that Hutchinson is telling "the story of cosmic evolution . . . as if it were a stunning play."

5. Rebecca Stott, *Darwin and the Barnacle: The Story of One Tiny Creature and History's Most Spectacular Scientific Breakthrough* (London: Faber and Faber, 2003), 43. Janet Browne notes that Jemmy behaved like "a stock character from stage and literature" (*Darwin: A Biography*, vol. 1, *Voyaging* [Princeton, N.J.: Princeton University Press, 1996], 237). Huxley, quoted in Chris Fleming and Jane Goodall, "Dangerous Darwinism," *Public Understanding of Science* 11 (2002): 261.

6. For example, Allardyce Nicoll lists the two-act comic drama *Lavater the Physiognomist and a Good Judge Too*, by an unknown author, which played at

Sadler's Wells in March 1848 (Lord Chamberlain Collection [hereafter LCP], catalogue number 35/3/48), and the farce *Mesmerism versus Galvanism* by George Beswick, which played at the Albert in spring 1845, in *A History of English Drama, 1660–1900*, vol. 4, *Early Nineteenth Century Drama, 1800–1850* (Cambridge: Cambridge University Press, 1955), 491, 267. *Lavater* was a "genuine and immense success" according to the *Morning Chronicle* (April 4, 1848). These plays are in addition to the numerous extravaganzas, equestrian melodramas, and other spectacular entertainments that proliferated during this period, well documented by Altick, *Shows of London*, 1978.

7. See, for example, Rae Beth Gordon, *Dances with Darwin, 1875–1919: Vernacular Modernity in France* (Farnham, Eng.: Ashgate, 2009).

8. Diana Donald and Jane Munro, *Endless Forms* (exhibition pamphlet for the exhibition presented at the Yale Center for British Art, New Haven, Conn., February–May 2009, and the Fitzwilliam Museum, Cambridge, Eng., June–October 2009), 14.

9. Peter Morton, *The Vital Science: Biology and the Literary Imagination, 1860–1900* (London: Unwin, 1984); Margot Norris, *Beasts of the Modern Imagination: Darwin, Nietzsche, Kafka, Ernst, and Lawrence* (Baltimore: Johns Hopkins University Press, 1985); Angelique Richardson, "The Life Sciences: 'Everybody Nowadays Talks About Evolution,'" in *A Concise Companion to Modernism*, ed. David Bradshaw (Oxford: Blackwell, 2003), 18.

10. Fleming and Goodall, "Dangerous Darwinism," 259.

11. Goodall, *Performance and Evolution in the Age of Darwin*, 6.

12. James Harrison, "Destiny or Descent? Responses to Darwin," *Mosaic* 14, no. 1 (1981): 114.

13. Ibid., 116.

14. Emile Zola, preface to *Thérèse Raquin*, in *Naturalism and Symbolism in European Theatre, 1850–1918*, ed. Claude Schumacher (Cambridge: Cambridge University Press, 1996), 71.

15. In France, "the first African village in a Paris zoological garden" was set up in 1877 (Gordon, *Dances with Darwin*, 67).

16. Sadiah Qureshi, *Peoples on Parade: Exhibitions, Empire, and Anthropology in Nineteenth-Century Britain* (Chicago: University of Chicago Press, 2011), and "Meeting the Zulus: Displayed Peoples and the Shows of London, 1853–79," in *Popular Exhibitions, Science and Showmanship, 1840–1910*, ed. Joe Kember, John Plunkett, and Jill A. Sullivan (London: Pickering and Chatto, 2012), 183–98. See also Bernth Lindfors, "Ethnological Show Business: Footlighting the Dark Continent," in *Freakery: Cultural Spectacles of the Extraordinary Body*, ed. Rosemarie Garland Thomson (New York: NYU Press, 1996), 207–18.

17. Gordon, *Dances with Darwin*, 67n.15, 95, 146–47 (referred to in one contemporary article as "Nubians").

18. Nigel Rothfels, *Savages and Beasts: The Birth of the Modern Zoo* (Baltimore: Johns Hopkins University Press, 2002), 114, 117.

19. Ibid., 117.

20. Ibid.

21. Charles Darwin, "Autobiography," in *Charles Darwin and T. H. Huxley: Autobiographies*, ed. Gavin de Beer (Oxford: Oxford University Press, 1983), 46.

22. Gordon, *Dances with Darwin*, 5, 83, 243.

23. Ralph O'Connor, *The Earth on Show: Fossils and the Poetics of Popular Science, 1802–1856* (Chicago: University of Chicago Press, 2007); Joe Kember, John Plunkett, and Jill A. Sullivan, eds., *Popular Exhibitions, Science and Showmanship, 1840–1910* (London: Pickering and Chatto, 2012); Lightman, *Victorian Popularizers of Science*; Iwan Rhys Morus, "'More the Aspect of Magic than Anything Natural': The Philosophy of Demonstration," in *Science in the Marketplace: Nineteenth-Century Sites and Experiences*, ed. Aileen Fyfe and Bernard Lightman (Chicago: University of Chicago Press, 2007), 122–39.

24. Peter J. Bowler, *The Non-Darwinian Revolution: Reinterpreting a Historical Myth* (Baltimore: Johns Hopkins University Press, 1988), 75.

25. Morus, "Worlds of Wonder," 815.

26. In 1894, the *Daily Telegraph* published a grave warning by a professor at the University of Geneva that bicycling will in the course of a thousand years render us unable to walk: "The theory of evolution leaves no doubt on the subject. The human feet will gradually get stunted and pine away," and we will look like "ugly apes" ("A Terrible Future for Cyclists," *Jackson's Oxford Journal*, September 22, 1894, 6).

27. In *The Earth on Show*, O'Connor reveals the theatricality of Victorian geology and its popularization, but in a narrative sense—the narrative techniques employed by geologists to dramatize for their readers the prehistoric earth and the dinosaurs. Apart from a fascinating discussion of Byron's unperformed play *Cain* (1821), there is no consideration of the depiction of landscape in melodrama and spectacle with regard to, and perhaps indirectly reflecting, the public interest in geological deep time.

28. O'Connor, *Earth on Show*, 386.

29. *Lord Byron's Cain, a Mystery: With Notes*, ed. Harding Grant (London: William Crofts, 1830), 2.i, 192.

30. There was more than one Astley's; the reference here is most likely to the Royal Amphitheatre at Westminster Bridge Road, whose most famous period was 1830 to 1841. See Allardyce Nicoll, *A History of Early Nineteenth Century Drama, 1800–1850* (Cambridge: Cambridge University Press, 1930), 1:223–25. For a succinct summary of key developments in Victorian theatre scholarship, see Tracy C. Davis and Peter Holland, "Introduction: The Performing Society," in *The Performing Century: Nineteenth-Century Theatre's History*, ed. Tracy C. Davis and Peter Holland (London: Palgrave Macmillan, 2007), 1–11.

31. Quoted in John R. Durant, "Innate Character in Animals and Man," in *Biology, Medicine and Society, 1840–1940*, ed. Charles Webster (Cambridge: Cambridge University Press, 2003), 157.

32. Ibid., 161. Durant notes that there are three definitions of ethology: the portrayal of character by mimicry, the science of ethics, and the science of character. This last has gained greatest currency.

33. Ibid., 164.

34. Ibid., 187.

35. Ibid., 181.

36. Julian Huxley, "Bird-Watching and Biological Science," *Auk* 33 (1916): 143–44, quoted in ibid.

37. Mike Hawkins, *Social Darwinism in European and American Thought, 1860–1945* (Cambridge: Cambridge University Press, 1997), 69.

38. Bernard Shaw, *Man and Superman*, in *The Complete Plays of Bernard Shaw* (London: Odhams Press, 1934), act 3, 379.

39. "The Cameleopard," *Morning Chronicle*, July 12, 1827.

40. "The Cameleopard," *Morning Post*, July 11, 1827. In 1832, Saint-Hilaire formally articulated the concept of teratology, casting the freak as a "pathological specimen of the terata [monstrosity]" (Rosemarie Garland Thomson, "Introduction: From Wonder to Error—A Genealogy of Freak Discourse in Modernity," in *Freakery*, ed. Garland Thomson, 4). Saint-Hilaire's observations of the occurrence of monstrosities in humans as compared with those in the lower animals shaped Darwin's own thinking about monsters. See Michael M. Chemers, *Staging Stigma: A Critical Examination of the American Freak Show* (London: Palgrave, 2008), 64.

41. "Cameleopard," *Morning Post*.

42. "Cameleopard," *Morning Chronicle*.

43. *Birds, Beasts and Fishes; or, Harlequin and Natural History*, written and produced by Nelson Lee for the City of London Theatre (1854), LCP, Add. 52951.

44. Established in 1828 for scientific study, the London Zoo opened to the public in 1847. The first orangutan arrived there in 1837. See Rothfels, *Savages and Beasts*.

45. "Christmas Novelties," *Era*, December 31, 1854, 11. The stage directions simply read: "A View of the Royal Zoological Gardens. With Birds, Beasts & Fishes" (Lee, *Birds, Beasts and Fishes*, 17).

46. "The Forthcoming Christmas Novelties," *Era*, December 24, 1854, 11.

47. John Perry, *James A. Herne: The American Ibsen* (Chicago: Nelson-Hall, 1978), 41.

48. Morus, "Worlds of Wonder," 816.

49. The extensive literature on freakery includes Rosemarie Garland Thomson, *Extraordinary Bodies: Figuring Disability in American Culture and Literature* (New York: Columbia University Press, 1996), and *Freakery*; Marlene

Tromp, ed., *Victorian Freaks: The Social Context of Freakery in Britain* (Columbus: Ohio State University Press, 2008); Nadja Durbach, *Spectacle of Deformity: Freak Shows and Modern British Culture* (Berkeley: University of California Press, 2010); Chemers, *Staging Stigma*; and Goodall, *Performance and Evolution in the Age of Darwin.*

50. See, for example, Garland Thomson, *Extraordinary Bodies* and *Freakery*; Tromp, ed., *Victorian Freaks*; Durbach, *Spectacle of Deformity*; Chemers, *Staging Stigma*; and Goodall, *Performance and Evolution in the Age of Darwin.*

51. Marlene Tromp and Karyn Valerius, "Introduction: Toward Situating the Victorian Freak," in *Victorian Freaks,* ed. Tromp, 1.

52. Heather McHold, "Even as You and I: Freak Shows and Lay Discourse on Spectacular Deformity," in *Victorian Freaks*, ed. Tromp, 23–24.

53. Gordon, *Dances with Darwin*, 271 and passim.

54. For example, in *History of Early Nineteenth Century Drama, 1800–1850*, Nicoll lists a pantomime in 1810 called *The Hottentot Venus; or, Harlequin in Africa* and a melodramatic entertainment called *Brazilian Jack; or, The Life of an Ape* in 1834. Caroline Radcliffe discusses specific instances of this cross-fertilization—theatre appropriating a popular exhibition and "remediating" it on the stage—in "*The Talking Fish*: Performance and Delusion in the Victorian Exhibition," in *Popular Exhibitions, Science and Showmanship, 1840–1910*, ed. Joe Kember, John Plunkett, and Jill A. Sullivan (London: Pickering and Chatto, 2012), 147.

55. Rebecca Stern, "Our Bear Women, Our Selves," in *Victorian Freaks,* ed. Tromp 200–233.

56. *The Missing Link*, unknown author, licensed March 11, 1893, LCP, no catalog number given. The piece is signed by W. H. Westwood, manager of Grand Theatre, Wolsall. Goodall notes the thriving genre of "monkey-man productions" in the 1840s and 1850s (*Performance and Evolution in the Age of Darwin*, 51). Other plays of the period bearing in their titles some allusion to missing links, apes, zoos, or savagery include *The Missing Link* (drama, 1886) by Hal Collier; *The Missing Link* (farce, 1894) by Arthur Shirley; *My Niece and My Monkey* (burlesque, 1876) by Henry Herman; *At the Zoo* (no date) by W. P. Ridge; *Buffalo Bill* (drama, 1887) by Gary Roberts; *Buffalo Bill; or, a Life in the Wild West* (drama, 1887) by H. J. Stanley, with Charles Hermann; *The Ourang Outang and His Double*; *or, The Runaway Monkey* (no date) by George Herbert Rodwell; *Zoo, a Musical Farce* (1875) with music by Arthur Sullivan; *The Zoo* (no date) by Benjamin Charles Stephenson; and *Aunt Chimpanzee* (musical farce, 1897) by Morton Williams, with music by Woodruffe. I am indebted to Tiziana Morosetti for compiling this list.

57. *Missing Link*, no page numbers.

58. Helen Hamilton Gardener, "The Moral Responsibility of Women in Heredity," in *Facts and Fictions of Life* (Boston: Fenno, 1895), 199.

59. Unnamed physiologist in *Deutsche Revue,* quoted in Gardener, "Moral Responsibility of Women," 179.

60. Shaw, *Man and Superman*, act 4, 400.

61. Bernard Lightman, "Scientists as Materialists in the Periodical Press: Tyndall's Belfast Address," in *Science Serialized: Representations of the Sciences in Nineteenth-Century Periodicals*, ed. Geoffrey Cantor and Sally Shuttleworth (Cambridge, Mass.: MIT Press, 2004), 202.

62. Ibid., 202.

63. Ibid., 221.

64. Carolyn Williams, *Gilbert and Sullivan: Gender, Genre, Parody* (New York: Columbia University Press, 2011), 50.

65. Ibid., 247–48. Already in the 1860s, long before the Darwinian Man song, Gilbert wrote about men as monkeys in verses for *La Vivandière, or True to the Corps!,* a burlesque of Donizetti's *Daughters of the Regiment.*

66. As Gordon notes, although Darwin never made this explicit connection himself, the idea of a common ancestry with the apes implied a sexual union at some point between humans and monkeys; according to Diana Snigurowicz, "The spectre of cross-species fertilization continued to haunt the social imaginary" (quoted in Gordon, *Dances with Darwin*, 68).

67. Williams, *Gilbert and Sullivan*, 249.

68. Rebecca Stott, "'Tennyson's Drift': Evolution in *The Princess,*" in *Darwin, Tennyson and Their Readers: Explorations in Victorian Literature and Science*, ed. Valerie Purton (London: Anthem, 2013), 14.

69. Stott, "'Tennyson's Drift,'" 25.

70. Ibid., 29–30.

71. Williams, *Gilbert and Sullivan*, 250.

72. Angelique Richardson, "Against Finality: Darwin, Mill and the End of Essentialism," *Critical Inquiry* 53, no. 4 (2011): 21–44.

73. *New York Times,* January 21, 1894 (accessed May 10, 2011). The New York *Daily Tribune,* January 21, 1894, called it "dull, disjointed, undramatic and hardly intelligible."

74. Robert Buchanan, *Charlatan*, LCP; no page numbers.

75. *Pall Mall Gazette*, October 14, 1890; *Times* (London), September 26, 1890, 7.

76. *Times*, September 26, 1890, 7.

77. *Theatre*, November 1, 1890.

78. *Times*, September 26, 1890, 7.

79. *Theatre*, November 1, 1890.

80. *Penny Illustrated Paper*, October 4, 1890, 210.

81. *Era*, September 27, 1890.

2. Confronting the Serious Side

1. Downing Cless, *Ecology and Environment in European Drama* (London: Taylor and Francis, 2010), 15.

2. John Perry, *James A. Herne: The American Ibsen* (Chicago: Nelson-Hall, 1978), 138.

3. James A. Herne, *Shore Acres and Other Plays* (London: Samuel French, 1928), 81, 87, 88.

4. Herbert Spencer, *First Principles* (1885), quoted in Perry, *James A. Herne*, 81.

5. Hamlin Garland, quoted in Perry, *James A. Herne*, 125.

6. Perry, *James A. Herne*, 188.

7. Herbert J. Edwards and Julie A. Herne, *James A. Herne: The Rise of Realism in the American Drama* (Orono: University of Maine Press, 1964), 113.

8. Helen Hamilton Gardener, "Woman as an Annex," in *Facts and Fictions of Life* (Boston: Fenno, 1895), 135.

9. J. Stanley Lemons, "Social Feminism," in *The Oxford Encyclopedia of Women in World History*, ed. Bonnie G. Smith (Oxford: Oxford University Press, 2008), 1:81.

10. Quoted in Perry, *James A. Herne*, 97.

11. Perry, *James A. Herne*, 97.

12. Herne, *Shore Acres*, act 1, 24.

13. Ibid., act 2, 75; act 1, 30.

14. Ibid., act 1, 29, 31.

15. Ibid., act 1, 21.

16. *Boston Evening Transcript*, quoted in Perry, *James A. Herne*, 234.

17. Herne, *Shore Acres*, act 4, 107, 109.

18. Perry, *James A. Herne*, 234.

19. Herne, *Shore Acres*, act 4, 107, 109.

20. Donald Pizer, "Herbert Spencer and the Genesis of Hamlin Garland's Critical System," *Tulane Studies in English* (1957): 157–58.

21. Mike Hawkins, *Social Darwinism in European and American Thought, 1860–1945* (Cambridge: Cambridge University Press, 1997), 34.

22. Ibid., 35.

23. "I left off writing a novel I was engaged upon, and gave most of my leisure to seeing plays and reading Herbert Spencer" (Doris Arthur Jones, *The Life and Letters of Henry Arthur Jones* [London: Gollancz, 1930], 34).

24. Letter dated February 19, 1878, in Doris Arthur Jones, *Life and Letters*, 39.

25. Henry Arthur Jones's *The Dancing Girl* (premiere January 1891) ran for 310 nights at the Haymarket and was revived by Herbert Beerbohm Tree at His Majesty's in 1909, though not to great success.

26. Steve Nicholson, *The Censorship of British Drama*, 4 vols. (Exeter, Eng.: University of Exeter Press).

27. Henry Arthur Jones, *The Dancing Girl*, act 4, 4, LCP 53466 G.

28. Henry Arthur Jones, *The Dancing Girl* (London: Samuel French, 1907).

29. Doris Arthur Jones, *Life and Letters*, 114.

30. Henry Arthur Jones, *Dancing Girl*, act 3, 13, LCP. These lines also appear in the Samuel French acting edition (1907).

31. James A. Herne's *Hearts of Oak* (1879) also features people setting out for "the Arctics."

32. Henry Arthur Jones, *The Case of Rebellious Susan*, in *Plays of Henry Arthur Jones*, ed. Russell Jackson (Cambridge: Cambridge University Press, 1982), act 3, 153.

33. Henry Arthur Jones, *Dancing Girl*, act 4, 8, LCP.

34. Henry Arthur Jones, quoted in Allardyce Nicoll, *A History of Late Nineteenth Century Drama, 1850–1900* (Cambridge: Cambridge University Press, 1949), 1:169.

35. Ibid., 172.

36. Cyril D. Darlington, *The Evolution of Man and Society* (London: Allen and Unwin, 1969), 678. In *Origin*, chapter 4 and the conclusion summarize Darwin's points about extinction.

37. Charles Darwin, *On the Origin of Species*, facsimile of the first edition, ed. Ernst Mayr (Cambridge, Mass.: Harvard University Press, 1964), 318.

38. Charles Darwin, *Origin of Species by Means of Natural Selection* (London: Murray, 1859), 431.

39. Charles Darwin, "Autobiography," in *Charles Darwin and T. H. Huxley: Autobiographies*, ed. Gavin de Beer (Oxford: Oxford University Press, 1983), 328.

40. De Curel's career includes several plays with evolution-related themes. He had an enduring interest in science; for instance, his play *The New Idol* (1895; performed by the Stage Society in March 1902) dealt with "the cult of science at the expense of human value," and his play *The Soul Gone Mad* (1919), "his only popular success," was a comedy comparing human and animal emotions. See *Encyclopaedia Britannica Online Academic Edition* (Encyclopaedia Britannica, 2014), s.v. "François, vicomte de Curel" (accessed July 12, 2013).

41. The play was written in October 1891 and premiered at the Théâtre Libre on November 29, 1892, with Claire played by Mlle. Berthe Bady and Antoine playing the duke. In 1897, *Les Fossiles* was considered by the Comédie française for its repertoire, and de Curel revised it to correct the weakness in how the revelation of paternity came about, which had seemed contrived; instead, he provided "la grande scène entre le duc et Robert, scène douloureuse" in act 3. See François de Curel, *Les Fossiles*, in *Théâtre complet* (Paris: Editions Georges Crès, 1920), 2:153. Fire at the theatre meant delay and an enforced move to the Odéon, and it

was May 21, 1900—during the Exposition—when the play was revived there by the Comédie française, "devant un public distrait, venu pour la grande kermesse et parfaitement incapable de prêter une attention soutenue à une œuvre âpre et violente. Aussi *les Fossiles* n'eurent-ils que 20 à 25 représentations. Ils n'ont jamais été repris depuis cette époque" (introduction, June 1919, 153).

42. De Curel, *Les Fossiles*, act 3.iii, 3.v.

43. Ibid., act 4, 253–54.

44. Ibid., 3.ii, 224–25; translation in Samuel Montefiore Waxman, *Antoine and the Théâtre Libre* (Cambridge, Mass.: Harvard University Press, 1926), 171.

45. L. K., "The Vaporings of Lovers" (review of the production of *L'amour brode* [*Love Embroiders*] in Paris), *New York Times*, November 26, 1893, 19.

46. *Henry James: Guy Domville: Play in Three Acts*, ed. Leon Edel (London: Rupert Hart-Davis, 1961), 1, 135.

47. Charles Darwin, *The Descent of Man, and Selection in Relation to Sex*, ed. James Moore and Adrian Desmond (London: Penguin, 2004), 143.

48. James, *Guy Domville*, act 1, 138.

49. Ibid., 139.

50. Ibid., 143 (emphasis added).

51. Gillian Beer, "Darwin and the Uses of Extinction," *Victorian Studies* 51, no. 2 (2009): 322.

52. Ibid.

53. De Curel, *Les Fossiles*, 2.ii, 202.

54. James, *Guy Domville*, 21.

55. Ibid., act 2, 140–42.

56. Ibid., 142.

57. Ibid., 157.

58. Thomas H. Huxley, "Evolution and Ethics," in *Collected Essays* (London, 1901), 9, 81–82, quoted in James Harrison, "Destiny or Descent? Responses to Darwin," *Mosaic* 14, no. 1 (1981): 112.

59. James, *Guy Domville*, act 1, 136.

60. Bernard Shaw, "The Drama's Laws," *Saturday Review*, January 12, 1895, in ibid., 205–7.

61. Arnold Bennett, "Fitful Beauty," *Woman*, January 16, 1895, in ibid., 216.

62. Herbert George Wells, "A Pretty Question," *Pall Mall Gazette*, January 7, 1895, in ibid., 211–12.

63. John Stokes, *In the Nineties* (London: Harvester Wheatsheaf, 1989), 116–43.

64. Thomas Malthus, *An Essay on the Principle of Population*, ed. George Thomas Bettany (London: Ward, Lock, 1890), viii.

65. Hawkins, *Social Darwinism*, 170.

66. Ibid., 171.

67. Gideon Lewis-Kraus, "It's Good to Be Alive," *London Review of Books*, February 9, 2012, 36.

68. Amy Cook provides a succinct overview of the key research on mirror neurons and its implications for theories of acting in "Interplay: The Method and Potential of a Cognitive Scientific Approach to Theatre," *Theatre Journal* 59, no. 4 (2007): 579–94. See also Bruce McConachie, "Falsifiable Theories for Theatre and Performance Studies," *Theatre Journal* 59, no. 4 (2007): 533–77.

69. Jane Goodall, *Performance and Evolution in the Age of Darwin: Out of the Natural Order* (London: Routledge, 2002), 7.

70. Joseph Roach, "Darwin's Passion: The Language of Expression on Nature's Stage," *Discourse* 13, no. 1 (1990–1991): 52, 53.

71. For discussion of Irving in *The Bells*, see, especially, Joseph R. Roach, *The Player's Passion: Studies in the Science of Acting* (Ann Arbor: University of Michigan Press, 1993). On mesmerism, see Jane R. Goodall, *Stage Presence* (London: Routledge, 2008), chap. 3.

72. Tom Gunning, "In Your Face: Physiognomy, Photography, and the Gnostic Mission of Early Film," in *The Mind of Modernism: Medicine, Psychology, and the Cultural Arts in Europe and America, 1880–1940*, ed. Mark S. Micale (Stanford, Calif.: Stanford University Press, 2004), 154.

73. Roach, *Player's Passion* and "Darwin's Passion"; Rose Whyman, *The Stanislavsky System of Acting* (Cambridge: Cambridge University Press, 2008); Lynn M. Voskuil, *Acting Naturally: Victorian Theatricality and Authenticity* (Charlottesville: University of Virginia Press, 2004).

74. Gunning, "In Your Face," 154.

75. Ibid.

76. Whyman, *Stanislavsky System*, 5.

77. Ibid., 6.

78. James M. Barrie, *What Every Woman Knows*, in *Peter Pan and Other Plays*, ed. Peter Hollindale (Oxford: Oxford University Press, 2008), act 4, i, 229.

79. Goodall, *Performance and Evolution in the Age of Darwin*, 177.

80. Bruce McConachie, review of *The Actor, Image, and Action: Acting and Cognitive Neuroscience*, by Rhonda Blair, *TDR: The Drama Review* 54, no. 2 (2010): 183.

81. Rae Beth Gordon, *Dances with Darwin, 1875–1919: Vernacular Modernity in France* (Farnham, Eng.: Ashgate, 2009), 2.

82. Ibid., 22.

83. John R. G. Turner, "Mimicry," in *Encyclopedia of Evolution*, ed. Mark D. Pagel, 2 vols. (Oxford: Oxford University Press, 2002), 1:734, 736.

84. Ibid., 732.

85. George Henry Lewes, *On Actors and the Art of Acting* (London: Smith, Elder, 1875), 103. Lord Henry Wotton in Oscar Wilde's *The Picture of Dorian Gray* says virtually the same thing: "We are no longer the actors, but the spectators of the play" (quoted in Voskuil, *Acting Naturally*, 19).

86. Susan Bassnett, "Eleonora Duse," in *Bernhardt, Terry, Duse: The Actress in Her Time,* ed. John Stokes, Michael R. Booth, and Susan Bassnett (Cambridge: Cambridge University Press, 1988), 166. On Duse, see also Arthur Symons, *Eleonora Duse* (London: Elkin Mathews, 1926); William Weaver, *Duse: A Biography* (London: Thames and Hudson, 1984); Laura M. Hansson, *Modern Women* (London, 1896); Eva Le Gallienne, *The Mystic in the Theatre: Eleonora Duse* (London: Bodley Head, 1966); Jeanne Bordeux [pseud.], *Eleonora Duse: The Story of Her Life* (London: Hutchinson, 1925); and Helen Sheehy, *Eleonora Duse* (New York: Knopf, 2003).

87. The article in the *Saturday Review* is reprinted in Bernard Shaw, *Our Theatres in the Nineties* (London: Constable, 1932), 13:148–54.

88. Charles Darwin, *The Expression of the Emotions in Man and Animals,* ed. Joe Cain and Sharon Messenger (London: Penguin, 2009), 286.

89. Gerhart Hauptmann, *Lonely People,* in *The Dramatic Works of Gerhart Hauptmann,* ed. Ludwig Lewisohn, vol. 3, *Domestic Dramas* (New York: Huebsch, 1922), act 2, 184; act 3, 236, 245; act 5, 304.

90. Maurice Maeterlinck, *The Betrothal, or The Blue Bird Chooses,* trans. Alexander Teixeira de Mattos (London: Methuen, 1919), 149.

91. Bernard Shaw, *Man and Superman,* in *The Complete Plays of Bernard Shaw* (London: Odhams Press, 1934), act 2, 355.

92. Roach, "Darwin's Passion," 54.

93. Quoted in Bassnett, "Eleonora Duse," 152.

94. Roach, "Darwin's Passion," 55.

95. Bassnett, "Eleonora Duse," 155.

96. Ibid.167.

97. Sos Eltis, *Acts of Desire: Women and Sex on Stage, 1800–1930* (Oxford: Oxford University Press, 2013). See also Voskuil, *Acting Naturally*; and John Stokes and Maggie Gale, eds., *The Cambridge Companion to the Actress* (Cambridge: Cambridge University Press, 2007).

98. Hansson, *Modern Women,* 106. Already in his early play *The Vikings at Helgeland* (1857), Ibsen asks for "repressed emotion" in many of his stage directions. See *The Vikings at Helgeland,* trans. William Archer (New York: Scribner, 1911), 61, 74, 94.

99. Elizabeth Robins, *The Mirkwater* (manuscript, Fales Collection, New York).

100. Hauptmann, *Lonely People,* act 3, 246.

101. Bassnett, "Eleonora Duse," 141–42.

102. Ibid., 145.

103. Ibid., 141.

104. Hugo Von Hoffmansthal, quoted in ibid., 154; Gay Gibson Cima, *Performing Women: Female Characters, Male Playwrights, and the Modern Stage* (Ithaca, N.Y.: Cornell University Press, 1993).

105. Quoted in Rhonda Blair, *The Actor, Image, and Action: Acting and Cognitive Neuroscience* (New York: Routledge, 2008), 40.

106. Scott A. Harmon, "Attention, Absorption and Habit: The Stanislavski System Reexamined as a Cognitive Process Using the 'Theatre of Consciousness' Model of Bernard Baars" (M.A. thesis, University of Illinois, 2010), 15. Harmon is referring to Bernard Baars, *In the Theater of Consciousness* (Oxford: Oxford University Press, 1997), and his concepts of the "suggestible state" and the "absorbed state."

107. Darwin, *Expression of the Emotions*, 286.

108. Rebecca Stott, *Darwin and the Barnacle: The Story of One Tiny Creature and History's Most Spectacular Scientific Breakthrough* (London: Faber and Faber, 2003), 47.

109. Darwin, *Expression of the Emotions*, 300.

110. Ibid., 308.

111. Thomas Burgess, *The Physiology or Mechanism of Blushing* (London: Churchill, 1839), 11, quoted in Maurice S. Lee, *Uncertain Chances: Science, Skepticism, and Belief in Nineteenth-Century American Literature* (Oxford: Oxford University Press, 2012), 159.

112. Darwin, *Expression of the Emotions*, 309.

113. Ibid., 307.

114. Ibid., 298.

115. Bassnett, "Eleonora Duse," 142.

116. Luigi Rasi, quoted in ibid., 142.

117. Darwin, *Expression of the Emotions*, 288.

118. Ibid., 291.

119. Ibid., 295–97.

120. Quoted in Bassnett, "Eleonora Duse," 168.

121. Ibid., 138.

122. Luigi Pirandello, "The Art of Duse," *Columbian Monthly* 1, no. 7 (1928), quoted in ibid., 124–25.

123. Adelaide Ristori, quoted in ibid., 137.

3. "On the Contrary!"

1. "Ibsen's Place in Letters," *New York Times*, May 24, 1906; Robert Ferguson, *Henrik Ibsen: A New Biography* (London: Cohen Books, 1996), 121, 261, 316, 349.

2. Work on Henrik Ibsen and Charles Darwin includes that of Asbjørn Aarseth, "Ibsen and Darwin: A Reading of *The Wild Duck*," *Modern Drama* 48, no. 1 (2005): 1–10; Mathias Clasen, Stine Slot Grumsen, Hans Henrik Hjermitslev, and Peter C. Kjærgaard, "Translation and Transition: The Danish Literary Response to Darwin," in *The Literary and Cultural Reception of Charles Darwin in Europe*, ed.

Thomas F. Glick and Elinor Shaffer (London: Bloomsbury, 2014), 3:103–27; Linn B. Konrad, "Father's Sins and Mother's Guilt: Dramatic Responses to Darwin," in *Drama, Sex and Politics*, ed. James Redmond, Themes in Drama, vol. 7 (Cambridge: Cambridge University Press, 1991), 137–49; Thore Lie, "The Introduction, Interpretation and Dissemination of Darwinism in Norway During the Period 1860–1890," trans. James Anderson, in *The Reception of Charles Darwin in Europe*, ed. Eve-Marie Engels and Thomas F. Glick (London: Continuum, 2008), 1:156–74; Tore Rem, "Darwin and Norwegian Literature," in *Literary and Cultural Reception of Charles Darwin,* ed. Glick and Shaffer, 160–80; Ross Shideler, *Questioning the Father: From Darwin to Zola, Ibsen, Strindberg and Hardy* (Stanford, Calif.: Stanford University Press, 1999); Eivind Tjønneland, "Darwin, J. P. Jacobsen og Ibsen," *Spring—Tidsskrift for moderne dansk litteratur* 13 (1998): 178–99, and "Repetition, Recollection and Heredity in Ibsen's *Ghosts*—The Context of Intellectual History," in *Ibsen on the Cusp of the Twenty-First Century: Critical Perspectives*, ed. Pål Bjørby, Alvhild Dvergsdal, and Idar Stegane (Fyllingsdalen, Norway: Alvheim and Eide, 2005); and Tamsen Wolff, *Mendel's Theatre: Heredity, Eugenics, and Early Twentieth-Century American Drama* (New York: Palgrave Macmillan, 2009).

3. Brian W. Downs, *Ibsen: The Intellectual Background* (Cambridge: Cambridge University Press, 1946), ix.

4. Michael Meyer, *Ibsen: A Biography* (New York: Doubleday, 1971), 814.

5. Ibid., 286.

6. Tjønneland, "Darwin, J. P. Jacobsen og Ibsen," 182 (my translation). I am indebted to the author for sending me this and other articles he has written on Ibsen and evolution.

7. Throughout this chapter, I give dates of plays' first publication; dates of first performances are listed on the repertoire database at ibsen.nb.no.

8. Jane R. Goodall, *Performance and Evolution in the Age of Darwin: Out of the Natural Order* (London: Routledge, 2002), 180.

9. Wolff, *Mendel's Theatre*, 31.

10. Clasen et al., "Translation and Transition," 105.

11. Inga-Stina Ewbank, "Dickens, Ibsen, and Cross-Currents," in *Anglo-Scandinavian Cross-Currents*, ed. Inga-Stina Ewbank, Olva Lausund, and Bjørn Tysdahl (London: Norvik Press, 1999), 301.

12. Brian Johnston, *Text and Supertext in Ibsen's Drama* (University Park: Pennsylvania State University Press, 1989), 27.

13. Ferguson, *Henrik Ibsen*, 349.

14. Ibid., 121.

15. "A toad. In the middle of a block of sandstone./In a fossil world. Just his head showing" (Henrik Ibsen, *Peer Gynt*, trans. Christopher Fry and Johan Fillinger [Oxford: Oxford University Press, 2009], act 4, 89).

16. Ferguson, *Henrik Ibsen*, 316.

17. Henrik Ibsen, speech presented at a banquet in Stockholm (September 24, 1897), in *Henrik Ibsen: Samlede Værker*, vol. 10, ed. Halvdan Koht and Jens B. Halvorsen (Copenhagen: Gyldendal, 1902), 516. This and all subsequent quotations from Ibsen's works, drafts, letters, and speeches in their original Dano-Norwegian are taken from *Samlede Værker* and from *Henrik Ibsen: Efterladte Skrifter*, ed. Halvdan Koht and Julias Elias (Copenhagen: Gyldendal, 1909); all of these can now be consulted on the online resource *Henrik Ibsens Skrifter* (http://ibsen.uio.no/forside.xhtml), which gives detailed annotations as well as facsimile versions of Ibsen's writings, although as yet only in Norwegian. Given space constraints, I provide English versions of only texts quoted and refer the reader to the English translations that were used during Ibsen's lifetime: *The Collected Works of Henrik Ibsen*, ed. William Archer, which includes *From Ibsen's Workshop: Notes, Scenarios, and Drafts of the Modern Plays*, trans. Arthur G. Chater, vol. 12 (New York: Scribner, 1911). In some cases, I give my own translations in preference to those in this edition. English translations of a selection of Ibsen's letters and speeches can be found in *Ibsen: Letters and Speeches*, ed. and trans. Evert Sprinchorn (New York: Hill and Wang, 1964), but there is still no English translation of Ibsen's complete letters and speeches. All translations throughout this chapter are mine except where indicated.

18. Clasen et al., "Translation and Transition," 125.

19. Robert Chambers, *Vestiges of the Natural History of Creation*, ed. James A. Secord (Chicago: University of Chicago Press, 1994), 386, quoted in Rebecca Stott, "'Tennyson's Drift': Evolution in *The Princess*," in *Darwin, Tennyson and Their Readers: Explorations in Victorian Literature and Science*, ed. Valerie Purton (London: Anthem, 2013), 31.

20. Henrik Ibsen to Georg Brandes, October 30, 1888, in Ibsen, *Letters and Speeches*, 271–73.

21. See, for example, David Quammen, *The Song of the Dodo* (London: Hutchinson, 1996); Errol Fuller, *The Dodo: Extinction in Paradise* (Charleston, Mass.: Bunker Hill, 2003); and Mark V. Barrow, Jr., *Nature's Ghosts* (Chicago: University of Chicago Press, 2009).

22. Henrik Ibsen to Sophus Schandorph, January 6, 1882, in Ibsen, *Letters and Speeches*, 201.

23. Henrik Ibsen, speech presented at the banquet of the Norwegian League for Women's Rights (May 26, 1898), in Ibsen, *Letters and Speeches*, 337.

24. Bernard Shaw, preface to *Getting Married*, in *Prefaces by Bernard Shaw* (London: Constable, 1934), 15.

25. Ibsen, *From Ibsen's Workshop*, 87; my translation differs from Chater's.

26. Henrik Ibsen, *Pillars of Society*, act 4, 408; Archer renders it as "no eyes for womanhood."

27. Ibsen, *Pillars*, act 4, 409; Archer renders it as "the spirits of Truth and Freedom—*these* are the Pillars of Society."

28. Ibsen, *From Ibsen's Workshop*, 185.

29. Shideler, *Questioning the Father,* 57. Shideler offers critical analyses of the Darwinian elements of some of Ibsen's plays, including *Pillars, Doll's House, Ghosts,* and *Hedda.* For him, these are the plays in which "environment and heredity in Darwin's randomly evolving nature serve as fundamental elements of plot" (60).

30. Ibid., 5.

31. Quoted in J. P. Jacobsen, "Menneskeslægtens Oprindelse," *Nyt Dansk Maanedsskrift* 2 (1871): 122, quoted in Clasen et al., "Translation and Transition," 110.

32. Ibsen, *From Ibsen's Workshop*, 487.

33. Ibid., 206.

34. Quoted in Downing Cless, *Ecology and Environment in European Drama* (New York: Routledge, 2010), 141.

35. Notice (1879?), in Ibsen, *Henrik Ibsen: Samlede Værker*, 18:364 (my translation).

36. Sigurd Ibsen (Ibsen's son), in *Human Quintessence*, refers to a "great harmony of things," echoing Haeckel's popular monism but perhaps also Ibsen's ideas on the synthesis of all laws into one law, all forms into one form. Without referring anywhere in his 300-page book to his father by name, Sigurd indirectly alludes to him in his wide-ranging musings on the nature of genius and what it means to be a great man. Certainly, the "power of anticipative synthesis is conspicuous in those who tower above others intellectually" (Sigurd Ibsen, *Human Quintessence*, trans. Marcia Hargis Janson [1911; London: Palmer, 1913], 294). This book was translated into many languages and was especially popular in America; in Eugene O'Neill's autobiographical play *Ah, Wilderness!* the hero names it as one of the progressive books that influenced him in his youth. Bergliot Ibsen (Sigurd's wife) claims that in America the book had the same standing as Friedrich Nietzsche's work in central Europe. See Bergliot Ibsen, *The Three Ibsens*, trans. Gerik Schjelderup (London: Hutchinson, 1951), 169.

37. Lie, "Introduction, Interpretation and Dissemination," 157. It is not clear whether Ibsen had first-hand knowledge of Chambers's *Vestiges of the Natural History of Creation* or Lyell's *Principles of Geology*.

38. Ibid., 171. See also Peter C. Kjærgaard, Niels Henrik Gregersen, and Hans Henrik Hjermitslev, "Darwinizing the Danes, 1859–1909," in *The Reception of Charles Darwin in Europe,* ed. Eve-Marie Engels and Thomas F. Glick (London: Continuum, 2008), 1:146–55. Ferguson notes the unique people and mechanisms in place that helped disseminate new ideas throughout Scandinavia (and put paid to the myth of Ibsen as the lone crusader struggling against narrow-minded Norwegian provincialism). See also the recent work of Narve Fulsås in this area, such as "Ibsen Misrepresented: Canonization, Oblivion, and the Need

for History," *Ibsen Studies* 11, no. 1 (2011): 3–20, and the invaluable commentary he provides on the database *Henrik Ibsens Skrifter*.

39. Downs, *Ibsen*, 162; Lie, "Introduction, Interpretation and Dissemination," 161–62.

40. Lie, "Introduction, Interpretation and Dissemination," 161.

41. Peter C. Asbjørnsen, "Darwins Nye Skabningslære [Darwin's New Theory of Creation]," *Budstikken* [*The Messenger*], February–March 1861, quoted in ibid., 161–62.

42. Henrik Ibsen, *Ghosts*, act 2, 225.

43. Richard Dawkins, *The Selfish Gene* (Oxford: Oxford University Press, 2006), 201. See also William W. Demastes, *Staging Consciousness: Theater and the Materialization of Mind* (Ann Arbor: University of Michigan Press, 2002), 78.

44. Meyer, *Ibsen*, 171.

45. Lie, "Introduction, Interpretation and Dissemination," 160.

46. Toril Moi argues that the finished text of the often-ignored *Emperor and Galilean* reflects "some of the most culturally contentious issues of its time, for example in its obsession with determinism. Darwinism surely contributed to the general preoccupation with the question, but so did the emerging disciplines of statistics and probability calculation" (*Henrik Ibsen and the Birth of Modernism* [Oxford: Oxford University Press, 2006], 193).

47. Oskar Seidlin, "Georg Brandes, 1842–1927," *Journal of History of Ideas* 3 (1942): 419.

48. Ibid., 419.

49. Ibid.

50. Alrik Gustafson, *Six Scandinavian Novelists: Lie, Jacobsen, Heidenstam, Selma Lagerlöf, Hamsun, Sigrid Undset* (Minneapolis: University of Minnesota Press, 1940), 81.

51. Ibid., 82.

52. Ibid., 83.

53. Quoted in ibid., 84.

54. Clasen et al., "Translation and Transition," 103–27; see also Tjønneland, "Darwin, J. P. Jacobsen og Ibsen," 178–99.

55. James Harrison, "Destiny or Descent?: Responses to Darwin," *Mosaic* 14, no. 1 (2002): 121.

56. Tjønneland proposes that this longing indicates that Ibsen's evolutionary vision may also have been shaped by the distinctly melancholic attitude to evolution in the Norwegian Darwin literature, that he had a "pessimistic understanding of evolution," shown not only through his characters but also by himself ("Darwin, J. P. Jacobsen og Ibsen," 186).

57. Johnston, *Text and Supertext*, 80. Johnston does not cite a source for this, and I have been unable to trace it. According to Narve Fulsås (in private

correspondence with author), there are no extant notes by Ibsen to *Emperor and Galilean*.

58. Tjønneland, "Darwin, J. P. Jacobsen og Ibsen," 189. In this paragraph, Haeckel characterizes the upward development of nature, of which humankind is the "highest triumph," and predicts that humans will "reach higher and higher spiritual perfection."

59. Cless, *Ecology and Environment,* 146.

60. Ludwig Lewisohn, introduction to *The Dramatic Works of Gerhart Hauptmann,* ed. Ludwig Lewisohn, vol. 3, *Domestic Dramas* (New York: Huebsch, 1922), vii.

61. Gerhart Hauptmann, *Lonely People,* in *Dramatic Works,* act 3, 255.

62. Ibid., act 2, 210. Shaw does this in *Man and Superman,* whose opening stage directions indicate the presence of a bust of Herbert Spencer (at whom Tanner will later stare "gloomily") and an "enlarged photograph" of T. H. Huxley.

63. Hauptmann, *Lonely People,* act 4, 274. It is also worth noting another possible connection between Ibsen and Haeckel, through the name Allmers, the protagonist of *Little Eyolf.* The German poet Hermann Allmers was a lifelong friend of Haeckel, and their travels in Italy in the 1850s and 1860s parallel Ibsen's own journey and residence there (where Haeckel spent his days in landscape painting and his nights in "dancing the tarantella"). See Robert J. Richards, *The Tragic Sense of Life: Ernst Haeckel and the Struggle over Evolutionary Thought* (Chicago: University of Chicago Press, 2008), 59–63.

64. Robert Brustein, *The Theatre of Revolt: Studies in Modern Drama from Ibsen to Genet* (Chicago: Dee, 1991), 12.

65. Ibid., 78.

66. Ibsen, *From Ibsen's Workshop,* 466.

67. Tjønneland, "Darwin, J. P. Jacobsen og Ibsen," 184.

68. Ibsen's titles often seem to allude to something familiar, but in unusual language that signals more complex meanings; for example, *A Doll's House* is actually "a doll home," a much more potent term indicating not a physical building ("house") or a little girl's plaything but the problematic meaning of "home" for modern women.

69. Ibsen, *From Ibsen's Workshop,* 344, 370 ("the dying mermaid on the dry—"); Wolff, *Mendel's Theatre,* 10.

70. Ibsen, *From Ibsen's Workshop,* 366. Knut Hamsun was intrigued by Ellida's state of being "a thing which isn't a person, not even a crazy person." In Ellida's first speech, "those words about 'human beings like sea creatures' express a state of mind which correlates with the one I have when I fall sensually in love with light. My blood seems to sense that I stand in a nervous relationship with the universe, with the elements. Some day perhaps—in the fullness of time—humans will stand to today's humans as today's humans stand to today's *protista,* beings which do not need to love some other being but which can love anything at all:

water, fire, air. You know, Ibsen does have flashes of genius"(Hamsun to Amalie Skram, in *Knut Hamsun: Selected Letters, 1879–98*, ed. Harald Naess and James Walter McFarlane [London: Norvik Press, 1990], 1:83).

71. Ibsen, *From Ibsen's Workshop,* 331.

72. Ibid., 364 (emphasis added).

73. Martin Puchner, *The Drama of Ideas: Platonic Provocations in Theater and Philosophy* (Oxford: Oxford University Press, 2010), 74.

74. *Era*, May 16, 1891, 10. Rose Meller played Ellida in this production at Terry's Theatre, London (May 1891); the translation was by Eleanor Marx-Aveling. Another reviewer called the play badly acted and scoffed at the "human mermaid" Ellida's "amphibious" nature. See "Flashes from the Footlights," *Licensed Victuallers' Mirror*, May 19, 1891, 238.

75. Ibsen, *From Ibsen's Workshop*, 377.

76. Charles Darwin, *On the Origin of Species*, facsimile of the first edition, ed. Ernst Mayr (Cambridge, Mass.: Harvard University Press, 1964), 140–41.

77. Ibsen, *From Ibsen's Workshop*, 331. The *Times* dismissed *The Lady from the Sea* in 1892: "Studies in morbid heredity are very well in a scientific treatise. On the stage, put forward as a public entertainment, they tend to perplex, irritate and repel, besides being useless for any practical purpose" (quoted in Meyer, *Ibsen*, 667).

78. Ibsen, *From Ibsen's Workshop*, 331.

79. Ferguson, *Henrik Ibsen*, xii.

80. For in-depth consideration of the domestication-versus-wildness motif in Ibsen, see, for example, Tjønneland, "Darwin, J. P. Jacobsen og Ibsen"; Aarseth, "Ibsen and Darwin"; and Rem, "Darwin and Norwegian Literature."

81. Tjønneland, "Darwin, J. P. Jacobsen og Ibsen," 187.

82. Ibsen, *From Ibsen's Workshop*, 185.

83. Max Nordau erroneously claimed that Ibsen actually "quotes Darwin" in his plays, in *Degeneration* (Lincoln: University of Nebraska Press, 1993), 350.

84. Downs, *Ibsen*, 165.

85. Ibsen, *From Ibsen's Workshop*, 331.

86. Henrik Ibsen, *The Lady from the Sea,* act 3, 255.

87. Peter J. Bowler, *The Non-Darwinian Revolution: Reinterpreting a Historical Myth* (Baltimore: Johns Hopkins University Press, 1988), 101.

88. Ibsen, *From Ibsen's Workshop,* 480.

89. Ibid. In the draft to *Little Eyolf,* the eleven-year-old Alfred/Eyolf is "slim, slight . . . and undersized, and looks somewhat delicate" (471).

90. Henrik Ibsen, *John Gabriel Borkman,* act 2, 258.

91. Wolff, *Mendel's Theatre*, 21.

92. Ibsen, *From Ibsen's Workshop*, 38.

93. For wider contextualization of this emphasis, see Elisabeth Gitter, "The Power of Women's Hair in the Victorian Imagination," *PMLA* 99, no. 5 (1984): 936–54; and Carol Hanbery MacKay, *Creative Negativity: Four Victorian Exemplars of the Female Quest* (Stanford, Calif.: Stanford University Press, 2001), 28–34.

94. Ibsen, *From Ibsen's Workshop,* 17.

95. I am indebted to Jane R. Goodall for suggesting to me this further implication of Hedda's hair.

96. Charles Darwin, *The Expression of the Emotions in Man and Animals,* ed. Joe Cain and Sharon Messenger (London: Penguin, 2009), 273–74.

97. Ibsen, *From Ibsen's Workshop,* 475.

98. Meyer, *Ibsen,* 540.

99. Henrik Ibsen, speech presented to the Swedish Society of Authors, Stockholm (April 11, 1898) (my translation). See also *Ibsen: Letters and Speeches,* 335.

100. H. A. E. Zwart, "The Birth of a Research Animal: Ibsen's *The Wild Duck* and the Origin of a New Animal Science," *Environmental Values* 9 (2000): 93, 100.

101. Cless, *Ecology and Environment,* 142–43.

102. Ibid., 143.

103. Ibsen, *From Ibsen's Workshop,* 492.

104. Margot Norris, *Beasts of the Modern Imagination: Darwin, Nietzsche, Kafka, Ernst, and Lawrence* (Baltimore: Johns Hopkins University Press, 1985), 16.

105. "I am a miner's son," explains Borkman, "and my father used sometimes to take me with him into the mines. The metal sings down there." He characterizes the hammer as setting the "metal" free because "it wants to come up into the light of day and serve mankind" (Henrik Ibsen, *John Gabriel Borkman,* act 2, 207–8).

106. Jessica H. Whiteside, "Wedges and Impacts: Darwin's Enduring Legacy," *Brown Medicine,* spring 2009, 2. I am indebted to James A. Secord for pointing me toward the wedge metaphor.

107. Darwin, *Origin of Species,* 67. Darwin includes the wedge metaphor in volume 2 of *Variation of Animals and Plants under Domestication* (London: Murray, 1868).

108. Whiteside, "Wedges and Impacts," 2.

109. Ibid.

110. Darwin, *Origin of Species,* 14–15.

111. Ibsen, *From Ibsen's Workshop,* 186.

112. Francis Galton, *Hereditary Genius: An Inquiry into Its Laws and Consequences* (London: Macmillan, 1869, 1892), 1.

113. Downs, *Intellectual Background,* 165.

114. Georg Brandes to the biologist Carl J. Salomonsen, quoted in Meyer, *Ibsen*, 389.

115. Wolff, *Mendel's Theatre*, 33.

116. Ibsen, *From Ibsen's Workshop*, 177.

117. Heather McHold, "Even as You and I: Freak Shows and Lay Discourse on Spectacular Deformity," in *Victorian Freaks: The Social Context of Freakery in Britain,* ed. Marlene Tromp (Columbus: Ohio State University Press, 2008), 24.

118. Ibsen, *From Ibsen's Workshop*, full draft of *A Doll's House*, act 1, 389.

119. Ibid., 390. Tjønneland notes Dr. Rank's "marked Darwinistic standpoint" in the draft, in "Darwin, J. P. Jacobsen og Ibsen," 183.

120. Tjønneland, "Darwin, J. P. Jacobsen og Ibsen," 184. Ibsen clearly mused on the concept of "base" or "lower" nature. In the draft of *The Pillars of Society*, the reverend Dr. Rörlund piously inveighs against "unaided human nature" (which finds things like dinner parties on a Sunday amusing) as something to be "overcome." There is some discussion between him and Dina of her "nature"—he says she must change it, but she feels that is just how she is. In the draft of *The Wild Duck*, Hjalmar, having discovered the truth about Gina, alludes to her "lower nature"(Ibsen, *From Ibsen's Workshop*, 35, 237).

121. Ibsen, *From Ibsen's Workshop*, 399.

122. Ibid., full draft of *A Doll's House*, act 1, 390–91 (emphasis added).

123. Ibsen, *From Ibsen's Workshop*, 186.

124. Zwart, "Birth of a Research Animal," 92.

125. Wolff, *Mendel's Theatre*, 36.

126. Gustafson, *Six Scandinavian Novelists*, 18.

127. Ibid. As Donald Mackenzie notes, "intellectual aristocracy" is a concept associated with the field of biometric eugenics headed by Karl Pearson and his followers in the late nineteenth and early twentieth centuries ("Sociobiologies in Competition," in *Biology, Medicine and Society, 1840–1940*, ed. Charles Webster [Cambridge: Cambridge University Press, 1981], 276).

128. Gustafson, *Six Scandinavian Novelists*, 19.

129. Bart Van Es, *Shakespeare in Company* (Oxford: Oxford University Press, 2013), 304–6.

130. Cless, *Ecology and Environment*, 140.

131. Ibsen, *From Ibsen's Workshop,* 331. A strong echo of this occurs in Shaw's play *On the Rocks* (1933), when Chavender proposes harnessing "power from the tides" (Bernard Shaw, *On the Rocks*, in *The Complete Plays of Bernard Shaw* [London: Odhams Press, 1934], act 2, 1199).

132. Ibsen, *From Ibsen's Workshop,* 309–10.

133. Ibid., 310.

134. Henry Arthur Jones, *The Corner Stones of Modern Drama* [lecture presented at Harvard University on October 31, 1906] (London: Chiswick Press, 1906), 6.

135. In *Notebook of Malte Laurids Brigge*, Rainer Maria Rilke paid tribute to Ibsen and described his works as wild animals: "Loneliest of men, withdrawn from all. . . . [The people who once opposed you now] carry your words about with them in the cages of their presumption, and exhibit them in the streets and excite them a little from their own safe distance: all those wild beasts of yours. When I first read you, they broke loose on me and assailed me in my wilderness—your desperate words" (quoted in Meyer, *Ibsen*, 816). This long passage is full of scientific allusions—microscope, test tubes, fossils, gradualism:

> You conceived the vast project of magnifying single-handed these minutiæ, which you yourself first perceived only in test-tubes, so that they should be seen of thousands, immense, before all eyes. Then your theatre came into being. You could not wait until this almost spaceless life, condensed into fine drops by the weight of centuries, should be discovered by the other arts, and gradually made visible to the few who, little by little, come together in their understanding and finally demand to see the general confirmation of these extraordinary rumours in the semblance of the scene opened before them. . . . You had to determine and record the almost immeasurable: the rise of half a degree in a feeling; the angle of refraction, read off at close quarters, in a will depressed by an almost infinitesimal weight; the slight cloudiness in a drop of desire, and the well-nigh imperceptible change of colour in an atom of confidence. All these: for of just such processes life now consisted, our life, which had slipped into us and had drawn so deeply in that it was scarcely possible even to conjecture about it any more. (Rilke, *Notebook*, trans. John Linton, 76–79, quoted in Meyer, *Ibsen*, 816–17)

This sounds like a climate more than a drama; in essence, Rilke describes Ibsen's drama in terms drawn from meteorology.

136. Henry James, "John Gabriel Borkman," *Harper's Weekly*, February 6, 1897, in Henry James, *The Scenic Art: Notes on Acting and the Drama, 1872–1901* ed. Allan Wade (London: Hart-Davis, 1949), 291.

137. Ibid.

138. Ibid., 292.

4. "Ugly . . . but Irresistible"

1. Sally Shuttleworth, "Demonic Mothers: Ideologies and Bourgeois Motherhood in the Mid-Victorian Era," in *Rewriting the Victorians,* ed. Linda M. Shires (London: Routledge, 1992), 37.

2. Sarah Blaffer Hrdy, "Motherhood," in *Encyclopedia of Evolution*, ed. Mark D. Pagel, 2 vols. (Oxford: Oxford University Press, 2002), 1:E64. See also Sarah

Blaffer Hrdy, *Mother Nature: Maternal Instincts and How They Shape the Human Species* (New York: Ballantine, 1999), and *The Woman That Never Evolved*, rev. ed. (Cambridge, Mass.: Harvard University Press, 1999).

3. Penny Farfan, *Women, Modernism, and Performance* (Cambridge: Cambridge University Press, 2004), 4.

4. Quoted in ibid., 5.

5. Jean Chothia, introduction to *The New Woman and Other Emancipated Woman Plays*, ed. Jean Chothia (Oxford: Oxford University Press, 1998), ix.

6. Mike Hawkins, *Social Darwinism in European and American Thought, 1860–1945* (Cambridge: Cambridge University Press, 1997), 252, 253. See also Hrdy, "Motherhood," E56–64.

7. Chothia, introduction to *New Woman*, ed. Chothia, ix. See also Sos Eltis, *Acts of Desire: Women and Sex on Stage, 1800–1930* (Oxford: Oxford University Press, 2013), chap. 6.

8. Angelique Richardson, "The Biological Sciences," in *A Companion to Modernist Literature and Culture,* ed. David Bradshaw and Kevin J. H. Dettmar (Oxford: Blackwell, 2006), 59.

9. See, for example, Cynthia Eagle Russett, *Sexual Science: The Victorian Construction of Womanhood* (Cambridge, Mass.: Harvard University Press, 1989), 43.

10. Andrew Sinclair, *The Emancipation of the American Woman* (1965), quoted in ibid., acknowledgments.

11. Quoted in Richardson, "Biological Sciences," 59. Allen fulminates against educating women to the extent that they no longer want to fulfill their biological function, as they will become "flat as a pancake and as dry as a broomstick" (60).

12. Charles Darwin, *The Descent of Man, and Selection in Relation to Sex,* ed. James Moore and Adrian Desmond (London: Penguin, 2004), 631, 629.

13. Quoted in Howard I. Kushner, "Suicide, Gender, and the Fear of Modernity in Nineteenth-Century Medical and Social Thought," *Journal of Social History* 26, no. 3 (1993): 472.

14. *Spectator*, April 7, 1894, quoted in *Plays by Henry Arthur Jones*, ed. and intro. Russell Jackson (Cambridge: Cambridge University Press, 1982), 17.

15. Hawkins, *Social Darwinism*, 161, 251.

16. Angelique Richardson, "Against Finality: Darwin, Mill and the End of Essentialism," *Critical Quarterly* 53, no. 2 (2011): 28. See also Elizabeth Grosz, *Time Travels: Feminism, Nature, Power* (Durham, N.C.: Duke University Press, 2005), 1–42.

17. Mona Caird, "Marriage," *Westminster Review*, August 1888, 186, in *A New Woman Reader*, ed. Carolyn Christensen Nelson (Peterborough, Ont.: Broadview, 2001), 185.

18. See, for example, Gilman's utopian novel *Herland* (1915), imagining an all-female civilization, and her nonfiction writing, such as *Women and Economics:*

A Study in the Economic Relation Between Men and Women as a Factor in Social Evolution (London: Putman, Small, Maynard, 1899).

19. Hawkins, *Social Darwinism*, 260–61.

20. Ibid., 261.

21. Quoted in Hrdy, *Woman That Never Evolved*, 12.

22. Hawkins, *Social Darwinism*, 263.

23. Helen Hamilton Gardener, "Environment: Can Heredity Be Modified?" in *Facts and Fictions of Life* (Boston: Fenno, 1895), 297.

24. Helen Hamilton Gardener, "Heredity in Its Relations to a Double Standard of Morals," in *Facts and Fictions of Life*, 201. In another essay, she draws on recent scientific findings that challenge the biometric basis for assuming women's inferiority, quoting a physiologist in the *Deutsche Revue* as saying that in fact the evidence points to women's superiority: "Darwin has demonstrated that female animals often *revert* to the masculine type, while the reverse seldom happens" (Helen Hamilton Gardener, "The Moral Responsibility of Woman in Heredity," in *Facts and Fictions of Life*, 179).

25. Hawkins, *Social Darwinism*, 256. See John Ruskin, *Sesame and Lilies: Three Lectures by John Ruskin* (London: George Allen, 1894).

26. Hrdy, *Woman That Never Evolved*, 10.

27. Hawkins, *Social Darwinism*, 257.

28. Ibid., 258 (original emphasis).

29. Eltis, *Acts of Desire*, 204.

30. Walter M. Gallichan, "The Equality of the Sexes," *Reynold's*, May 27, 1894.

31. Ibid.

32. Ibid. (emphasis added).

33. Charles Darwin, *The Expression of the Emotions in Man and Animals*, ed. Joe Cain and Sharon Messenger (London: Penguin, 2009), 197, 80.

34. Ibid., 80, 197.

35. Henry Arthur Jones, *The Physician* (London: Chiswick Press, 1897), act 1, 18.

36. Harry W. Paul, *Henri de Rothschild, 1872–1947* (Farnham, Eng.: Ashgate, 2011). For a thorough documentation of these developments in America, see Janet Golden, *A Social History of Wet-Nursing in America: From Breast to Bottle* (Cambridge: Cambridge University Press, 1996); for Britain, see Valerie A. Fildes, *Breasts, Bottles and Babies* (Edinburgh: Edinburgh University Press, 1986). More generally, see Valerie A. Fildes, *Wet Nursing: A History from Antiquity to the Present* (Oxford: Blackwell, 1988); and Vanessa Maher, *The Anthropology of Breast-Feeding: Natural Law or Social Construct* (Oxford: Berg, 1992).

37. Grant Allen, "Woman's Place in Nature," *Forum* 7 (1899), quoted in Russett, *Sexual Science*, 43.

38. Darwin, *Descent of Man*, 533.

39. Ibid., 535.

40. Ibid., 535–36.

41. Ibid., 537 (emphasis added).

42. Elisabeth Badinter, *Le Conflit, la femme et la mère* (Paris: Flammarion, 2010), and *The Myth of Motherhood: An Historical Overview of the Maternal Instinct*, trans. Roger DeGaris (London: Souvenir Press, 1981). See also Hrdy, *Woman That Never Evolved* and *Mother Nature* (whose arguments Badinter fundamentally misinterprets).

43. Rae Beth Gordon, *Dances with Darwin, 1875–1910: Vernacular Modernity in France* (Farnham, Eng.: Ashgate, 2009), 4.

44. John Perry, *James A. Herne: The American Ibsen* (Chicago: Nelson-Hall, 1978); Dorothy S. Bucks and Arthur H. Nethercot, "Ibsen and Herne's *Margaret Fleming*: A Study of the Early Ibsen Movement in America," *American Literature* 17, no. 4 (1946): 311–33; Donald Pizer, "The Radical Drama in Boston, 1889–1891," *New England Quarterly* 31, no. 3 (1958): 361–74; Bernard Hewitt, "'Margaret Fleming' in Chickering Hall: The First Little Theatre in America?" *Theatre Journal* 34, no. 2 (1982): 165–71; Herbert J. Edwards and Julie A. Herne, *James A. Herne: The Rise of Realism in the American Drama* (Orono: University of Maine Press, 1964); Arthur Hobson Quinn, introduction to *Margaret Fleming*, in *Representative American Plays: From 1767 to the Present Day*, ed. Arthur Hobson Quinn, 7th ed. (New York: Appleton-Century-Crofts, 1957), 515–18.

45. "There is not, he declares [in an article in the *Arena* in November 1890], and there never has been, a literary institution which can be called the American Drama" (Dion Boucicault, "The Future of American Drama," quoted in *Review of Reviews* 1 [January–June 1890]: 585).

46. Jean Chothia, *Forging a Language: A Study of the Plays of Eugene O'Neill* (Cambridge: Cambridge University Press, 1979), 23.

47. Gerald Bordman, *The Oxford Companion to American Theatre* (Oxford: Oxford University Press, 1984), 461.

48. Theodore Hatlen, "'Margaret Fleming' and the Boston Independent Theatre," *Educational Theatre Journal* 8, no. 1 (1956): 18. The play was also staged for one matinee in New York (December 9, 1891), where it was "an unequivocal failure," and in Chicago (July 7–16, 1892), where it received mixed reviews despite Herne's concession to the producer to "soften" the play by allowing Philip and Margaret to be reconciled in the end. A further New York production, in April 1894, fared no better. Then, sixteen years after its Boston premiere (though unfortunately after Herne's death), the play finally had theatrical success in Chicago in January 1907 in a production starring Chrystal Herne (James and Katharine Herne's daughter)—an unexpected critical and popular success so great that the directors wanted to take it to New York after an extended Chicago run. But Chrystal Herne had already signed a three-year contract with another

company; so just at the play's "great moment . . . there was no one to grasp it" (Edwards and Herne, *James A. Herne*, 69–73).

49. This distinguished audience recalls the "gathering of literary, social, and scientific luminaries such as only Boston could assemble" for the debates in 1860 between Louis Agassiz and Asa Gray over Darwin's theory of natural selection. Though American assimilation of Darwin's ideas may have been slower than British, Boston was an important center for debate and discussion of them. See Russett, *Darwin in America*, 9.

50. Hamlin Garland, "On the Road with James A. Herne," *Century Magazine*, August 1914, 577, quoted in Hatlen, "'Margaret Fleming' and the Boston Independent Theatre," 18 (emphasis added).

51. Edwards and Herne, *James A. Herne*, 54. Boucicault's *The Octoroon* (1859) can hardly be dismissed as mere "sentimental melodrama"; while certainly melodramatic in many ways, it is also a politically engaged drama that tackles the issues of slavery and racism and in so doing provides a bracing antidote to the often-saccharine and reductive stage adaptations of *Uncle Tom's Cabin* that were popular throughout the last few decades of the nineteenth century and well into the twentieth.

52. Bucks and Nethercot, "Ibsen and Herne's *Margaret Fleming*," 327–30.

53. Edwards and Herne, *James A. Herne*, 161nn.11, 13.

54. Despite the introduction of better copyright laws for dramatic authors in this period, "Herne did not publish his plays, fearing that they would be performed without his consent" (Arthur Hobson Quinn, "Act III of James A. Herne's *Griffith Davenport*," *American Literature* 24, no. 3 [1952]: 330, prefatory note).

55. Gary A. Richardson, for example, clearly bases his detailed plot summary and extended discussion of the play on the reconstruction, but only later mentions in passing that there are two versions, in *American Drama from the Colonial Period Through World War I: A Critical History* (New York: Twayne, 1993), 195.

56. See, for example, Donald Pizer, "An 1890 Account of Margaret Fleming," *American Literature* 27, no. 2 (1955): 264–67, which reprints and discusses Hamlin Garland's detailed plot synopsis of the play.

57. *Boston Evening Transcript*, July 8, 1890.

58. Tice L. Miller, "Plays and Playwrights: Civil War to 1896," in *The Cambridge History of American Theatre*, vol. 2, *1870–1945*, ed. Don Wilmeth and Christopher W. E. Bigsby (Cambridge: Cambridge University Press, 1999), 254.

59. Perry, *James A. Herne*, 191.

60. James A. Herne, *Margaret Fleming*, rev. and ed. Mrs. James A. Herne, in *Representative American Plays*, ed. Quinn, 544. The play was also published in Myron Matlaw, ed., *Nineteenth-Century American Plays* (New York: Applause

Books, 1967), with an introduction discussing the differences between the original (lost) version and Katharine Herne's reconstruction. Perry also lays out these differences in *James A. Herne*, 190–92.

61. Edwards and Herne, *James A. Herne*, 164n.36.

62. Bucks and Nethercot, "Ibsen and Herne's *Margaret Fleming*," 317.

63. Herne, *Margaret Fleming*, 540 (emphasis added). Apparently, this breast-feeding scene was Katharine's idea, according to Edwards and Herne, *James A. Herne*, 164.

64. Marc Robinson offers an insightful and persuasive analysis of this moment in light of Herne's innovative realist aesthetics in *The American Play, 1787–1900* (New Haven, Conn.: Yale University Press, 2009), 117–25, arguing that this moment in particular exemplifies the play's "hesitating realism" (124).

65. *New York Spirit of the Times*, May 9, 1891, quoted in Perry, *James A. Herne*, 158.

66. Thomas Russell Sullivan, quoted in ibid., 156.

67. *New York Spirit of the Times*, May 9, 1891, quoted in ibid., 158.

68. Perry, *James A. Herne*, 158.

69. Ibid., 183.

70. Introduction to Wilkie Collins, *Hide and Seek* (Oxford: Oxford University Press, 1993), xvii.

71. Ibid., 81.

72. Lillian Nayder, *Unequal Partners: Charles Dickens, Wilkie Collins, and Victorian Authorship* (Ithaca, N.Y.: Cornell University Press, 2002), 77.

73. For particularly potent evidence that public breast-feeding is a contentious issue in the United States, see *New York Times*, October 21, 2003, and June 7, 2005.

74. Darwin, *Descent of Man*, 462.

75. Interspecies nursing has a long history, as Fildes notes in *Breasts, Bottles and Babies*, 272–73.

76. Bucks and Nethercot, "Ibsen and Herne's *Margaret Fleming*," 18.

77. William Winter, quoted in ibid., 318.

78. Don Wilmeth, ed., *The Cambridge Guide to American Theatre*, 2nd ed. (Cambridge: Cambridge University Press, 2007), 426.

79. Herne, *Margaret Fleming*, 526–27.

80. Ibid., 527.

81. Quoted in Pizer, "An 1890 Account of Margaret Fleming," 266.

82. Richardson, *American Drama*, 190.

83. Ibid., 193. There are many theatrical variations on this theme, most notably Rachel Crothers's *A Man's World* (1910), in which Frank, a woman who like Margaret has adopted a child not her own, rejects her lover because of what he

has done to the fallen woman. She honors the dead mother by chastising the man who ruined her and in so doing indicates that she finds fulfillment in her adopted son rather than in sexual love.

84. *Boston Post,* May 5, 1891, 4, quoted in Perry, *James A. Herne,* 158.

85. Hatlen, "'Margaret Fleming' and the Boston Independent Theatre," 19.

86. Matthew Arnold, "The French Play in London," *Nineteenth Century*, August 1879, 243, quoted in John Stokes, *Resistible Theatres: Enterprise and Experiment in the Late Nineteenth Century* (London: Elek Books, 1972), 4.

87. Sadiah Qureshi, in discussion with the author, April 2012. Sometimes spectators could pay extra to arrange private viewings of these display peoples in their homes, where conditions would have been more conducive to breast-feeding.

88. First performed in February 1901 (for a three-week run) and again in early February 1907, though no performance date is given. See James B. Sanders, *André Antoine, directeur a l'Odéon* (Paris: Minard, 1978). There is no record of performance in England, although the *Athenaeum* mentions *Les Remplaçantes* as slated for a "season of French plays at the Avenue" in 1904.

89. Margaret Wiley, "Mother's Milk and Dombey's Son," *Dickens Quarterly* 13, no. 4 (1996): 225.

90. Darwin, *Descent of Man*, 143.

91. Brieux, *Les Remplaçantes*, 8 (stage directions).

92. Jeffrey D. Mason, "'Affront or Alarm': Performance, the Law and the 'Female Breast' from Janet Jackson to *Crazy Girls*," *New Theatre Quarterly* 21, no. 2 (2005): 179.

93. Ibid., 192n.7.

94. Anne Varty, "The Rise and Fall of the Victorian Stage Baby," *New Theatre Quarterly* 21 (2005): 218.

95. Claire Tomalin, quoted in ibid., 218.

96. Ibid.

97. Ibid., 228.

98. Perry, *James A. Herne*, 159, 170–71.

99. Quoted in ibid., 53.

100. James A. Herne, *Hearts of Oak*, act 5, 314, in James A. Herne, *Shore Acres and Other Plays*, rev. and ed. Mrs. James A. Herne (New York: French, 1928).

101. Ibid., act 3, 284–85.

102. *Boston Evening Transcript*, quoted in Perry, *James A. Herne*, 244.

103. Herne, *Shore Acres*, act 4, 112.

104. Ibid., act 2, 73.

105. Ibid., act 1, 20.

106. Varty, "Rise and Fall," 218–29.

107. Alex Roe, correspondence with the author, September 4, 2010. My thanks to Alex Roe for sharing these insights with me.

108. Ibid., September 2, 2010.

109. Martin Denton, review of *Margaret Fleming*, September 24, 2007, indie theater now, http://nytheatre.com/Review/martin-denton-2007-9-24-margaret-fleming (accessed October 28, 2014). Note that Denton does not mention the breast-feeding.

110. Sally Mitchell, *Daily Life in Victorian England* (Westport, Conn.: Greenwood Press, 1996), 125. See also Fildes, *Wet Nursing.*

111. Henrik Ibsen, *From Ibsen's Workshop: Notes, Scenarios, and Drafts of the Modern Plays*, trans. Arthur G. Chater (New York: Scribner, 1911), 12:174.

112. Ian Hacking, *Representing and Intervening* (Cambridge: Cambridge University Press, 1983), 131.

113. See, especially, Shuttleworth, "Demonic Mothers," 31–51.

114. *The Interlopers* was staged on September 15, 1913, at the Royalty Theatre, London.

115. H. M. Harwood, *The Supplanters* (London: Benn, 1926), 85.

116. Ibid., 84.

117. Ibid. In Gerhart Hauptmann's play *Lonely People* (1891), Kitty puts her husband above her child: "I believe I could sooner give the baby up than you" (Gerhart Hauptmann, *Lonely People,* in *The Dramatic Works of Gerhart Hauptmann,* ed. Ludwig Lewisohn, vol. 3, *Domestic Dramas* [New York: Huebsch, 1922], act 2, 222).

118. Anne Varty, *Children and Theatre in Victorian Britain* (London: Palgrave Macmillan, 2008), 15. See also Marah Gubar, "The Drama of Precocity: Child Performers on the Victorian Stage," in *The Nineteenth-Century Child and Consumer Culture,* ed. Dennis Denisoff (Farnham, Eng.: Ashgate, 2008), 63–78. Hawkins confirms that for the social Darwinists: "Women and children occupy the same position as 'savages' in the scale of evolution" (*Social Darwinism*, 34).

119. See, for example, Golden, *Social History of Wet-Nursing in America.*

120. Wiley, "Mother's Milk and Dombey's Son," 217.

121. Golden, *Social History of Wet-Nursing in America*, 127.

122. Ibid., 98.

123. Ibid., 102.

124. Sally Shuttleworth, "Tickling Babies: Gender, Authority, and 'Baby Science,'" in *Science in the Nineteenth-Century Periodical: Reading the Magazine of Nature* ed. Geoffrey Cantor et al. (Cambridge: Cambridge University Press, 2004), 201. As Darwin's biographers have observed, Darwin experimented frequently on his own babies, for instance, deliberately startling them into crying to note their instinctive physiological responses. See, for example, Janet Browne, *Charles Darwin: Voyaging*, vol. 1 of *A Biography* (London: Pimlico, 2003); and

Sally Shuttleworth, *The Mind of the Child: Child Development in Literature, Science, and Medicine, 1840–1900* (Oxford: Oxford University Press, 2010), especially chapter 12, "Experiments on Babies."

125. Joseph Jacob, quoted in Shuttleworth, "Tickling Babies," 205.

126. Shuttleworth, "Tickling Babies," 209.

127. Ibid., 206.

128. Herne, *Margaret Fleming*, 541.

129. Kerry Powell, *Women and Victorian Theatre* (Cambridge: Cambridge University Press, 1997), 136.

130. Clement Scott, "Tares," *Theatre*, March 1, 1888, 152. He also notes that Beringer herself admits drawing on Gustav Freitag's *Graf Waldemar* (1850) for inspiration.

131. Ibid., 153.

132. *Tares: A Social Problem in Three Acts* [no author given] (London: Miles, 1887). Printed as manuscript, for private circulation only. Lord Chamberlain Play Collection, 9.

133. Ibid., 10.

134. Ibid., 11.

135. Ibid., 39.

136. Ibid., 60–62.

137. Ibid., 62.

138. Ibid., 63.

139. Ibid., 65.

140. Oscar Wilde to unidentified correspondent, February 23, 1983, in *The Complete Letters of Oscar Wilde*, ed. Merlin Holland and Rupert Hart-Davis (London: Fourth Estate, 2000), 353. I am indebted to Sos Eltis for pointing this letter out to me.

141. Powell, *Women and Victorian Theatre*, 136.

142. Quoted in Powell, *Women and Victorian Theatre*, 138. The play has been temporarily removed from the Lord Chamberlain collection in the British Library for conservation due to its fragile condition and can therefore not be consulted at this time. Crackanthorpe's correspondence with Lady Randolph Churchill shows that she repeatedly tried to have *The Turn of the Wheel* published in the *Anglo-Saxon Review*; the Churchill papers indicate that it was rejected, even though she persevered for nearly a year. A novel called *The Turn of the Wheel* by Crackanthorpe's son Hubert (who committed suicide in 1896, in his mid-twenties) was published in 1901, edited by Blanche, but its plot bears little resemblance to Powell's description of Crackanthorpe's play; certainly, there is no rejection of a baby.

143. Francine du Plessix Gray, foreword to Badinter, *Myth of Motherhood*, xii–xiii.

144. Garland, "On the Road," 577, quoted in Hatlen, "'Margaret Fleming' and the Boston Independent Theatre," 19.

145. "Oh come," Janet says to her parents, "I could have got along quite well without a father if it comes to that." She says her father has never done anything for her or her sister except "try and prevent us from doing something we wanted to do" (St. John Hankin, *The Last of the De Mullins*, in *The Dramatic Works of St John Hankin*, intro. John Drinkwater [London: Secker, 1912], 123).

146. Shuttleworth, "Demonic Mothers," 39–40. See also Nurse's comment to Juliet: "Were not I thine only nurse, I would say thou hadst sucked wisdom from thy teat" (*Romeo and Juliet*, I.iii, in *William Shakespeare: The Complete Works*, ed. Alfred Harbage [New York: Viking, 1969], 864).

147. Hrdy, "Motherhood, " E57.

148. Perry, *James A. Herne*, 186–87.

149. Ibid., 188.

150. Helen Hamilton Gardener, "Woman as an Annex," in *Facts and Fictions of Life*, 132.

151. Ibid., 130.

152. Ibid., 132.

153. Ibid., 138.

154. Gardener, "Moral Responsibility of Women," 169.

155. Ibid., 153–55.

156. Ibid., 177–78.

157. Ibid., 183.

158. Hankin, *Last of the De Mullins,* act 3, 126.

159. Eltis, *Acts of Desire*, 179.

160. Hankin, *Last of the De Mullins,* act 3, 126.

161. Eltis, *Acts of Desire,* 179n.68.

162. Hankin, *Last of the De Mullins,* act 3, 127.

163. Janet's impulse may be eugenic—co-opting "women's desire for the reproductive good of the race" (Eltis, *Acts of Desire*, 180)—but it could also be a case of desire for desire's sake and for the fulfillment of the maternal urge, not necessarily to improve the race through its genetic stock.

164. Max Beerbohm, *Last Theatres, 1904–1910* (London: Hart-Davis, 1970), 414.

165. Ibid., 413–15; originally reviewed December 12, 1908.

166. Sophie Treadwell, *Constance Darrow*, in *Broadway's Bravest Woman: Selected Writings of Sophie Treadwell*, ed. Jerry Dickey and Miriam López-Rodriguez (Carbondale: Southern Illinois University Press, 2006), 107. Dickey points out that the play was retitled *The High Cost* and copyrighted in 1911 ("Treadwell the Dramatist," 72).

167. Dickey, "Treadwell the Dramatist," 73.

168. Ibid.
169. Ibid.

5. Edwardians and Eugenicists

1. August Strindberg, "To the Heckler," in *Selected Essays*, ed. and trans. Michael Robinson (Cambridge: Cambridge University Press, 1996), 159, quoted in Tamsen Wolff, *Mendel's Theatre: Heredity, Eugenics, and Early Twentieth-Century American Drama* (New York: Palgrave Macmillan, 2009), 231n.75.

2. Michael Meyer, *Strindberg: A Biography* (London: Secker and Warburg, 1985), 30, 32.

3. Robert Brustein, *The Theatre of Revolt: Studies in Modern Drama from Ibsen to Genet* (Chicago: Dee, 1991), 87.

4. Ibid., 100, 113.

5. Michael Robinson, introduction to Strindberg, *Selected Essays*, 12.

6. August Strindberg, "The Death's Head Moth," in Strindberg, *Selected Essays*, 150.

7. August Strindberg, "Indigo and the Line of Copper," in Strindberg, *Selected Essays*, 158.

8. Meyer, *Strindberg*, 485, 492.

9. In addition to Strindberg, *Selected Essays*, see Sue Prideaux, *Strindberg: A Life* (New Haven, Conn.: Yale University Press, 2012).

10. Strindberg, "To the Heckler," 159.

11. Robinson, editorial comment in a footnote to Strindberg, "The Mysticism of World History," in Strindberg, *Selected Essays*, 267.

12. August Strindberg to Torsten Hedlund, October 25, 1895, in Strindberg, *Selected Essays*, 12–13.

13. Strindberg, "Mysticism of World History," 207.

14. August Strindberg, "Whence We Have Come," in Strindberg, *Selected Essays*, 248, 108–9.

15. August Strindberg, "In the Cemetery," in Strindberg, *Selected Essays*, 142.

16. Strindberg, "Death's Head Moth," 154.

17. Strindberg, "Mysticism of World History,"190.

18. Wolff, *Mendel's Theatre*, 17.

19. Ibid., 20; chapter 1 gives extended analysis of Strindberg.

20. Marvin Carlson, "Ibsen, Strindberg, and Telegony," *PMLA* 100, no. 5 (1985): 774–82.

21. See, especially, Ross Shideler, *Questioning the Father: From Darwin to Zola, Ibsen, Strindberg and Hardy* (Stanford, Calif.: Stanford University Press, 1999).

22. Ivan Alekseevich Bunin, *About Chekhov: The Unfinished Symphony* (Evanston, Ill.: Northwestern University Press, 2007), xxi. On Chekhov's interaction with various popular notions about seeing, particularly those relevant to his work in medicine, see also Michael Finke, *Seeing Chekhov: Life and Art* (Ithaca, N.Y.: Cornell University Press, 2005)

23. Bunin, *About Chekhov*, xxi.

24. Downing Cless, *Ecology and Environment in European Drama* (New York: Routledge, 2010), 15.

25. Ibid., 147.

26. Rae Beth Gordon, *Dances with Darwin, 1875–1910: Vernacular Modernity in France* (Farnham, Eng.: Ashgate, 2008), 244.

27. Ibid., 247.

28. Michel Pharand, *Bernard Shaw and the French* (Gainesville: University Press of Florida, 2000), 85.

29. Ibid., 91.

30. *Maternité* (1903) was translated as *Maternity* by Mrs. Shaw and published along with *Damaged Goods* (trans. John Pollock) and *The Three Daughters of M. Dupont* (trans. St. John Hankin) in *Three Plays by Brieux* (New York: Brentano, 1911). This volume also includes a new version of *Maternity* translated by John Pollock. In his preface to *Three Plays by Brieux*, Shaw linked sexuality to creativity: "Sex is a necessary and healthy instinct; and its nurture and education is one of the most important uses of all art; and, for the present at all events, the chief use of the theatre." Four years later, Shaw wrote another defense of Brieux, "The Play and Its Author"—two thousand words on *La Femme Seule* (1912), which Charlotte Shaw translated as *Woman on Her Own* (produced by the Actresses Franchise League, December 8–13, 1913). Shaw says in this preface that Brieux treats prostitution "much more disturbingly" than Shaw himself did in *Mrs. Warren's Profession,* which (although staged independently in 1902) was not granted a license for public performance until 1924.

31. Quoted in Pharand, *Bernard Shaw*, 89. In 1910, Bennett predicted, accurately, that "nothing can keep Brieux's plays alive . . . because they are false to life" (96–97).

32. Julian B. Kaye, *Bernard Shaw and the Nineteenth Century Tradition* (Norman: University of Oklahoma Press, 1958), 155.

33. William Archer to Bernard Shaw, August 1903, in *Bernard Shaw: Collected Letters,* ed. Dan E. Laurence, vol. 2, *1898–1910* (London: Max Reinhardt, 1972), 356.

34. Bernard Shaw, preface to *Back to Methuselah*, in *Prefaces by Bernard Shaw* (London: Constable, 1934), 523.

35. Ibid., 505.

36. Bernard Shaw to Charles Rowley, February 11, 1907, in *Collected Letters,* 2:672; in a letter to Julie Moore (October 15, 1909), he likewise refers to "the

chapter of accidents called Natural Selection" (873), a phrase that recurs in his other writings on evolution.

37. J. D. Bernal, "Shaw the Scientist," in *G.B.S. 90: Aspects of Bernard Shaw's Life and Work,* ed. S. Winsten (London: Hutchinson, 1946), 94, 96.

38. Bernard Shaw, *Back to Methuselah*, part 2, in *The Complete Plays of Bernard Shaw* (London: Odhams Press, 1934), 880. Unless otherwise noted, all quotations from Shaw's plays are from this edition.

39. Ishrat Lindblad, "Creative Evolution and Shaw's Dramatic Art: With Special Reference to *Man and Superman* and *Back to Methuselah*" (Ph.D. diss., Uppsala University, 1971), 10.

40. Bernard Shaw to Augustin Hamon, January 9, 1907, in *Collected Letters,*2:670. I am indebted to Sos Eltis for pointing this letter out.

41. Shaw, preface to *Back to Methuselah*, 492. For a consideration of Schopenhauer's interpretation of evolution of species in terms of a theatrical analogy, see Kaye, *Bernard Shaw and the Nineteenth Century Tradition*, 116.

42. Wolff, *Mendel's Theatre*, 49.

43. Although Shaw "admitted to appropriating from him [Bergson] the expressions *évolution créatrice* and *élan vital*" (Pharand, *Bernard Shaw*, 246), his ideas about creative evolution had been fermenting for at least four years before Bergson's terminology became known to him, as *Man and Superman* makes clear.

44. George Levine, *Darwin Loves You: Natural Selection and the Re-enchantment of the World* (Princeton, N.J.: Princeton University Press, 2006), 187.

45. J. L. Wisenthal, *Shaw's Sense of History* (Oxford: Oxford University Press, 1988), 132. See also J. L. Wisenthal, *The Marriage of Contraries: Bernard Shaw's Middle Plays* (Cambridge, Mass.: Harvard University Press, 1974).

46. Bernard Shaw, *Man and Superman*, act 1, 340, 342. Ramsden is described in the opening stage directions as "an Evolutionist from the publication of the Origin of Species."

47. Ibid., act 1, 340. In the Don Juan in Hell scene, the Devil is "enormously less vital than the woman" (act 3, 372). John Osborne complained in a letter to the *Guardian* that Shaw made women into mere "bullies"; my thanks to Michael Billington for pointing this out.

48. Shaw, *Man and Superman*, act 4, 403.

49. Ibid., act 1, 346.

50. Ibid., act 3, 384, 378. See also Shaw's *Misalliance*, one long theatrical discussion of female sexual selection; in one scene, for example, Hypatia declares "I'll catch you" as "she dashes off in pursuit" of Percival, who has "bolted" from her (626), and in the end she suggests to her father, "Papa: buy the brute for me" (640).

51. Bernard Shaw, *You Never Can Tell,* act 2, 197.

52. Ibid.

53. Shaw, *Misalliance*, 639.

54. Shaw, *Man and Superman,* act 1, 337. As with so many of his contemporaries, Shaw uses "race" and "species" interchangeably. See, for example, *Man and Superman*, act 4, 392.

55. Ibid., act 3, 378. Shaw also acknowledges the self-sacrifice implicit in childbirth as woman must "risk her life to create another life" (act 1, 343).

56. A full consideration of Shaw's women as the Life Force can be found in the section "Woman: The Biological Agent of Evolution" in Lindblad, "Creative Evolution and Shaw's Dramatic Art." He notes how frequently Shaw's plays characterize women in animal similes; thus, Ann Whitefield is a "boa-constrictor," "grizzly bear," and "seabird" devouring the "scrap of fish" that is Tanner.

57. Shaw, *Man and Superman*, act 3, 375.

58. Ibid., 379.

59. Ibid., 378. On the wider cultural resonance of the megatherium, see Gowan Dawson, "'Like a Megatherium Smoking a Cigar': Darwin's *Beagle* Fossils in Nineteenth-Century Popular Culture," in *Darwin, Tennyson and Their Readers: Explorations in Victorian Literature and Science*, ed. Valerie Purton (London: Anthem, 2013), 81–96. The megatherium, which crops up again in *Back to Methuselah*, is a "scrapped experiment," a favorite motif of Shaw's, and his invocation of this beast is, like so much of his thinking on evolution, redolent of earlier theories of evolutionary already being displaced.

60. Shaw, *Back to Methuselah*, part 2, 886.

61. Ibid., 888. Shaw's faith in Creative Evolution endured; in the preface to *On the Rocks* (1933), for example, he calls it his "religion."

62. Strindberg, "Mysticism of World History," 219.

63. Ibid., 220.

64. Shaw, *Misalliance*, 611.

65. Shaw, *Back to Methuselah*, part 2, 885.

66. James Harrison, "Destiny or Descent? Responses to Darwin," *Mosaic* 14, no. 1 (2002): 114–15.

67. Shaw, *Back to Methuselah*, part 2, 866–67; Alfred Russel Wallace, "Evolution and Character,"*Fortnightly Review*, January 1908.

68. Shaw, preface to *Back to Methuselah*, 490.

69. Peter J. Bowler, *The Eclipse of Darwinism: Anti-Darwinian Evolution Theories in the Decades Around 1900* (Baltimore: Johns Hopkins University Press, 1992).

70. Peter J. Bowler, *Evolution: The History of an Idea* (Berkeley: University of California Press, 2009), 270.

71. Bernard Shaw, *The Shewing-Up of Blanco Posnet* (produced by the Stage Society in its 1909/1910 season) and *The Adventures of the Black Girl in Her Search for God* (1932) both admit that the Life Force makes mistakes on the road to perfection, as witness disease, evil, and so on.

72. Shaw, *Back to Methuselah*, part 5, 948.

73. Ibid., 958.

74. Bernard Shaw, *Too True to Be Good: A Political Extravaganza*, act 2, 1150.

75. Lindblad, "Creative Evolution and Shaw's Dramatic Art," 52.

76. Peter J. Bowler, *Science for All: The Popularization of Science in Early Twentieth-Century Britain* (Chicago: University of Chicago Press, 2009), 43.

77. Jeff Wallace, "T. H. Huxley, Science and Cultural Agency," in *Darwin, Tennyson and Their Readers*, ed. Purton, 154.

78. *Back to Methuselah* had its world premiere in 1922 at the New York Theatre Guild. In October 1921, one part of it (part 5, the final section) had been produced at the Court Theatre, to generally positive responses. The play in its entirety was first produced in England in 1924.

79. Lindblad, "Creative Evolution and Shaw's Dramatic Art," 48.

80. Shaw, *Back to Methuselah*, part 5, 938.

81. Wisenthal, *Shaw's Sense of History*, 134.

82. Shaw, *Man and Superman*, act 3, 375.

83. Shaw, *Back to Methuselah*, part 5, 962.

84. Maurice Valency, *The Cart and the Trumpet: The Plays of George Bernard Shaw* (New York: Oxford University Press, 1973), 362.

85. Louis Crompton, *Shaw the Dramatist* (London: Allen and Unwin, 1971), 183–84.

86. Shaw, *Back to Methuselah*, part 3, 905.

87. Shaw, preface to *Back to Methuselah*, 489.

88. Pharand, *Bernard Shaw*, 93.

89. Lindblad, "Creative Evolution and Shaw's Dramatic Art," 114 (emphasis added).

90. Bernard Shaw, preface to *Getting Married*, in *Prefaces*, 16.

91. Lindblad, "Creative Evolution and Shaw's Dramatic Art," 86.

92. Jill Davis, "Women Defined: The New Woman and the New Life," in *The New Woman and Her Sisters: Feminism and Theatre 1850–1914*, ed. Viv Gardner and Susan Rutherford (London: Prentice-Hall, 1992), 27–29.

93. Sos Eltis, *Acts of Desire: Women and Sex on Stage, 1800–1930* (Oxford: Oxford University Press, 2013), 177.

94. Kaye, *Bernard Shaw and the Nineteenth-Century Tradition*, vii.

95. Bernal, "Shaw the Scientist," 93. Yet, far from dismissing it, Bernal wrote that the preface to *Back to Methuselah* "should form part of every biological student's education because it shows better than any other single piece of writing both the social origins and the social effects of Darwin's teaching" (96).

96. *The Journals of Thornton Wilder, 1939–61*, ed. Donald Gallup (New Haven, Conn.: Yale University Press, 1985), 89.

97. Shaw, *Too True to Be Good*, act 3, 1157.

98. Ibid., act 1, 1132.

99. Maria Grazia Gregori, "Movement as a Metaphor of Time: An Interview with Luca Ronconi," trans. Bruna Tortorella (manuscript, 2002), kindly given to me by Pino Donghi. Ronconi is specifically referring to playwrights engaging with science; he has directed several science-related theatre productions, including *Infinities* (to which his comments specifically refer) and *Biblioetica*. See also Kirsten Shepherd-Barr, *Science on Stage: From Doctor Faustus to Copenhagen* (Princeton, N.J.: Princeton University Press, 2006).

100. *The Admirable Crichton* was produced at the Duke of York's Theatre (November 4, 1902), under the management of Charles Frohman and the direction of Dion Boucicault; it ran for 828 performances.

101. Hubert Henry Davies, *Mrs. Gorringe's Necklace*, in *The Plays of Hubert Henry Davies*, 2 vols. (London: Chatto and Windus, 1921), quoted in Hugh Walpole, introduction, 1:xviii.

102. Davies, *Mrs. Gorringe's Necklace*, 73–74.

103. Wilson Barrett and Louis N. Parker, *Man and His Makers*, LCP Add 53692 B. The play was performed at the Lyceum in October 1899.

104. Ibid., act 2, 10.

105. Ibid., act 4, 2.

106. *Era*, October 14, 1899; *Lloyds Weekly*, October 8, 1899, 13.

107. *Morning Post*, October 9, 1899, 6.

108. *Pall Mall Gazette*, October 9, 1899, 3.

109. Max Beerbohm, *Last Theatres, 1904–1910* (London: Hart-Davis, 1970), 535.

110. *The Mollusc* premiered at Wyndham's on May 12, 1903.

111. "Plays of the Month," *Play Pictorial* 53, no. 320 (1928): 4.

112. Hubert Henry Davies, *The Mollusc*, in *Late Victorian Plays, 1890–1914*, ed. George Rowell (London: Oxford University Press, 1968), act 1, 174. The reference to the Germans here is worth noting, perhaps indicating their status as recognized leaders in the biological sciences and proponents of morphological studies as opposed to the geneticists led by Bateson.

113. Davies, *Mollusc*, 322.

114. "Plays of the Month," 4.

115. Percival P. Howe, *Dramatic Portraits* (London: Secker, 1913), 215–16.

116. *Athenaeum*, October 26, 1907, 527.

117. Walpole, introduction to *Plays of Hubert Henry Davies*, xx.

118. Max Beerbohm, *Saturday Review*, November 2, 1907, in *Last Theatres*, 392, 332.

119. Rebecca Stott, *Darwin and the Barnacle: The Story of One Tiny Creature and History's Most Spectacular Scientific Breakthrough* (London: Faber and Faber, 2003), 10.

120. Ibid., 248–49, 246.

121. Ibid., 246.

122. Penny Farfan, "Comic Form and Conservative Reaction: Hubert Henry Davies's *The Mollusc*," *Journal of Dramatic Theory and Criticism* 18, no. 1 (2003): 45–56.

123. Beerbohm, *Last Theatres*, 332.

124. *Doormats* was staged at Wyndham's (October 3, 1912), a high-profile production starring Gerald du Maurier as Noel Gale and Nina Boucicault as Josephine. The *Playgoer and Society Illustrated* ran a twenty-two-page feature on the play in its December 1912 issue.

125. Wolff, *Mendel's Theatre*, 59.

126. Hubert Henry Davies, *Doormats*, in *Plays of Hubert Henry Davies*, vol. 2, act 3, 201 (emphasis in original).

127. Ibid., act 3, 200.

128. Ibid., act 3, 206.

129. Ibid., act 3, 200 (emphasis in original).

130. Howe, *Dramatic Portraits*, 216, 218.

131. Daniel J. Kevles, "Genetics in the United States and Great Britain, 1890–1930," in *Biology, Medicine and Society, 1840–1940,* ed. Charles Webster (Cambridge: Cambridge University Press, 1981), 206–8.

132. Ibid., 203–4.

133. Ibid., 212.

134. Quoted in Donald Mackenzie, "Sociobiologies in Competition," in *Biology, Medicine and Society*, ed. Webster, 267.

135. Ibid., 269.

136. Quoted in ibid., 282.

137. *Bookman*, March 1922, 277.

138. Ibid.

139. Ted Bain, "St John Hankin," in *British Playwrights, 1880–1956: A Research and Production Sourcebook,* ed. William W. Demastes and Katherine E. Kelly (Westport, Conn.: Greenwood Press, 1996), 207–8.

140. William H. Phillips, *St. John Hankin: Edwardian Mephistopheles* (Rutherford, N.J.: Fairleigh Dickinson University Press, 1979), 53.

141. Ibid., 46.

142. Ibid., 51, 50. Jean Chothia mentions "plays that concerned themselves with Natural Selection" such as Strindberg's *Miss Julie*, Granville-Barker's *The Marrying of Ann Leete* and *The Voysey Inheritance*, and Hankin's *The Last of the De Mullins* (introduction to *The New Woman and Other Emancipated Woman Plays*, ed. Jean Chothia [Oxford: Oxford University Press, 1998], xxiv).

143. Phillips, *St. John Hankin*, 59, 61.

144. The play was performed on February 14, 1918, at the New Theatre, London; it was published in 1922 by Heinemann, though the copyright page gives the date 1917.

145. I have been unable to find evidence of Edward F. Benson's book as source material for Arthur Wing Pinero's play, though this would be entirely plausible. Benson also employs terms associated with evolution elsewhere, as in his highly successful *Dodo* (1983), reflecting their absorption into popular discourse.

146. Arthur Wing Pinero, *The Freaks: An Idyll of Suburbia* (London: Heinemann, 1922), 37.

147. *Tatler*, March 13, 1918. In "Claude Shepperson (An Appreciation)," Alfred Noyes discusses Shepperson's "infallible instinct for beauty," which was his "distinguishing characteristic" as an artist; he eschewed "the ugliness of modernity" (*Catalogue of the Memorial Exhibition of Works by the Late Claude A. Shepperson* [London: Ernest Brown and Phillips, Leicester Galleries], Exhibition 332 [March–April 1922], 7, John Johnson Collection, Bodleian Library, Oxford).

148. Penny Griffin, *Arthur Wing Pinero and Henry Arthur Jones* (London: Macmillan, 1991), 162.

149. Arthur Pinero, March 19, 1918, in *The Collected Letters of Sir Arthur Wing Pinero,* ed. J. P. Wearing (Minneapolis: University of Minnesota Press, 1974), 269–70.

150. Pinero, *Freaks*, 43.

151. Ibid., 45.

152. Ibid., 103.

153. Ibid., 100–101.

154. Ibid., 107–8.

155. Christine C. Ferguson, "Elephant Talk: Language and Enfranchisement in the Merrick Case," in *Victorian Freaks: The Social Context of Freakery in Britain*, ed. Marlene Tromp (Columbus: Ohio State University Press, 2008), 115.

156. For example, the script indicates that the play is set "before the War—those far-off days when, in our ignorance, small troubles seemed great, and minor matters important." Pinero thought of the play as wartime amusement: "The little piece is simple in subject and treatment, and has no higher aim than to amuse—which I take to be the function of the theatre at the present moment. There are more ways than one of trying to be amusing, you will say. I must hope *The Freaks* will not be judged as falling into the lowest category" (Arthur Pinero, January 25, 1918, in *Collected Letters*, 269). See also Griffin, *Arthur Wing Pinero and Henry Arthur Jones*, 161.

157. Nadja Durbach, *Spectacle of Deformity: Freak Shows and Modern British Culture* (Oxford: Oxford University Press, 2010), 175.

158. Ibid., 176.

159. "The Re-Appearance of Sir Arthur Pinero," *Saturday Review,* February 23, 1918, 156.

160. The sensational and shocking film *Freaks* (directed by Tod Browning, MGM, 1932) has a similar story but is far more explicit in the suggestion of sexual encounters between freaks and nonfreaks.

161. Arthur Pinero to Louis E. Shipman, January 7, 1918, in *Collected Letters*, 269. His deprecation of the play carries on to his biographer, who calls it an "inconsequential comedy" (John Dawick, *Pinero: A Theatrical Life* [Niwot: University Press of Colorado, 1993], 344).

162. Griffin, *Arthur Wing Pinero and Henry Arthur Jones*, 159.

163. Michael M. Chemers, *Staging Stigma: A Critical Examination of the American Freak Show* (New York: Palgrave Macmillan, 2008), 4.

164. Ibid., 4.

165. Francis Galton, "Eugenics: Its Definition, Scope and Aims" (1904), in *The Fin de Siecle: A Reader in Cultural History, c. 1880–1900*, ed. Sally Ledger and Roger Luckhurst (Oxford: Oxford University Press, 2000), 332–33.

166. Charles Darwin, "Autobiography," in *Charles Darwin and T. H. Huxley: Autobiographies*, ed. Gavin de Beer (Oxford: Oxford University Press, 1983), 22.

167. Wolff, *Mendel's Theatre*. On the ubiquity of the eugenics movement, see Daniel J. Kevles, *In the Name of Eugenics: Genetics and the Uses of Human Heredity* (New York: Knopf, 1985); Geoffrey R. Searle, "Eugenics and Class," in *Biology, Medicine and Society*, ed. Webster, 214–42; Angelique Richardson, *Love and Eugenics in the Late Nineteenth Century: Rational Reproduction and the New Woman* (Oxford: Oxford University Press, 2003); and William Greenslade, *Degeneration, Culture, and the Novel* (Cambridge: Cambridge University Press, 1994), which among other things reminds us of the respectability eugenics once enjoyed (Winston Churchill was a proponent). See also Marius Turda, *Modernism and Eugenics* (London: Palgrave Macmillan, 2010). Early promoters of eugenics with particular relevance for my discussion include Clémence Royer, the first French translator of *On the Origin of Species,* and Helen Hamilton Gardener, popular lecturer and feminist writer whose work influenced James A. Herne, as discussed in previous chapters.

168. Searle, "Eugenics and Class," 238 (emphasis added).

169. Ibid., 230, 239–40. Searle notes that "the Fabians never joined the Eugenics Society" (230).

170. Pharand, *Bernard Shaw*, 93.

171. Bernard Shaw, preface to *Misalliance*, in *Prefaces*, 51.

172. Searle, "Eugenics and Class," 231.

173. Adam Neave, *Woman and Superwoman: A Comedy of 1963 in Three Acts* (London: Griffiths, 1914), 10.

174. Ibid., act 2, 36.

175. Ibid., act 1, 18.

176. *Bookman*, December 1914, 106.

177. H. M. Harwood, *The Supplanters* (London: Benn, 1926), 12. In Susan Glaspell's play *Springs Eternal* (1943), a character utters a sarcastic jibe at the heart of Harwood's criticism of Beatrice: "I've always heard that a woman who

never had children of her own knows just how to bring them up" (*Susan Glaspell: The Complete Plays*, ed. Linda Ben-Zvi and J. Ellen Gainor [Jefferson, N.C.: McFarland, 2010], 380).

178. Harwood, *Supplanters*, 15. In *The Return of the Prodigal*, Eustace declares: "England is covered with hospitals for the incurably diseased and asylums for the incurably mad. If a tenth of the money were spent on putting such people out of the world, and the rest were used in preventing the healthy people from falling sick, and the sane people from starving, we should be a wholesomer nation" (St. John Hankin, *The Return of the Prodigal* [New York: Samuel French, 1907], 4:97).

179. Harwood, *Supplanters*, 16. In 1893, Helen Hamilton Gardener argued similarly against erecting more and more "charitable and eleemosynary institutions" to house the unfit when positive eugenics would prevent their birth in the first place ("Environment: Can Heredity Be Modified?" in *Facts and Fictions of Life* [Boston: Fenno, 1895], 298–99).

180. Karl Pearson, *Tuberculosis, Heredity and Environment* (1912), quoted in Searle, "Eugenics and Class," 225.

181. M. J. S. Hodge, "England", in *The Comparative Reception of Darwinism*, ed. Thomas F. Glick (Chicago: University of Chicago Press, 1988) 17.

182. Harwood, *Supplanters*, 17.

183. Ibid., 11.

184. Ibid., 17 (emphasis in original).

185. Brian Harrison, "Women's Health and the Women's Movement," in *Biology, Medicine and Society*, ed. Webster, 61.

186. Harwood, *Supplanters*, 19.

187. Elizabeth Baker, *Bert's Girl: A Comedy in Four Acts*, Contemporary British Dramatists, vol. 56 (London: Benn, 1927), act 1, 12. The play was produced at the Court Theatre, under Sir Barry Jackson, in March 1927. I am indebted to Sos Eltis for drawing it and others by Baker to my attention.

188. Ibid., act 1, 30; act 3, 75.

189. Ibid., act 2, 43.

190. Ibid., act 2, 47–48.

191. Ibid., act 2, 53.

192. Ibid., act 2, 48–49.

193. Ibid., act 2, 51.

194. Ibid., act 3, 66.

195. Ibid., act 3, 62.

196. Ibid., act 3, 67.

197. Ibid., act 4, 98.

198. Ibid., act 4, 99.

199. Pinero's *The Mind-the-Paint Girl* likewise revolves around breeding and puts the burden of producing fitter humans on women because men (especially

upper-class ones) are inadequate. One character wonders why lately so many "weedy" young men have married sturdy, "keen-witted young women full of the joy of life, with strong frames, beautiful hair and fine eyes, and healthy pink gums and big white teeth" (Arthur Wing Pinero, *The Mind-the-Paint Girl* [London: Heinemann, 1912], 1:20). The play slyly suggests that in their liaisons with upper-class men the actresses are doing eugenic work; they will be "the salvation of the aristocracy in this country and the long run!" (act 1, 21, and repeated in act 4, 234).

6. Reproductive Issues

1. *Review of Reviews*, January–June 1904, 615. See also Pat Jalland, *Women, Marriage and Politics, 1860–1914* (Oxford: Oxford University Press, 1986), 133–88.

2. Sos Eltis, *Acts of Desire: Women and Sex on Stage, 1800–1930* (Oxford: Oxford University Press, 2013), 210.

3. Henrik Ibsen, *From Ibsen's Workshop: Notes, Scenarios, and Drafts of the Modern Plays*, trans. Arthur G. Chater (New York: Scribner, 1911), 12:294.

4. Eltis, *Acts of Desire*, 175.

5. Jalland, *Women, Marriage and Politics*, 144–45.

6. Ibid., 145.

7. The actress Eleonora Duse was hit hard by menopause at age fifty, "shipwrecked by a natural process shared by all women, yet one no playwright had ever written about" (Helen Sheehy, *Eleonora Duse* [New York: Knopf, 2003], 250).

8. John Holmes, *Darwin's Bards: British and American Poetry in the Age of Evolution* (Edinburgh: Edinburgh University Press, 2009), 187.

9. St. John Hankin, *The Last of the De Mullins* (London: Fifield, 1909), act 3, 124.

10. Ibid., act 3, 125.

11. Stanley Houghton, *Hindle Wakes*, in *Late Victorian Plays, 1890–1914*, ed. George Rowell (London: Oxford University Press, 1968), act 3, 503.

12. Hankin shares Bernard Shaw's "sense of the life force, the biological imperative underlying human activity and relationship . . . [and] is alert to the contemporary attention to Darwinian ideas of the survival of the fittest" (Jean Chothia, introduction to *The New Woman and Other Emancipated Woman Plays* [Oxford: Oxford University Press, 1998], xxv). Hankin knew Brieux's *Three Daughters of M Dupont* well and probably wrote an unpublished essay on it in 1904. See William H. Phillips, *St. John Hankin: Edwardian Mephistopheles* (Rutherford, N.J.: Fairleigh Dickinson University Press, 1979), 86, 82. The essay is called "The Propagandist as Playwright," and it lauds Brieux's "masterpiece."

13. See, especially, Joel H. Kaplan and Sheila Stowell, *Theatre and Fashion: Oscar Wilde to the Suffragettes* (Cambridge: Cambridge University Press, 1994).

14. Tracy C. Davis, *Actresses as Working Women: Their Social Identity in Victorian Culture* (London: Routledge, 1991), 138.

15. Gillian Beer, "Systems and Extravagance: Darwin, Meredith, Tennyson," in *Darwin, Tennyson and Their Readers: Explorations in Victorian Literature and Science,* ed. Valerie Purton (London: Anthem, 2013), 140.

16. Ibid., 147–48.

17. Alfred Russel Wallace, "Human Selection," *Fortnightly Review*, September 1890, 329, http://people.wku.edu/charles.smith/wallace/S427.htm (accessed October 10, 2013).

18. Ibid.

19. George Levine, *Darwin Loves You: Natural Selection and the Re-enchantment of the World* (Princeton, N.J.: Princeton University Press, 2006), 201.

20. Ibid., 177, 178. See also Peter J. Bowler, *Evolution: The History of an Idea* (Berkeley: University of California Press, 2009), 315.

21. As Levine points out, "virtually nobody" believed in it in Darwin's lifetime or, indeed, until very recently: "The scientific community found it impossible to credit the idea that the female could have had much to do with evolutionary development." The deep cultural "hostility" to the idea of female choice makes Darwin's theory of sexual selection all the more "thrilling, . . . inventive and productive." It also makes Darwin "a kind of ideological hero in spite of himself" (*Darwin Loves You*, 189, 200).

22. Charles Webster, ed., *Biology, Medicine and Society, 1840–1940* (Cambridge: Cambridge University Press, 1981), 3.

23. Dorothy Brandon, *Wild Heather*, Lord Chamberlain Play collection, 1917/17. The play was produced at the Gaiety Theatre, Manchester (August 1917) and later transferred to London's Strand Theatre. I am indebted to Sos Eltis for bringing this play to my attention; see also her discussion of it in *Acts of Desire*, 208.

24. Bernard Shaw, *Misalliance*, in *The Complete Plays of Bernard Shaw* (London: Odhams Press, 1934), 610.

25. Throughout *The Descent of Man*, Darwin frequently refers to the male of a species having "special organs of prehension for holding" the female securely while mating (*The Descent of Man, and Selection in Relation to Sex*, ed. James Moore and Adrian Desmond [London: Penguin, 2004], 241).

26. Martha Vicinus, *Suffer and Be Still: Women in the Victorian Age* (Bloomington: Indiana University Press, 1972).

27. Marie Stopes to Alfred Sutro, November 1927, quoted in Esther Beth Sullivan, "*Vectia*, Man-Made Censorship, and the Drama of Marie Stopes," *Theatre Survey* 46, no. 1 (2005): 85.

28. Marjorie Strachey, "Women and the Modern Drama," *The Englishwoman*, May 1911, quoted in Maria DiCenzo, "Feminism, Theatre Criticism, and the Modern Drama," *South Central Review* 25, no. 1 (2008): 51.

29. Cicely Hamilton, *Life Errant* (London: Dent, 1935), 55.

30. See, for example, Brian Harrison, "Women's Health and the Women's Movement in Britain: 1840–1940," in *Biology, Medicine and Society*, ed. Webster, 64.

31. The autobiographical basis for this can be found in Ruth Hall, *Marie Stopes: A Biography* (London: Virago, 1977).

32. Marie Stopes, *A Banned Play and a Preface on the Censorship* (London: Bale, Danielsson, 1926), 113. See also Eltis, *Acts of Desire*, 206–7.

33. Sullivan, "*Vectia*," 84.

34. Stopes, *Banned Play*, 143 (emphasis in original).

35. Sullivan, "*Vectia*," 80.

36. Elizabeth Baker, *Partnership: A Comedy in Three Acts* (New York: French, 1921), 38. The play also suggests an environmentalist stance in depicting Fawcett as anti-urban and back-to-nature; being on the hills at daybreak on a spring morning, he says, shows you real color, and "a human being then seems an intruder, and you step softly as if not sure of your place on earth. Your cocksuredness gets rubbed" (61).

37. H. M. Harwood, *The Supplanters* (London: Benn, 1926), 80.

38. Bernard Shaw, *Getting Married*, in *Complete Plays*, 551.

39. Harwood, *Supplanters*, 86.

40. Ibid., 113.

41. *Saturday Review*, September 20, 1913, 360.

42. *Academy*, September 27, 1913, 401.

43. Harwood, *Supplanters*, 68.

44. See the discussion of marriage's enduring popularity in Eltis, *Acts of Desire*, 201–2; and Martin Pugh, *We Danced All Night: A Social History of Britain Between the Wars* (New York: Vintage, 2009), 145.

45. Eltis, *Acts of Desire*, 201–2, citing Pugh, *We Danced All Night*, 160.

46. Ibid., 202.

47. Darwin, *Descent of Man*, 656.

48. Ibid.

49. Ibsen, *From Ibsen's Workshop*, 484.

50. Ibid., 485.

51. Ibid., 497.

52. Ibid., 493.

53. Ibid., 496.

54. Ibid., 495.

55. Elizabeth Robins, *The Silver Lotus*, act 1, 4 (in copy of typescript, Fales Collection, New York, kindly sent to me by Joanne E. Gates); each act begins at p. 1.

56. Joanne E. Gates, *Elizabeth Robins, 1862–1952: Actress, Novelist, Feminist* (Tuscaloosa: University of Alabama Press, 1994), 88. Angela V. John, Robins's

other biographer, does not comment on the play, in *Elizabeth Robins: Staging a Life* (London: Routledge, 1995).

57. Robins, *Silver Lotus*, act 1, 7.

58. Ibid., act 1, 13.

59. Ibid., act 2, 6.

60. Ibid., act 1, 21.

61. Ibid., act 1, 24.

62. Ibid., act 2, 8.

63. Ibid., act 2, 9.

64. Ibid., act 2, 30.

65. Ibid., act 2, 27.

66. Ibid., act 2, 34.

67. Ibid., act 2, 8.

68. Gates, *Elizabeth Robins*, 89.

69. Ibid., 90.

70. Though neither of Robins's biographers mentions this coincidence, it is possible that Robins had known about or even seen James A. Herne's play, since she was working in Boston where it was performed, establishing her acting career at the time. In January 1889, *Drifting Apart* was in Boston and was seen and admired by the young Hamlin Garland (discussed at length in chapter 4), whose brother Franklin was an actor who lived with him in Boston, suggesting a possible further connection to Robins. Playwrights continued to mine the idea of hereditary addiction— for example, in Wilson Barrett and Louis N. Parker's *Man and His Makers* (1899). An amateur science enthusiast, Sir Henry Faber, declares that his would-be son-in-law comes from a long line of drunkards and drug addicts and that in their conceptions of heredity, "Darwin, Henley, Spencer, Nisbet, Nordau, Zola and Lombroso are all perfectly right!" (Licensing copy of *Man and His Makers,* licensed October 26, 1899 [LC Add 53692 B], act 1, 3).

71. The play was an adaptation of a Swedish story by Elin Ameen called "Befriad" (Freed), and under Swedish law, a woman guilty of infanticide was not sentenced to death but to imprisonment. William Archer states in his introduction to *Alan's Wife* that Robins showed him the Swedish magazine *Ur Dagens Krönika* of January 1891 containing "Befriad" ([Florence Bell and Elizabeth Robins,] *Alan's Wife, a Dramatic Study in Three Scenes. First Acted at the Independent Theatre in London*, ed. Jacob T. Grein [London: Henry, 1893], xi). Unpublished letters by Robins and Bell from 1892 indicate that their working title for the adaptation was "Mother's Hands," Fales Collection, New York University Library.

72. Kirsten E. Shepherd-Barr, "'It Was Ugly': Maternal Instinct on Stage at the Fin de Siècle," *Women: A Cultural Review* 23, no. 2 (2012): 216–34.

73. Henrik Ibsen, *Samlede Digterverker* (Kristiania [Oslo]: Gyldendal, 1922), 5:378 (my translation).

74. Ibsen, *From Ibsen's Workshop*, 439.

75. Ibid., 521.

76. Allardyce Nicoll, *A History of English Drama, 1660–1900,* vol. 4, *Early Nineteenth Century Drama, 1800–1850* (Cambridge: Cambridge University Press, 1955), and vol. 5, *Late Nineteenth Century Drama, 1850–1900* (Cambridge: Cambridge University Press, 1959). *The Cataract of the Ganges! or, The Rajah's Daughter* opened at Drury Lane on October 27, 1823.

77. Betsy Bolton, "Saving the Rajah's Daughter: Spectacular Logic in Moncrieff's *Cataract of the Ganges*," *European Romantic Review* 17, no. 4 (2006): 481.

78. Josephine McDonagh, *Child Murder and British Culture, 1720–1900* (Cambridge: Cambridge University Press, 2008), 138.

79. Ibid., 138–39.

80. Quoted in ibid., 139.

81. Ibid., 125.

82. Nicoll lists an unknown author's drama called *Infanticide* at Queen's in 1831, in *History of English Drama*, 4:482.

83. Review of *Alexandra, Bristol Mercury*, March 11, 1893, 8.

84. Review of *Jeanie Deans, Aberdeen Weekly Journal*, May 30, 1899, 4. Though many theatrical adaptations of Scott's novel were made, the only one called *Jeanie Deans* was by Dion Boucicault (New York, 1860). See Ernest Reynolds, *Early Victorian Drama, 1830–1870* (Cambridge: Heffer, 1936), 140.

85. Review of *The Scarlet Dye, Era*, February 9, 1887, 15.

86. Arthur B. Walkley, review of *Alan's Wife, The Speaker*, May 6, 1893, quoted in Archer, introduction to [Bell and Robins,] *Alan's Wife*, xxix.

87. Darwin, *Descent of Man,* 659.

88. Ibid., 644, 65.

89. Ibid., 141. Darwin relies on others' reports in many of these instances.

90. Ibid., 297.

91. Ibid., 660.

92. Glenn Hausfater and Sarah Blaffer Hrdy, eds., *Infanticide: Comparative and Evolutionary Perspectives* (New Brunswick, N.J.: Aldine Transaction Press, 2008), xi. The editors argue that "the evolutionary importance of infanticide and related phenomenon [*sic*]" have been "heretofore overlooked," and that "infanticide has only recently come to be regarded as a biologically significant phenomenon" (xi, xiii).

93. Grein, editor's preface to [Bell and Robins,] *Alan's Wife*, vi.

94. Archer, introduction to [Bell and Robins,] *Alan's Wife*, xlvi. Only years later were Robins and Bell revealed as the authors/adapters; they had elected anonymity most likely because they feared that if their gender were known, it would damage the response to the play even more. Even Grein's preface goes so far as to refer to the play (twice) as one of the finest and truest tragedies "ever written by a modern Englishman." He further adds, "I am not able to divulge the name of the

author, which, in deference to my solemn promise to Miss Robins [who was playing the main character], I have not even endeavoured to ascertain" (viii). Archer refers to the author as "he" throughout his lengthy introduction.

95. John, *Elizabeth Robins*, 127.

96. Archer, introduction to [Bell and Robins,] *Alan's Wife*, xiii (emphasis added). Not showing the infanticide would not necessarily have lessened its intensity for the audience. In the Finnish playwright Minna Canth's play *Anna-Liisa* (1895), the act of infanticide has taken place four years before the play begins, but it is vividly recalled by the protagonist, who was fifteen at the time and is now engaged to a man who knows nothing of her past. The infanticide is described so graphically that it is arguably far more uncomfortable for the audience than Jean's swift crime, as Anna-Liisa dwells at length on the baby's physicality, whereas Jean's act is oddly more symbolic than real for an audience that never sees the baby itself.

97. Katherine E. Kelly, "*Alan's Wife*: Mother Love and Theatrical Sociability in London of the 1890s," *Modernism/Modernity* 11, no. 3 (2004): 550. In her memoir *Both Sides of the Curtain* (1940), Robins recalls that the work of aspiring female playwrights in the 1890s went into "an unmarked grave," and that women in theatre generally were subjected to "slavery . . . the unworthy bondage of the *successful* as well as the unsuccessful women of the stage" (quoted in Kerry Powell, *Women and Victorian Theatre* [Cambridge: Cambridge University Press, 1997], 158). In an unpublished sequel to these memoirs, *Whither and How*, Robins is even more forceful in her condemnation of 1890s theatre with regard to women. She quotes her own diary kept at the time, in which she imagined a theatre made up of "an association of workers" rather than monopolized by any individual such as George Alexander, Herbert Beerbohm Tree, or Henry Irving (Robins, quoted in Powell, *Women and Victorian Theatre*, 159). Her founding in 1891 of the Joint Management with Marion Lea was a step toward realizing this dream. See Kirsten Shepherd-Barr, *Ibsen and Early Modernist Theatre, 1890–1900* (Westport, Conn.: Greenwood Press, 1997).

98. McDonagh, *Child Murder*, 179.

99. John, *Elizabeth Robins*, 163.

100. Kelly, "*Alan's Wife*," 558.

101. McDonagh, *Child Murder*, 180.

102. Sheila Stowell, *A Stage of Their Own*, quoted in Penny Farfan, *Women, Modernism, and Performance* (Cambridge: Cambridge University Press, 2004), 4.

103. Elizabeth Robins, *Discretion* (manuscript, Fales Collection).

104. John, *Elizabeth Robins*, 163.

105. Susan Carlson, "Conflicted Politics and Circumspect Comedy: Women's Comic Playwriting in the 1890s," in *Women and Playwriting in Nineteenth-Century*

Britain, ed. Tracy C. Davis and Ellen Donkin (Cambridge: Cambridge University Press, 1999), 266.

106. Grein, editor's preface to [Bell and Robins,] *Alan's Wife,* viii.

107. In *Anna-Liisa,* the baby-killing mother likewise refuses to speak, though this is temporary. Her husband-to-be complains, "You can hear what they're accusing you of, but you don't/say a word." She becomes lifeless, immobile, "like a statue. . . . The poor thing can't see or hear anything any more" (Minna Canth, *Anna-Liisa,* in *Portraits of Courage: Plays by Finnish Women,* ed. S. E. Wilmer [Helsinki: Helsinki University Press, 1997], 77, 79, 73–74).

108. Julie Holledge, *Innocent Flowers: Women in the Edwardian Theatre* (London: Virago, 1981), 44.

109. Archer, introduction to [Bell and Robins,] *Alan's Wife,* xii.

110. Sally Shuttleworth, "Demonic Mothers: Ideologies and Bourgeois Motherhood in the Mid-Victorian Era," in *Rewriting the Victorians,* ed. Linda M. Shires (London: Routledge, 1992), 36.

111. Archer, introduction to [Bell and Robins,] *Alan's Wife,* xxiv.

112. McDonagh, *Child Murder,* 180.

113. In this regard, the translation of the original title "Befriad" becomes absolutely central. In a letter to Robins as they worked on their translation/adaptation, Bell wrote: "Let me implore you to call it *Set Free.* . . . The last sentence of the play ends w. it. do *do* say you like it" (Florence Bell to ER, November 9, 1892, Fales Collection).

114. Florence Bell and Elizabeth Robins, *Alan's Wife,* in *New Woman Plays,* ed. Linda Fitzsimmons and Viv Gardner (London: Methuen, 1991), 10–11.

115. Angelique Richardson, "The Life Sciences: 'Everybody Nowadays Talks About Evolution,'" in *A Concise Companion to Modernism,* ed. David Bradshaw (Oxford: Blackwell, 2003), 18. See also Cynthia Eagle Russett, *Sexual Science: The Victorian Construction of Womanhood* (Cambridge, Mass.: Harvard University Press, 1989).

116. For a critique of this trend, see Anna Farkas, "Between Orthodoxy and Rebellion: Women's Drama in England, 1890–1918" (Ph.D. diss., University of Oxford, 2010).

117. ER to Florence Bell, probably November 1892, Fales Collection.

118. Michael Benedikt and George E. Wellwarth, eds., *Modern French Theatre: The Avant-Garde, Dada and Surrealism, an Anthology of Plays* (New York: Dutton, 1964), xvi.

119. Guillaume Apollinaire, *Les Mamelles de Tiresias,* in ibid., 70.

120. Guillaume Apollinaire, preface to *Les Mamelles de Tiresias,* in ibid., 58.

121. Apollinaire, *Les Mamelles de Tiresias,* 80.

122. Susan Glaspell, *Bernice,* in *Susan Glaspell: The Complete Plays,* ed. Linda Ben-Zvi and J. Ellen Gainor (Jefferson, N.C.: McFarland, 2010), act 2, 103.

123. *Chains of Dew* was performed in April 1922, though probably written around 1920. See Susan Glaspell, *Chains of Dew*, in *Complete Plays*, 125.

124. Ibid. Note Emma Goldman's interest in theatre as in the previously mentioned lectures published as *The Social Significance of the Modern Drama* (Boston: Badger, 1914).

125. Eltis notes that an earlier draft of the play submitted to the Lord Chamberlain for licensing ended "not with Vida identifying herself as particularly suited to political activism because of her childlessness, but rather with a wider identification of herself with all vulnerable women" (*Acts of Desire*, 171). The line indicating this was deleted in this licensing copy.

126. Mary Papke, *Susan Glaspell: A Research and Production Sourcebook* (Westport, Conn.: Greenwood Press, 1992), 73.

127. Barbara Ozieblo and Jerry Dickey, *Susan Glaspell and Sophie Treadwell: American Modernist Women Dramatists* (London: Routledge, 2008), 38.

128. Michael Billington gave *Chains of Dew* four of five stars and called the play "astonishing" and "amazing," in "Glaspell Shorts," *Guardian*, April 9, 2008, www.theguardian.com/stage/2008/apr/09/theatre (accessed October 12, 2013).

129. Patricia Knight, "Women and Abortion in Victorian and Edwardian England," *History Workshop* 4 (1977): 57–68; Leslie Reagan, *When Abortion Was a Crime: Women, Medicine, and Law in the United States, 1867–1973* (Berkeley: University of California Press, 1997).

130. Eltis, *Acts of Desire*, 208. Although the play was refused a license initially, Barker presented it privately in two performances sponsored by the Stage Society at the Imperial Theatre in November 1907, playing Trebell. The play was finally licensed in 1920 and performed in 1936, though in a rewritten version. See Eric Salmon, *Granville Barker: A Secret Life* (London: Heinemann Educational Books, 1983). Salmon's study provides a detailed comparison of passages from both versions of the play.

131. Salmon, *Granville Barker*, 143. Salmon argues that *Waste* is about "the clash, at the very centre of life, of the natural forces of creation with the natural forces of destruction, which bafflingly co-exist in one life, in all life" (142).

132. Ibid., 143. Harwood indirectly caused the revision of the play; he offered to produce it at Ambassador's Theatre, so it was resubmitted to the Censor in 1920, though not performed then despite being given a license.

133. Granville Barker, *Waste*, quoted in Salmon, *Granville Barker*, 159; he gives no citation for this excerpt from the play.

134. Elizabeth Robins, *Votes for Women!* in *New Woman*, ed. Chothia, act 3, 198.

135. Eltis, *Acts of Desire*, 170.

136. *Abortion* was rejected by the Provincetown Players and not performed in O'Neill's lifetime; though published in *Ten "Lost" Plays of Eugene O'Neill,* ed. Bennett Cerf (London: Cape, 1964), it was virtually unknown until staged (without sets or costumes) during the O'Neill Festival in 1999 to mark the reopening of the Provincetown Playhouse in Greenwich Village. As with *Chains of Dew*, the play was a revelation to critics and audiences, a "stunning surprise . . . an emotional roller coaster of a play" (Les Gutman, review of *Abortion*, *CurtainUp*, www.curtainup.com/oneillreport.html [accessed October 13, 2009]). Tennessee Williams adapted this dramatic situation in *Sweet Bird of Youth*, in which a "townie" youth gets a rich girl pregnant, she has a botched abortion, and her father viciously avenges her death. However, O'Neill frames abortion within an evolutionary discourse that is absent from the Williams play.

137. Eugene O'Neill, *Abortion* (1914), 7, http://www.eoneill.com/texts/abortion/contents.htm (accessed November 13, 2009).

138. Ibid., 8.

139. Robert M. Dowling, *Critical Companion to Eugene O'Neill* (New York: Facts on File, 2009), 1:25.

7. Midcentury American Engagements with Evolution

Carol Bird, "Enter the Monkey Man," *Theatre Magazine*, 1922, quoted in Erika Rundle, "The Hairy Ape's Humanist Hell: Theatricality and Evolution in O'Neill's 'Comedy of Ancient and Modern Life,'" *Eugene O'Neill Review* 30 (2008): 51. This is Bird's reaction on seeing *The Hairy Ape*, a play in whose "theatricalised ape-man figure" Bird recognized "a zeitgeist of post–WWI American drama" (55).

1. Peter Middleton, "Poetry, Physics, and the Scientific Attitude at Mid-Century," *Modernism/Modernity* 21, no. 1 (2014): 147–68.

2. Paul Lifton, "*Vast Encyclopedia*": *The Theatre of Thornton Wilder* (Westport, Conn.: Greenwood Press, 1995), 43.

3. Peter J. Bowler, *Evolution: The History of an Idea* (Berkeley: University of California Press, 2009), 316. See also Peter J. Bowler, *Science for All: The Popularization of Science in Early Twentieth-Century Britain* (Chicago: University of Chicago Press, 2009).

4. Bowler, *Evolution*, 316.

5. Woody Guthrie's novel *House of Earth* (1947; New York: HarperCollins, 2013), for example, conveys this sense of optimism, "part of a broad philosophical vision of man's place in nature" at this time, according to Gerald Mangan, "Adobe Solutions," *Times Literary Supplement*, March 15, 2013.

6. Kirsten Shepherd-Barr, *Science on Stage: From Doctor Faustus to Copenhagen* (Princeton, N.J.: Princeton University Press, 2006), 61–69.

7. Michael Valdez Moses, "'Saved from the Blessings of Civilization': John Ford, the West, and American Vernacular Modernism" (paper presented at the Northern Modernisms seminar, University of Birmingham, March 15, 2013).

8. Christopher W. E. Bigsby, introduction to *Plays by Susan Glaspell,* ed. Christopher W. E. Bigsby (Cambridge: Cambridge University Press, 1987).

9. Michael Billington, review of *Alison's House, Guardian,* October 11, 2009, and "Glaspell Shorts," *Guardian,* April 9, 2008, www.theguardian.com /stage/2008/apr/09/theatre (accessed October 12, 2013).

10. Tamsen Wolff, *Mendel's Theatre: Heredity, Eugenics, and Early Twentieth-Century American Drama* (New York: Palgrave Macmillan, 2009), 8.

11. Mary Papke, "Susan Glaspell's Naturalist Scenarios of Determinism and Blind Faith," in *Disclosing Intertextualities: The Stories, Plays, and Novels of Susan Glaspell,* ed. Martha C. Carpentier and Barbara Ozieblo (Amsterdam: Rodopi, 2006), 25. Papke argues that in her mature novels, Glaspell interrogates monism, replacing it with her own brand of "impressionistic epistemology" (26), but the plays remain strongly influenced by Ernst Haeckel.

12. Barbara Ozieblo, *Susan Glaspell: A Critical Biography* (Chapel Hill: University of North Carolina Press, 2000), 286n.16. Glaspell's hagiography of George Cram Cook, *The Road to the Temple* (London: Benn, 1926), reveals the profound influence of Charles Darwin on Cook, ground through monistic and Spencerian mills, and shows how deeply Cook in turn shaped Glaspell's views throughout her most creative period.

13. Quoted in Ozieblo, *Susan Glaspell,* 38; some of this paper Glaspell later included in *Road to the Temple.*

14. Linda Ben-Zvi, *Susan Glaspell: Her Life and Times* (New York: Oxford University Press, 2005), 309. *Bernice* was produced in London at the Gate Theatre on October 30, 1925. The *Era* called it a play of "rare beauty" (quoted in Mary Papke, *Susan Glaspell: A Research and Production Sourcebook* [Westport, Conn.: Greenwood Press, 1992], 42).

15. Susan Glaspell, *Bernice,* in *Susan Glaspell: The Complete Plays,* ed. Linda Ben-Zvi and J. Ellen Gainor (Jefferson, N.C.: McFarland, 2010), act 2, 103.

16. Ibid., act 2, 105.

17. Ibid., act 3, 109.

18. Papke, "Susan Glaspell's Naturalist Scenarios," 26. See also J. Ellen Gainor, *Susan Glaspell in Context: American Theater, Culture, and Politics, 1915–48* (Ann Arbor: University of Michigan Press, 2001), 123.

19. Susan Glaspell, *Inheritors,* in *Complete Plays,* 191. As the editors note, in this scene the extended discussion of evolutionary ideas oscillates between Darwinism and social Darwinism.

20. Ibid., 191–92.

21. Ibid., 192.

22. Ibid.

23. Ibid., 193.

24. Quoted in Barbara Ozieblo and Jerry Dickey, *Susan Glaspell and Sophie Treadwell: American Modernist Women Dramatists* (London: Routledge, 2008), 50.

25. Wolff, *Mendel's Theatre*, 230n.55.

26. Glaspell, *Inheritors*, 196. *Inheritors* is Glaspell's reworking of her story from 1919, "Pollen." See Papke, "Susan Glaspell's Naturalist Scenarios," 27. The play predates Barbara McClintock's pioneering work in maize cytogenetics by almost a decade (she was taking her first course in genetics in 1921, while studying for her BSc in botany at Cornell—the period when Glaspell was working plant genetics into *Inheritors* and *The Verge*).

27. "A Brief Biography," in Glaspell, *Complete Plays*, 3.

28. Susan Glaspell, *The Outside*, in *Complete Plays*, 59. *The Outside* was produced by the Provincetown Players at the Playwrights' Theatre, New York (December 28, 1917 to January 3, 1918). The play emulates J. M. Synge's and Maurice Maeterlinck's symbolist dramas but adds a pointedly feminist slant.

29. Antoine, quoted in Ben-Zvi, *Susan Glaspell*, 95–96. Maeterlinck became the greatest influence on her, especially his monism. He won the Nobel Prize in 1911, and while Glaspell was in Paris, his opera *Monna Vanna* and his play *The Bluebird* were staged.

30. Susan Glaspell, *Close the Book*, in *Complete Plays*, 46.

31. Glaspell, *Inheritors*, 188.

32. Susan Glaspell, *Springs Eternal*, in *Complete Plays*, 357.

33. Glaspell, *Inheritors*, 188–89. The premiere was by the Provincetown Players, Playwrights' Theatre, New York (March 21 to April 10, 1921; additional performances on April 13 and 16). The first British production was at Liverpool Repertory Theatre (September 25, 1925), transferring for an extended run to London's Everyman Theatre on December 28, 1925. Papke writes that the Liverpool production was called a "triumph" in a report wired to the *New York Times* (*Susan Glaspell*, 52).

34. Susan Kingsley Kent, *Gender and Power in Britain, 1640–1990* (London: Routledge, 1999), chap. 12.

35. *The Verge* was performed by the Provincetown Players, Playwrights' Theatre, New York (November 14 to December 1, 1921; thirty-eight performances). Theatre Guild then took over the production, moving it uptown to the Garrick Theatre on West Thirty-fifth Street for special matinee performances (December 6–16, 1921), but it proved unprofitable there. It then returned to the Provincetown Players for a two-week run. See Papke, *Susan Glaspell*, 56.

36. Susan Glaspell, *The Verge*, in *Complete Plays*, 227; Alfred North Whitehead, *The Concept of Nature* (Cambridge: Cambridge University Press, 1920).

37. Margot Norris, *Beasts of the Modern Imagination: Darwin, Nietzsche, Kafka, Ernst, and Lawrence* (Baltimore: Johns Hopkins University Press, 1985), 1–2.

38. Ozieblo, *Susan Glaspell,* 189.

39. Glaspell, *The Verge*, act 1, 230.

40. Ibid., act 1, 246.

41. Jörg Thomas Richter, "Generating Plants and Women: Intersecting Conceptions of Biological and Social Mutations in Susan Glaspell's *The Verge* (1921)," in *Making Mutations: Objects, Practices, Contexts,* ed. Luis Campos and Alexander Von Schwerin (Munich, Max Planck Institutes, 2010), 71–84, http://www.mpiwg-berlin.mpg.de/Preprints/P393.PDF (accessed May 16, 2013). See also Kristina Hinz-Bode, *Susan Glaspell and the Anxiety of Expression: Language and Isolation in the Plays* (Jefferson, N.C.: McFarland, 2006), 172.

42. Wolff, *Mendel's Theatre,* 113.

43. August Strindberg to Torsten Hedlund, October 25, 1895, in *Selected Essays by August Strindberg,* ed. and trans. Michael Robinson (Cambridge: Cambridge University Press, 1996), 13.

44. Steve Bottoms, "Building on the Abyss: Susan Glaspell's *The Verge* in Production," *Theatre Topics* 8, no. 2 (1998): 127–47.

45. Glaspell, *Road to the Temple*, 154.

46. Wolff, *Mendel's Theatre*, 121. In addition, the tower in the final act has a possible precedent in James A. Herne's lighthouse in *Shore Acres* that prompted much critical attention at the time, as discussed previously.

47. Hugh Elliott, introduction to Jean-Baptiste Lamarck, *Zoological Philosophy*, trans. Hugh Elliott (1914; repr., New York: Hafner, 1963), xviii.

48. Steve Jones, *Darwin's Island: The Galapagos in the Garden of Eden* (London: Little, Brown, 2009).

49. Charles Darwin, "Autobiography, May 31, 1876," in *Charles Darwin and T. H. Huxley: Autobiographies*, ed. Gavin de Beer (Oxford: Oxford University Press, 1983), 75–80.

50. Ibid., 80.

51. Bowler, *Evolution,* 268.

52. Charles Darwin, *The Descent of Man, and Selection in Relation to Sex*, ed. James Moore and Adrian Desmond (London: Penguin, 2004), 63.

53. George Levine, *Darwin and the Novelists: Patterns of Science in Victorian Fiction* (Chicago: University of Chicago Press, 1988), 239.

54. Peter J. Bowler, *The Non-Darwinian Revolution: Reinterpreting a Historical Myth* (Baltimore: Johns Hopkins University Press, 1988), 121.

55. Michael M. Chemers, *Staging Stigma: A Critical Examination of the American Freak Show* (London: Palgrave, 2008), 65.

56. Ibid., 66.

57. See, for example, Mark Pagel, *Wired for Culture: The Natural History of Human Cooperation* (London: Allen Lane, 2012).

58. See, for example, Martin Nowak, with Roger Highfield, *Supercooperators: The Mathematics of Evolution, Altruism and Human Behavior (Or Why We Need Each Other to Succeed)* (Edinburgh: Canongate, 2012).

59. Katharine Cockin, *Edith Craig* (London: Continuum, 1998), 150.

60. Ben-Zvi, *Susan Glaspell*, 309.

61. Ibid., 306. As mentioned, Emma Goldman had already published *The Social Significance of the Modern Drama* (Boston: Badger, 1914), based on her lectures on that topic.

62. "Experiment in Drama," *New Statesman* 24, no. 599 (1924): 22.

63. Papke, *Susan Glaspell*, 66–67.

64. Quoted in Elizabeth Sprigge, *Sybil Thorndike Casson* (London: Gollancz, 1971), 171.

65. James Harrison, "Destiny or Descent? Responses to Darwin," *Mosaic* 14, no. 1 (1981): 114.

66. Sybil Thorndike, "Religion and the Stage," in *Affirmations: God in the Modern World* (London: Benn, 1928), 6–7, 11. She also expresses the view that "in each of us is contained *in embryo* every other human being" (23), hence the actor "is identified with all men" (25).

67. Sprigge, *Sybil Thorndike Casson*, 172. Over forty years later, Thorndike could still quote by heart some of Claire's speeches. See also Jonathan Croall, *Sybil Thorndike: A Star of Life* (London: Haus, 2008), 188; and Sheridan Morley, *Sybil Thorndike: A Life in the Theatre* (London: Weidenfeld and Nicolson, 1977).

68. Excerpted in Croall, *Sybil Thorndike*, 188–89.

69. John C. Trewin, *Sybil Thorndike* (London: Rockliff, 1955), 53.

70. See, especially, Richard J. Hand and Michael Wilson, *Grand-Guignol: The French Theatre of Horror* (Exeter, Eng.: University of Exeter Press, 2002), and *London's Grand Guignol and the Theatre of Horror* (Exeter, Eng.: University of Exeter Press, 2007).

71. Wolff, *Mendel's Theatre*, 139.

72. Edy Craig was connected to all the major "formal and technical experiments in art theatre by male theatre practitioners at this time," such as Edward Gordon Craig (her brother) at the Moscow Art Theatre in 1912, Lugne-Poë and Jacques Copeau in France, Max Reinhardt in Germany, and Adolphe Appia's ideas about lighting. See Katharine Cockin, *Edith Craig (1869–1947): Dramatic Lives* (London: Cassell, 1998), 121.

73. Sprigge, *Sybil Thorndike Casson*, 139.

74. Croall, *Sybil Thorndike*, 190.

75. Julie Holledge, *Innocent Flowers: Women in the Edwardian Theatre* (London: Virago, 1981), 146–50. Virginia Woolf reviewed the London production in

1925 of *The Verge* in the *New Statesman* (along with Desmond McCarthy); so did Cicely Hamilton in *Time and Tide*. See Cockin, *Edith Craig*, 127.

76. Croall, *Sybil Thorndike,* 189.

77. Ibid., 189–90.

78. Ibid., 190.

79. Cockin, *Edith Craig*, 150.

80. Ozieblo, *Susan Glaspell*, 270.

81. Glaspell, *Springs Eternal*, 357.

82. Ibid., 361.

83. Ibid., 402.

84. Ibid., 405.

85. Ibid., 360.

86. Ibid., 365.

87. Veronica Makowsky, "Susan Glaspell and Modernism," in *The Cambridge Companion to American Women Playwrights,* ed. Brenda Murphy (Cambridge: Cambridge University Press, 1999), 49.

88. Bottoms, "Building on the Abyss," 137.

89. Ozieblo and Dickey, *Susan Glaspell and Sophie Treadwell*, 13. Similarly, Ben-Zvi and Gainor say that much of Glaspell's work is a "curious blend of traditional and progressive perspectives" (Glaspell, *Complete Plays*, 37).

90. Ibid., 13.

91. Bertolt Brecht, "Short Organon for the Theatre," in *Brecht on Theatre,* ed. Marc Silberman, Steve Giles, and Tom Kuhn (London: Bloomsbury, 2014), 229; his essay on Shaw is also in this edition ("Three Cheers for Shaw," 28–31).

92. See, for example, Bertolt Brecht, "The Street Scene" ("how thoroughly he has to imitate"), in *Brecht on Theatre*, 176–77.

93. Astrid Oesmann, *Staging History: Brecht's Social Concepts of Ideology* (Albany: SUNY Press, 2005), 26.

94. Bertolt Brecht, "Notes on the Opera *Rise and Fall of the City of Mahagonny,*" in *Brecht on Theatre*, 65.

95. Oesmann, *Staging History,* 20–21.

96. Ibid., 17.

97. Downing Cless, *Ecology and Environment in European Drama* (London: Routledge, 2010), 169.

98. Lifton, "*Vast Encyclopedia,*" 35–36.

99. Ibid., 43.

100. *The Journals of Thornton Wilder, 1939–61,* ed. Donald Gallup (New Haven, Conn.: Yale University Press, 1985), 22.

101. Ibid., 21. *The Skin of Our Teeth* was finished on January 1, 1942; after tryouts in New Haven and Baltimore, it opened in New York at the Plymouth

Theatre on November 18, 1942. "The period from January 1942, when Wilder finished *The Skin of Our Teeth,* until May 1943, when he left for active military duty in North Africa, was one of the most active in Wilder's life" ("Thornton Wilder: January 1942 to May 1943," in *The Letters of Gertrude Stein and Thornton Wilder*, ed. Edward M. Burns and Ulla E. Dydo, with William Rice [New Haven, Conn.: Yale University Press, 1996], 377). He was working on finding a producer for his play and writing scripts for two army training films and for Alfred Hitchcock's *Shadow of a Doubt.*

102. *Journals of Thornton Wilder,* 26.

103. Thornton Wilder, *The Skin of Our Teeth*, in *Three Plays* (New York: Avon, 1957), xii.

104. Jackson R. Bryer, ed., *Conversations with Thornton Wilder* (Jackson: University Press of Mississippi, 1992), 75.

105. Wilder, *Skin of Our Teeth*, act 1, 91.

106. Byron's "cosmic drama" *Cain* (1821) retold the story of Cain and Abel from Cain's point of view; it also explored the theme of catastrophism as a way of explaining gaps in the fossil record and "played fast and loose with the illustrious name of Georges Cuvier" (Ralph O'Connor, *The Earth on Show: Fossils and the Poetics of Popular Science, 1802–1856* [Chicago: University of Chicago Press, 2007], 7).

107. Wilder, *Skin of Our Teeth*, act 1, 72.

108. Ibid., act 3, 37.

109. Ibid., act 3, 128.

110. Ibid., act 1, 83.

111. Lifton, "*Vast Encyclopedia,*" 43. Haeckel is not mentioned in Lifton's book.

112. Ibid., 43.

113. Lifton acknowledges a possible influence from Shaw's Creative Evolution on Wilder's concept of nature's workings but says it is much more likely that Goethe's idea of *ewig Wirkende*, the "perpetually achieving force at work in the universe," shaped his thinking about evolution, as Wilder envisions an optimistic nature that is ever striving to improve. He had a "faith in nature's benevolence" (Lifton, "*Vast Encyclopedia,*" 41).

114. Wilder had been reading *Finnegans Wake*, and he stated this in his preface to *Three Plays* in response to a scandal surrounding accusations of plagiarism. Christopher W. E. Bigsby notes clear parallels between the two works, including the Antrobuses and the Earwickers, Lily Sabina and Lily Kinsella, and the fact that both "move their characters through different historical periods," but sees this not as plagiarism but homage (*A Critical Introduction to Twentieth-Century American Drama, 1900–1940* [Cambridge: Cambridge University Press, 1982],

269–70). See also Gilbert A. Harrison, *The Enthusiast: A Life of Thornton Wilder* (New Haven, Conn.: Ticknor and Fields, 1983), chap. 18.

115. Four plays were written by e. e. (Edward Estlin) cummings, each one utterly different from the other in mode. If there is one unifying idea, it is the role of the artist in society. They are also notable for their evolutionary ideas. *Santa Claus* (1946) is, like Brecht's *Life of Galileo*, informed by the development of the atomic bomb, and it is clearly against science: it is "an American version of a Faust story" in which Santa wears a death mask (given to him by Death) and is called "Science" and is persuaded by Death that all people really want in this age is knowledge without understanding. Santa/Science becomes a salesman. See Richard S. Kennedy, *Dreams in the Mirror: A Biography of E. E. Cummings* (New York: Liveright, 1980), 407–8. Kennedy does not refer at all to *Anthropos*, and does not mention Wilder.

116. Bigsby, *Critical Introduction*, 270.

117. *Journals of Thornton Wilder*, 37. His journal also indicates that he excised early material dealing with "natural history . . . the arch of the natural world that surrounds us" (38).

118. Lifton, "*Vast Encyclopedia*," 42.

119. Wilder, *Skin of Our Teeth*, act 2, 95.

120. Ibid., act 2, 95–96.

121. Lifton, "*Vast Encyclopedia*," 42.

122. Wilder, *Skin of Our Teeth*, act 2, 115–18.

123. Ibid., act 2, 98–99.

124. Ibid., act 2, 114.

125. The play opened on September 8, 1949, at St. Martin's Theatre, London, and ran for forty-three performances; it was produced by Michael Macowan. See William Konkle, "J. B. Priestly (1894–1984)," in *British Playwrights, 1880–1956: A Research and Production Sourcebook*, ed. William Demastes and Katherine E. Kelly (Westport, Conn.: Greenwood Press, 1996), 332.

126. J. B. Priestley, *Summer Day's Dream*, in *Plays Two* (London: Oberon, 2004), 2.1, 175–76.

127. Tom Priestley, introduction to Priestley, *Plays Two*, 10.

128. Priestley, *Summer Day's Dream*, 1.1, 107.

129. Ibid., 2.1, 157.

130. Ibid., 1.1, 120.

131. Ibid., 1.1, 132.

132. Tom Priestley, introduction to Priestley, *Plays Two*, 11.

133. Priestley, *Summer Day's Dream*, 2.2, 185.

134. Maggie B. Gale, *J. B. Priestley* (London: Routledge, 2008), 119.

135. Ibid., 56.

136. Bowler, *Non-Darwinian Revolution*, 162.

137. Eugene O'Neill, *The First Man*, in *Eugene O'Neill: Complete Plays, 1920–1931,* ed. Travis Bogard (New York: Library of America, 1988), act 1, 59. For a discussion of this play, see, for example, Travis Bogard, *Contour in Time: The Plays of Eugene O'Neill* (New York: Oxford University Press, 1972), 150.

138. Rundle, "Hairy Ape's Humanist Hell," 50. *The Hairy Ape* was first produced at the Playwrights Theater in New York in March 1922, receiving good reviews, and transferring quickly to Broadway in April at the Plymouth Theater for 127 performances.

139. Ibid., 56.

140. Ibid., 50.

141. Eugene O'Neill, *The Hairy Ape*, in *The Plays of Eugene O'Neill* (New York: Modern Library, 1941), 1:39.

142. Bogard, *Contour in Time*, 242.

143. Rundle, "Hairy Ape's Humanist Hell," 54.

144. Eugene O'Neill to Kenneth Macgowan, December 24, 1921, quoted in Bogard, *Contour in Time*, 241.

145. Peter Egri, "'Belonging' Lost: Alienation and Dramatic Form in Eugene O'Neill's *The Hairy Ape*," in *Critical Essays on Eugene O'Neill*, ed. James J. Martine (Boston: Hall, 1984), 108.

146. Bogard, *Contour in Time*, 248.

147. Rundle, "Hairy Ape's Humanist Hell," 51.

148. Bogard, *Contour in Time,* 251.

149. Ibid., 250.

150. Ibid., 252.

151. Rundle analyzes this connection in detail in "Hairy Ape's Humanist Hell," 72.

152. Kenneth Macgowan, *The Theatre of Tomorrow* (New York: Boni and Liveright, 1921), quoted in Bogard, *Contour in Time,* 243–44.

153. *Inherit the Wind* played at the National Theater, New York (April 21, 1955 to June 22, 1957) for a total of 806 performances. It won several Tony awards in 1956.

154. Jerome Lawrence and Robert E. Lee, *Inherit the Wind* (New York: Bantam Books, 1955), 115.

155. Edward J. Larson, *Summer for the Gods: The Scopes Trial and America's Continuing Debate Over Science and Religion* (New York: Basic Books, 1997), 243.

156. Quoted in ibid., 243.

157. Stephen Jay Gould, *Hen's Teeth and Horse's Toes* (New York: Norton, 1983), 270, 273, quoted in ibid., 245.

158. Ullica Segerstråle, "Neo-Darwinism," in *Encyclopedia of Evolution*, ed. Mark D. Pagel, 2 vols. (Oxford: Oxford University Press, 2002), 2:809.

8. Beckett's "Old Muckball"

1. See, for example, Dan Rebellato, *1956 and All That* (London: Routledge, 1999); and John Russell Taylor, *Anger and After* (London: Methuen, 1962). *Waiting for Godot* premiered in Paris in 1953 and in London in 1955 and (after its American premiere in Florida) had its Broadway premiere in April 1956. *Look Back in Anger* premiered in London in May 1956.

2. Ruby Cohn, *Samuel Beckett: The Comic Gamut* (New Brunswick, N.J.: Rutgers University Press, 1962), 3, quoted in Linda Ben-Zvi, introduction to *Beckett at 100: Revolving It All*, ed. Linda Ben-Zvi and Angela Moorjani (Oxford: Oxford University Press, 2008), 5. The phrase "so various, so beautiful, so new" is from Matthew Arnold's poem "Dover Beach" (published 1861).

3. Downing Cless, *Ecology and Environment in European Drama* (London: Routledge, 2011), 171.

4. Ben Brantley, "When a Universe Reels, a Baryshnikov May Fall," *New York Times*, December 19, 2007.

5. James Knowlson and John Pilling, *Frescoes of the Skull: The Later Prose and Drama of Samuel Beckett* (London: Calder, 1979), 94.

6. Samuel Beckett, *Waiting for Godot*, in *Samuel Beckett: The Complete Dramatic Works* (London: Faber and Faber, 1986), act 2, 57, hereafter *CDW*. "Muckheap" also occurs in *Rough for Theatre I*, though in the sense not of the earth but as something "we're heading for" (*Samuel Beckett: Collected Shorter Plays* [London: Faber and Faber, 2006], 69, hereafter *CSP*).

7. Steven Connor, "Beckett's Atmospheres," in *Beckett After Beckett*, ed. Stanley E. Gontarski and Anthony Uhlmann (Gainesville: University Press of Florida, 2006), 57.

8. *The Theatrical Notebooks of Samuel Beckett*, vol. 1, *Waiting for Godot*, ed. Dougald McMillan and James Knowlson (New York: Grove Press, 1993), xiv. This elemental thinking is already evident in the notes on philosophy that he took in the early 1930s.

9. Beckett, *Waiting for Godot*, act 2, 60.

10. Chris J. Ackerley, "Samuel Beckett and Science," in *A Companion to Samuel Beckett*, ed. Stanley E. Gontarski (Oxford: Blackwell, 2011), 144.

11. Ibid., 162.

12. Paul Davies, "Beckett from the Perspective of Ecocriticism," in *Beckett After Beckett*, ed. Gontarski and Uhlmann, 72.

13. Ibid., 74.

14. Cyril D. Darlington, *The Evolution of Man and Society* (London: Allen and Unwin, 1969), 673.

15. Cless, *Ecology and Environment*, 15.

16. Joseph Roach, "'All the Dead Voices': The Landscape of Famine in *Waiting for Godot*," in *Land/Scape/Theater*, ed. Elinor Fuchs and Una Chaudhuri (Ann Arbor: University of Michigan Press, 2002), 85.

17. Ibid.

18. Ibid., 88.

19. Ibid., 88–89.

20. Samuel Beckett, *Happy Days*, in *CDW*, act 2, 161. The words *muck*, *muckball*, and *muckheap* all appear frequently in Beckett's prose and theatre; see, for example, Molloy's "But it's a change of muck. And if all muck is the same muck that doesn't matter, it's good to have a change of muck, to move from one heap to another a little further one, from time to time" (Samuel Beckett, *Molloy*, in *Three Novels: Molloy, Malone Dies, The Unnamable* [New York: Grove Press, 2005], 41).

21. Samuel Beckett, *Murphy* (New York: Grove Press, 2011), 73, and *Happy Days*, act 1, 150.

22. Ruby Cohn, *A Beckett Canon* (Ann Arbor: University of Michigan Press, 2001), 263.

23. Connor, "Beckett's Atmospheres," 52.

24. Ibid., 54.

25. Samuel Beckett, "Whoroscope" notebook, 63, Beckett Collection, University of Reading.

26. Matthew Feldman, *Beckett's Books: A Cultural History of Samuel Beckett's "Interwar Notes"* (New York: Continuum, 2006), 1.

27. Shane Weller, foreword to ibid., viii.

28. Ibid., 14.

29. Rita Felski, "'Context Stinks!'" *New Literary History* 42, no. 4 (2011): 573.

30. Weller, foreword to Feldman, *Beckett's Books*, ix.

31. Ibid.

32. Ibid., x.

33. Karim Mamdani, "Conclusion: Beckett in Theses," in *Beckett/Philosophy*, ed. Matthew Feldman and Karim Mamdani (Sophia, Bulgaria: University Press St. Klimrnt Ohridski, 2012), 312. "Interlocutor" is David Addyman's term, in "'Speak of Time, Without Flinching . . . Treat of Space with the Same Easy Grace': Beckett, Bergson and the Philosophy of Space," in ibid., 68–88.

34. Jane R. Goodall, *Performance and Evolution in the Age of Darwin: Out of the Natural Order* (London: Routledge, 2002), 7.

35. Linda Ben-Zvi, "Biographical, Textual and Historical Origins," in *Palgrave Advances in Samuel Beckett Studies*, ed. Lois Oppenheim (London: Palgrave Macmillan, 2004), 133–53.

36. Stanley E. Gontarski notes that the original French, "Les gens sont des cons," does not have the Darwinian implications of the translation, in "A Centenary of

Missed Opportunities: A Guide to Assembling an Accurate Volume of Samuel Beckett's Dramatic 'Shorts,'" *Modern Drama* 54, no. 3 (2011): 364.

37. Frederik N. Smith, "Dating the Whoroscope Notebook," *Journal of Beckett Studies* 3, no. 1 (1993), manuscript in typescript, Beckett Collection, University of Reading. Smith does not elaborate on what the latter reference might be.

38. Ibid., 6n.3.

39. Samuel Beckett to Tom MacGreevy, August 4, 1932, in *The Letters of Samuel Beckett*, vol. 1, *1929–1940*, ed. Martha Dow Fehsenfeld and Lois More Overbeck (Cambridge: Cambridge University Press, 2009), 111.

40. Ibid. On the marginalia in Beckett's copy of *On the Origin of Species*, see Dirk Van Hulle, "Notebooks and Other Manuscripts," in *Samuel Beckett in Context*, ed. Anthony Uhlmann (Cambridge: Cambridge University Press, 2013), 421; and Dirk Van Hulle and Mark Nixon, *Samuel Beckett's Library* (Cambridge: Cambridge University Press, 2013), 7.

41. Charles Darwin, *On the Origin of Species*, facsimile of the first edition, ed. Ernst Mayr (Cambridge, Mass.: Harvard University Press, 1964), 11–12.

42. Ibid., 144.

43. Michael Beausang, "*Watt*: Logique, démence, aphasie," in *Beckett avant Beckett: Essais sur les premières œuvres*, ed. Jean-Michel Rabaté (Paris: Accents, 1984), 153–72.

44. The word *whimsical* also occurs early in Beckett's "Whoroscope" notebook.

45. Valerie Purton, ed., *Darwin, Tennyson and Their Readers: Explorations in Victorian Literature and Science* (London: Anthem, 2013), xvi.

46. Beckett, quoted in Feldman, *Beckett's Books*, 71.

47. Ibid., 72.

48. See, for example, Erik Tonning, "'I Am Not Reading Philosophy': Beckett and Schopenhauer," in *Beckett/Philosophy*, ed. Feldman and Mamdani, 44–67.

49. Dirk Van Hulle, "'Eff it': Beckett and Linguistic Skepticism," in ibid., 225.

50. Ibid., 224.

51. Ibid., 225.

52. Peter Fifield, "'Of Being—or Remaining': Beckett and Early Greek Philosophy," in *Beckett/Philosophy*, ed. Feldman and Mamdani, 89–109, and personal correspondence with the author, May 8, 2013.

53. Fifield, "'Of Being—or Remaining,'" 98.

54. Adrian Desmond and James Moore, *Darwin: The Life of a Tormented Evolutionist* (London: Penguin, 2009), 255.

55. Jessica Whiteside, "Wedges and Impacts: Darwin's Enduring Legacy," *Brown Medicine*, spring 2009, 2.

56. Desmond and Moore, *Darwin*, 346.

57. John Pilling, "Dates and Difficulties in Beckett's *Whoroscope* Notebook," in *The Beckett Critical Reader: Archives, Theories and Translations*, ed. Stanley E.

Gontarski (Edinburgh: Edinburgh University Press, 2012), 87. Ackerley summarizes the key references to rocks in Beckett's work in "Samuel Beckett and Science," 145. See also "Geology," in *The Grove Companion to Samuel Beckett: A Reader's Guide to His Works, Life, and Thought*, ed. Christopher J. Ackerley and Stanley E. Gontarski (New York: Grove, 2004), 220.

58. "Geology," 219–20.

59. There is also a link with Shaw, whose wise Ancient in *Back to Methuselah* characterizes the earth in a poetic vision that likewise uses the dead fish as metaphor for vast geological changes: "In the hardpressed heart of the earth, where the inconceivable heat of the sun still glows, the stone lives in fierce atomic convulsion, as we live in our slower way. When it is cast out to the surface it dies like a deep-sea fish: what you see is only its cold dead body" (Bernard Shaw, *Back to Methuselah*, in *The Complete Plays of Bernard Shaw* [London: Constable, 1931], part 5, 958).

60. Goodall, *Performance and Evolution in the Age of Darwin*, 7. See also Jane Goodall, "Popular Culture," in *Samuel Beckett in Context*, ed. Gontarski and Uhlmann, 289–98.

61. William Greenslade, *Degeneration, Culture and the Novel, 1880–1940* (Cambridge: Cambridge University Press, 2010), 201–4.

62. Jane Goodall, personal correspondence with author, January 20, 2012.

63. Andrew Gibson, *Samuel Beckett* (London: Reaktion, 2010), 90–92.

64. Ibid., 90, 92. Feldman argues that Beckett's psychosomatic expression has been overstated by scholars, in *Beckett's Books*, 86.

65. Pietro Corsi, "Jean-Baptiste Lamarck: From Myth to History," in *Transformations of Lamarckism: From Subtle Fluids to Molecular Biology*, ed. Snait B. Gissis and Eva Jablonka (Cambridge, Mass.: MIT Press, 2011), 20.

66. Beausang, "*Watt*," 160. My thanks to Richard Parish for help with this translation.

67. Samuel Beckett, *Watt*, ed. Christopher Ackerley (London: Faber, 2009).

68. Charles Darwin, *On the Origin of Species*, chap. 7, 208 (emphasis added), Darwin Online, http://darwin-online.org.uk/content/frameset?itemID=F373&viewtype=text&pageseq=1 (accessed March 28, 2013).

69. Beausang, "*Watt*," 163.

70. Ibid., 164.

71. Feldman notes that "uterine tropes" abound in Beckett's works, in *Beckett's Books*, 107.

72. Samuel Beckett, *All That Fall*, in *CSP*, 36.

73. John P. Mahaffy, *Descartes* (Edinburgh: Blackwood, 1901), quoted in Francis Doherty, "Mahaffy's Whoroscope," in *Beckett Critical Reader*, ed. Gontarski, 21.

74. Samuel Beckett, quoted in Feldman, *Beckett's Books*, 86.

75. Beckett, *All That Fall*, 16, 31. The Rooneys speculate, "Can hinnies procreate, I wonder?" (36), a question that brings Mr. Rooney up sharp. They discuss this. "Aren't they barren, or sterile?" (37). Their childlessness makes this emphasis on sterility especially significant.

76. Quoted in Cohn, *Beckett Canon*, 233.

77. Beckett, *All That Fall*, 17.

78. Beckett, *Waiting for Godot*, 50, 86.

79. Ibid., act 2, 84.

80. Samuel Beckett, *Krapp's Last Tape*, in *CSP*, 59.

81. Beckett, *Rough for Theatre I*, 70.

82. Gontarski, "Centenary of Missed Opportunities," 380.

83. James Harrison, "Destiny or Descent? Responses to Darwin," *Mosaic* 14, no. 1 (1981): 119.

84. Beckett, *Rough for Theatre I*, 72.

85. John Calder, *The Philosophy of Samuel Beckett* (London: Calder, 2001), 72, quoted in Cless, *Ecology and Environment*, 213.

86. Cohn, *Beckett Canon*, 179.

87. Jane Goodall, "Lucky's Energy," in *Beckett After Beckett*, ed. Gontarski and Uhlmann, 190.

88. Quoted in Gontarski, "Centenary of Missed Opportunities, 362.

89. T. H. Huxley, quoted in Gillian Beer, "Darwin and the Uses of Extinction," *Victorian Studies* 51, no. 2 (2009): 329.

90. Joseph Roach, "Darwin's Passion: The Language of Expression on Nature's Stage," *Discourse: Journal for Theoretical Studies in Media and Culture* 13, no. 1 (1990): 56.

91. Katharine Worth, *Samuel Beckett's Theatre: Life Journeys* (Oxford: Clarendon Press, 2001), 22–47.

92. Kathryn White, *Beckett and Decay* (London: Continuum, 2009).

93. Roach, "'All the Dead Voices,'" 92.

94. Beckett, *Waiting for Godot*, act 1, 39.

95. Samuel Beckett, *Rough for Theatre II*, in *CSP*, 71, 82.

96. Beckett, *Waiting for Godot*, act 1, 24. The term is also mentioned explicitly in *How It Is*: "I was young I clung on to the species we're talking of the species the human."

97. Beckett, *Waiting for Godot*, act 1, 28–29; act 2, 78.

98. Ibid., act 2, 74.

99. Beckett, *All That Fall*, 23. Note that Beckett refers to "this wretched planet" in a letter to Robert Pinget in 1966. See Lois More Overbeck, "Audience of Self/Audience of Reader," *Modernism/Modernity* 18, no. 4 (2011): 723.

100. Beckett, *Waiting for Godot*, act 2, 79–80.

101. Knowlson and Pilling, *Frescoes of the Skull*, 104.

102. Ulrika Maude, "Pavlov's Dogs and Other Animals in Samuel Beckett," in *Beckett and Animals*, ed. Mary Bryden (Cambridge: Cambridge University Press, 2013), 82.

103. Beckett, *Happy Days*, act 2, 166.

104. Knowlson and Pilling, *Frescoes of the Skull*, 95–96.

105. Beckett, *Happy Days*, act 1, 151.

106. Ibid., 152.

107. Ibid., 153. Pilling points out that Beckett directs this section of the text to be delivered at a very slow pace, compared with the deliberately fast and garbled section recalling her episode with the mouse, indicating that the lines about adaptation should be unmistakable and clear. See Knowlson and Pilling, *Frescoes of the Skull*, 107.

108. Beckett, *Happy Days*, act 1, 154.

109. Ibid., 153.

110. Ibid., act 2, 161.

111. Ibid., 166.

112. Franz Kafka, "A Report to an Academy," in *In the Penal Settlement: Tales and Short Prose Works*, trans. Willa Muir and Edwin Muir (London: Secker and Warburg, 1949), 169–81.

113. Margot Norris, *Beasts of the Modern Imagination: Darwin, Nietzsche, Kafka, Ernst, and Lawrence* (Baltimore: Johns Hopkins University Press, 1985), 67; full discussion of Kafka's story, 66–72. See also Shane Weller, "Forms of Weakness: Animalisation in Kafka and Beckett," in *Beckett and Animals*, ed. Bryden, 13.

114. In a recent stage version starring Kathryn Hunter as the "creature stranded between a human present and a simian past," he subtly conveys "dark messages about humanity's bestiality." His initial capture and captivity were so brutal that "'I'm still overcome with such an aversion to human beings, I can barely stop myself from retching.'" Ultimately, his "forced march up the evolutionary scale" has left him isolated (Charles Isherwood, "A Captive of Human Nature" [review of *Kafka's Monkey*], *New York Times*, April 4, 2013). So convincingly did Hunter portray Red Peter's "ape-ness" that "if you squinted, you would probably believe that the creature onstage was a real chimpanzee in white tie and tails." *Kafka's Monkey* was adapted by Colin Teevan from Kafka's "A Report to an Academy" for the Young Vic and first performed there by Kathryn Hunter in 2009; the New York production in 2013 was for Theatre for a New Audience.

115. Becket, *Happy Days*, act 1, 147. In his notes to the German production of *Happy Days* in September 1971, Beckett gave detailed movements for Willie in his final moments on mound, which he must clutch (a term used twice), not looking at Winnie until he collapses and lies completely motionless in this

position until the end of the play (MS 1396/4/10, 5, Beckett Collection, University of Reading).

116. Feldman, *Beckett's Books*, 106.

117. Maude, "Pavlov's Dogs and Other Animals," 89.

118. Beckett, *Happy Days*, act 1, 145–46.

119. Knowlson and Pilling, *Frescoes of the Skull*, 97.

120. See, for example, Yoshiki Tajiri, "Beckett, Coetzee and Animals," in *Beckett and Animals*, ed. Bryden, 37.

121. Knowlson and Pilling, *Frescoes of the Skull*, 96.

122. John Bolin, *Beckett and the Modern Novel* (Cambridge: Cambridge University Press, 2013), 56–59.

123. Overbeck, "Audience of Self/Audience of Reader," 722.

124. See, for example, Bernard Shaw's discussion of it in the "Paley's Watch" section of his preface to *Back to Methuselah*, in *Prefaces by Bernard Shaw* (London: Constable, 1934), 495–96.

125. Beckett, *Rough for Theatre II*, 81.

126. Ibid., 88.

127. Cohn refers to the birds as thrushes (*Beckett Canon*, 242), but Beckett's term in the original French is *pinson* (chaffinch). See Samuel Beckett, *Pas, suivi de quatre esquisses* (Paris: Éditions de Minuit, 1978), 60. His detailed description indicates that "they are probably blue-capped parrot finches" (Linda Ben-Zvi, "Beckett's 'Necessary' Cat[s]," in *Beckett and Animals*, ed. Bryden, 137n.22).

128. Beckett, *Rough for Theatre I*, 67, 69.

129. Knowlson and Pilling, *Frescoes of the Skull*, 229.

130. Fifield, "'Of Being—or Remaining,'" 97.

131. Beer, "Darwin and the Uses of Extinction," 323.

132. Ibid., 324.

133. "These human beings are the last of their species" (Cohn, *Beckett Canon*, 226).

134. Knowlson and Pilling, *Frescoes of the Skull*, 93.

135. Ibid., 94.

136. Cohn, *Beckett Canon*, 227.

137. Ibid.

138. Ibid.

139. Ibid., 231, referring to Robert Haerdter, *Samuel Beckett inszeniert das "Endspiel"* (Frankfurt: Suhrkamp, 1969), 50.

140. Roach, "'All the Dead Voices,'" 89, citing Beckett, *Theatrical Notebooks*, 1:109.

141. Ibid., 90.

142. Brian Boyd, *On the Origin of Stories: Evolution, Cognition, and Fiction* (London: Belknap, 2009).

143. Cohn, *Beckett Canon*, 264.

144. Mike Hawkins, *Social Darwinism in European and American Thought* (Cambridge: Cambridge University Press, 1997), 30.

145. Knowlson and Pilling, *Frescoes of the Skull*, 34.

146. David Wheatley, "'Quite Exceptionally Anthropoid': Species Anxiety and Metamorphosis in Beckett's Humans and Other Animals," in *Beckett and Animals*, ed. Bryden, 61. On Beckett's antieugenic stance, see also Gibson, *Samuel Beckett*.

147. Samuel Beckett, *Eleuthéria*, trans. Barbara Wright (London: Faber and Faber, 1996), 44–45.

148. Knowlson and Pilling, *Frescoes of the Skull*, 25. In *Krapp's Last Tape*, Krapp follows a gnostic, even "specifically Manichean tradition," emphasizing sexual abstinence and refusal to marry, "so as not to play the Creator's game" (86–87). Feldman likewise sees "this old muckball" as evidence of the thoroughgoing gnosticism of the play, in *Beckett's Books*, 4.

149. Marius Buning, "Eleuthéria Revisited" (lecture presented in Spain, December 2, 1997), http://samuel-beckett.net/Eleutheria_Revisited.html (accessed January 5, 2012).

150. David Bradby calls *Les Mamelles de Tiresias* a "famous predecessor" to *Godot*, but he does not say why, in *Beckett: Waiting for Godot*, Plays in Production (Cambridge: Cambridge University Press, 2001), 54. James Knowlson notes that *Eleuthéria* indicates Beckett's attitudes toward the theatre of the past, parodying "many features of traditional plays and experiments," yet he makes no mention of Apollinaire's *Mamelles* as a possible precedent, in *Damned to Fame: The Life of Samuel Beckett* (New York: Grove Press, 2004), 363–64.

151. Guillaume Apollinaire, *Les Mamelles de Tiresias*, in *Modern French Theatre: The Avant-Garde, Dada and Surrealism, an Anthology of Plays*, ed. Michael Benedikt and George E. Wellwarth (New York: Dutton, 1964), 81.

152. Davies, "Beckett from the Perspective of Ecocriticism," 74.

153. Chris Ackerley, "'Despised for the Obviousness': Samuel Beckett's Dogs," in *Beckett and Animals*, ed. Bryden, 186–87.

Epilogue

1. Richard Bean, *The Heretic* (London: Oberon, 2011), act 3, 58.

2. *Him* ends with a freak show scene, complete with a hunchback barker and eight freaks, including a missing link and a fat lady. It's Me's "psychic fantasy" as she gives birth, and the barker pulls back the curtain to reveal Me holding her newborn baby.

3. Jill Lepore, "The Odyssey: Robert Ripley and His World," *New Yorker*, June 3, 2013, 62–65. Ripley insisted, "I only deal in oddities, not freaks," and made the distinction that "an oddity is a high-class freak."

4. Tony Kushner, jacket copy for *Venus* (New York: Theatre Communications Group, 1996). Another play apparently dealing with this character is *Hottentot Venus* (1981) by Brian Rotman, but I have been unable to locate a published copy of it.

5. Kate Kellaway, review of *Totem* in *The Observer*, January 9, 2011, consulted online at http://www.theguardian.com/stage/2011/jan/09/cirque-du-soleil-totem-review on January 23, 2012.

6. Michael Billington, review of *Totem*, *Guardian*, January 6, 2011, consulted online at http://www.theguardian.com/stage/2011/jan/06/cirque-du-soleil-totem-review on January 23, 2012.

7. Lyn Gardner, review of *Totem*, *Guardian*, January 8, 2012, http://www.theguardian.com/stage/2012/jan/08/totem-review (accessed January 23, 2012).

8. Francis Galton, preface to *Hereditary Genius* (London: Macmillan, 1892).

9. Nessa Carey, *The Epigenetics Revolution* (London: Icon, 2011), 3.

10. Jonathan Hodgkin, "Thin and Fat from the Start," review of *The Epigenetic Revolution,* by Nessa Carey, *TLS*, May 4, 2012, 24.

11. See, for example, Edward J. Steele, *Lamarck's Signature: How Retrogenes Are Changing Darwin's Natural Selection Paradigm* (Reading, Mass.: Perseus, 1998).

12. Jean-François Peyret and Alain Prochiantz, *Ex Vivo/In Vitro* (2011), carrousel 5, 27, http://theatrefeuilleton2.net/spectacles/ex-vivo-in-vitro/ (accessed November 29, 2011).

13. Ibid., 28.

14. Philip Ball, *Unnatural: The Heretical Idea of Making People* (London: Bodley Head, 2011). As Gillian Beer points out, in the Frankenstein story, which is, like Faust, one of the most powerful myths around a negative image of the scientist, there is creation without growth: "He is a creature denied the experience of growth. He is fabricated as if he were a machine, but out of organic bits and pieces. There is a gap between concept and material" (*Darwin's Plots: Evolutionary Narrative in Darwin, George Eliot, and Nineteenth-Century Fiction* [Cambridge: Cambridge University Press, 2000], 103).

15. See, for example, Ric Knowles, ed., "Interspecies Performance," special issue, *Theatre Journal* 65, no. 3 (2013): Ric Knowles, "Editorial Comment: Interspecies Performance," i–v; Una Chaudhuri, "Bug Bytes: Insects, Information, and Interspecies Theatricality," 321–34; Courtney Ryan, "Playing with Plants," 335–53; Susan Nance, "Game Stallions and Other 'Horseface Minstrelsies' of the American Turf," 355–72; and Jennifer Parker-Starbuck, "Animal Ontologies and Media Representations: Robotics, Puppets, and the Real of War Horse," 373–93.

16. Una Chaudhuri, "Animal Geographies: *Zooësis* and the Space of Modern Drama," in *Performing Nature: Explorations in Ecology and the Arts*, ed. Gabriella Giannachi and Nigel Stewart (Oxford: Lang, 2005), 103–4.

17. Katharine Worth, "Edward Albee: Playwright of Evolution," in *Essays on Contemporary American Drama,* ed. Hedwig Bock and Albert Wertheim (Munich: Hueber, 1981), 33–53. See also Toby Zinman, *Edward Albee* (Ann Arbor: University of Michigan Press, 2008), 1–9, 79–86.

18. Charles Isherwood, "A Captive of Human Nature" [review of *Kafka's Monkey*], *New York Times*, April 4, 2013. *Kafka's Monkey*, adapted by Colin Teevan, based on "A Report to an Academy," by Franz Kafka; directed by Walter Meierjohann. A Young Vic production (premiered 2009), presented by Theater for a New Audience, Jeffrey Horowitz, artistic director, in association with Baryshnikov Arts Center. At the Jerome Robbins Theater, Baryshnikov Arts Center, New York.

19. Downing Cless, *Ecology and Environment in European Drama* (London: Routledge, 2010), 16.

20. Julie Hudson, "'If You Want to Be Green Hold Your Breath': Climate Change in British Theatre," *New Theatre Quarterly* 28, no. 3 (2012): 261.

21. Gideon Lewis-Kraus, "It's Good to Be Alive," *London Review of Books*, February 9, 2012, 36.

22. William Flesch, "Acting Together," review of *Theatre and Mind*, by Bruce McConachie, *TLS*, September 20, 2013, 29.

23. Ibid.

Index